Progress in Mathematics
Volume 303

Series Editors
Hyman Bass
Joseph Oesterlé
Yuri Tschinkel
Alan Weinstein

Angel Cano
Juan Pablo Navarrete
José Seade

Complex Kleinian Groups

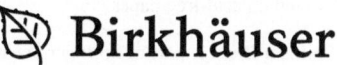 Birkhäuser

Angel Cano
José Seade
Instituto de Matemáticas
Universidad Nacional Autónoma de México
Unidad Cuernavaca
Cuernavaca, Morelos
Mexico

Juan Pablo Navarrete
Facultad de Matemáticas
Universidad Autónoma de Yucatán
Mérida, Yucatán
Mexico

ISBN 978-3-0348-0480-6   ISBN 978-3-0348-0481-3 (eBook)
DOI 10.1007/978-3-0348-0481-3
Springer Basel Heidelberg New York Dordrecht London

Library of Congress Control Number: 2012951146

Mathematics Subject Classification (2010): 37F, 22E40, 32M05, 32Q45

Printed on acid-free paper

Springer Basel is part of Springer Science+Business Media (www.springer.com)

Ferran Sunyer i Balaguer (1912–1967) was a self-taught Catalan mathematician who, in spite of a serious physical disability, was very active in research in classical mathematical analysis, an area in which he acquired international recognition. His heirs created the Fundació Ferran Sunyer i Balaguer inside the Institut d'Estudis Catalans to honor the memory of Ferran Sunyer i Balaguer and to promote mathematical research.

Each year, the Fundació Ferran Sunyer i Balaguer and the Institut d'Estudis Catalans award an international research prize for a mathematical monograph of expository nature. The prize-winning monographs are published in this series. Details about the prize and the Fundació Ferran Sunyer i Balaguer can be found at

http://ffsb.iec.cat/EN

**This book has been awarded the
Ferran Sunyer i Balaguer 2012 prize.**

The members of the scientific commitee of the 2012 prize were:

Alejandro Adem
  University of British Columbia

Núria Fagella
  Universitat de Barcelona

Joseph Oesterlé
  Université de Paris VI

Joan Verdera
  Universitat Autònoma de Barcelona

Alan Weinstein
  University of California at Berkeley

Ferran Sunyer i Balaguer Prize winners since 2003:

Preface

The purpose of this monograph is to lay down the foundations of the theory of complex Kleinian groups, a concept and a name introduced by José Seade and Alberto Verjovsky in the late 1990s, though their origin traces back to classical work by Henri Poincaré, Emile Picard, Georges Giraud and many others. This brings together several important areas of mathematics, as for instance classical Kleinian group actions, complex hyperbolic geometry, chrystallographic groups and the uniformization problem for complex manifolds. Each of these is in itself a fascinating area of mathematics, with a vast literature, both classical and modern. In fact, real and complex hyperbolic geometry are indeed at the very roots of the theory of complex Kleinian groups and therefore we have devoted the first two chapters of this work to giving a fast overview of these rich areas of mathematics.

A classical Kleinian group is a discrete group of conformal automorphisms of the Riemann sphere $\mathbb{S}^2$, acting on the sphere with a nonempty region of discontinuity. Since the Riemann sphere is biholomorphic to the complex projective line $\mathbb{P}^1_{\mathbb{C}}$, and the orientation preserving conformal automorphisms of $\mathbb{S}^2$ are exactly the elements in $\mathrm{PSL}(2, \mathbb{C})$, one has that in this dimension, the classical Kleinian groups can be regarded too as being groups of holomorphic automorphisms of $\mathbb{P}^1_{\mathbb{C}}$.

When going into higher dimensions, there is a dichotomy: Should we look at conformal automorphisms of the n-sphere $\mathbb{S}^n$?, or should we look at holomorphic automorphisms of the complex projective space $\mathbb{P}^n_{\mathbb{C}}$? These two theories are different because in higher dimensions, neither are conformal maps always holomorphic, nor are holomorphic maps necessarily conformal. In the first case we are talking about groups of isometries of real hyperbolic spaces, an area of mathematics where there is a rich body of knowledge thanks to the contributions of people like Ahlfors, Thurston, Margulis, Sullivan, Mostow, Kapovich, McMullen and many others. In the second case we are talking about an area of mathematics that still is in its childhood, and its study is the theme of this work. Complex Kleinian groups are discrete subgroups of $\mathrm{PSL}(n + 1, \mathbb{C})$, the group of holomorphic automorphisms of $\mathbb{P}^n_{\mathbb{C}}$, having a nonempty invariant set where the action is properly discontinuous.

The group $\mathrm{PU}(n, 1)$ of holomorphic isometries of complex hyperbolic n-space consists of the elements in $\mathrm{PSL}(n+1, \mathbb{C})$ that preserve a ball, while the affine group $\mathrm{Aff}(\mathbb{C}^n) \cong \mathrm{GL}(n, \mathbb{C}) \ltimes \mathbb{C}^n$ consists of the elements in $\mathrm{PSL}(n + 1, \mathbb{C})$ that leave invariant a given projective hyperplane. These are two very important subgroups of PSL$(n + 1, \mathbb{C})$, but there are others, as for instance all the Lorentz groups PU(p, q) with $p + q = n + 1$, the groups coming via twistor theory, Schottky type groups, and many others. Thus we see that the study of complex Kleinian groups is at the very heart of complex geometry.

An important difference with the classical case springs from the notion of the limit set. We know that if we consider a Kleinian group G acting on the n-sphere $\mathbb{S}^n$, then its limit set Λ is the set of accumulation points of the orbits. Its complement Ω is the region of discontinuity; this is the maximal region where the action is properly discontinuous, and it is also the equicontinuity set of the family of transformations defined by the group action. The same definitions and properties apply to discrete subgroups of PU$(n, 1)$, essentially because when one looks at the action of isometry groups in real or complex hyperbolic space, one has the convergence property, in Misha Kapovich's language. Yet, even in the case of subgroups of PU$(n, 1)$, when we look at the action on the whole space $\mathbb{P}^n_{\mathbb{C}}$ and not only on the unit ball, there is not a well-defined notion of the limit set. There are actually several possible definitions of the limit set, each with its own properties and characteristics, and their study is one of the main features of this monograph.

Another important point to notice is that, in complex dimension 1, we have Sullivan's dictionary (highly enriched by McMullen and others) between the theory of Kleinian groups and the study of iterates of holomorphic functions of the Riemann sphere. There is currently much interesting work being done on iteration theory in several complex variables, and it is to be expected that there should be plenty of analogies (or perhaps a dictionary, to some extent) between iteration theory of endomorphisms of $\mathbb{P}^n_{\mathbb{C}}$ and the theory of complex Kleinian groups. Recent work by W. Barrera, A. Cano, J.- P. Navarrete and others, points in this direction, but there is still a lot to be understood.

We finish this preface by saying that the theory of complex Kleinian groups is a rich area of mathematics that is waiting to be explored. We believe that anyone who absorbs the material in this book, will get plenty of ideas and insights about interesting questions and further lines of research.

Angel Cano,
Juan Pablo Navarrete, and
José Seade.
Cuernavaca and Mérida, México, Spring of 2012.

Contents

Introduction

Kleinian groups were introduced by Henri Poincaré in the 1880s as the monodromy groups of certain 2nd order differential equations on the complex plane $\mathbb{C}$. These are, classically, discrete subgroups of $PSL(2, \mathbb{C})$, the group of holomorphic automorphisms of the complex projective line $\mathbb{P}^1_{\mathbb{C}}$, which act on this space with nonempty region of discontinuity. Equivalently, these can be regarded as groups of conformal automorphisms of the sphere $\mathbb{S}^2$, or as groups of (orientation preserving) isometries of the hyperbolic 3-space.

Kleinian groups have played for decades a major role in several fields of mathematics, as for example in Riemann surfaces and Teichmüller theory, automorphic forms, holomorphic dynamics, conformal and hyperbolic geometry, 3-manifolds theory, iteration theory of rational maps, etc.

Much of the theory of Kleinian groups has been generalised to conformal Kleinian groups in higher dimensions (also called *Möbius* or *hyperbolic* Kleinian groups), i.e., discrete groups of conformal automorphisms of the sphere $\mathbb{S}^n$. We refer to [105, 106] for clear accounts on conformal Kleinian groups.

Also, D. P. Sullivan's dictionary gives remarkable relations between classical Kleinian groups and the iteration theory of rational maps on $\mathbb{P}^1_{\mathbb{C}}$. Many interesting results about the dynamics of rational maps on $\mathbb{P}^1_{\mathbb{C}}$ in the last decades have been motivated by the dynamics of Kleinian groups (see for instance the references to Sullivan's and McMullen's work in the bibliography). Further striking relations in this sense have been also obtained by M. Lyubich, Y. Minsky and others.

On the other hand, the iteration theory of rational maps is being generalised to higher dimensions by various authors, like J. F. Fornæss, E. Bedford, J. Smillie, N. Sibony, amongst others, obtaining many interesting results about the dynamics of rational endomorphisms of $\mathbb{P}^n_{\mathbb{C}}$ (see [51] for a clear account on the subject).

It is thus natural to ask: what about the other side of the dictionary, in holomorphic dynamics, in higher dimensions? I.e., what about discrete groups of automorphisms of $\mathbb{P}^n_{\mathbb{C}}$? That is the subject we explore in this monograph.

As we will see in the sequel, this includes the theory of discrete groups of isometries in both, real and complex hyperbolic geometry, as well as discrete complex affine groups. There are several other ways and sources from which complex Kleinian groups arise, and we discuss some of these throughout the text.

This monograph originates in articles about *complex Kleinian groups* by A. Verjovsky, J. Seade, J. P. Navarrete and A. Cano (see references in the bibliography), enriched through the work of many authors who have studied and written about real and complex hyperbolic geometry, and also about holomorphic projective structures on complex manifolds. We have particularly profited from W. Goldman's excellent book on complex hyperbolic geometry, as well as John Parker's articles cited in our bibliography. Misha Kapovich's work has also been very helpful and highly inspiring. These and many other works have played a significant role for us while writing this monograph; throughout the text we indicate the most significant references on each topic, and each chapter begins with an introduction that includes references for further reading.

A complex Kleinian group means a discrete group of automorphisms of $\mathbb{P}^n_{\mathbb{C}}$ which acts within a nonempty region of discontinuity. When the group acts on $\mathbb{P}^n_{\mathbb{C}}$ preserving a ball, then it is conjugate to a subgroup of $PU(n,1)$ and we are in the framework of complex hyperbolic geometry; the groups one gets in this way are called *complex hyperbolic Kleinian groups*. If the group acts on $\mathbb{P}^n_{\mathbb{C}}$ preserving a projective hyperplane $\mathbb{P}^{n-1}_{\mathbb{C}}$, then we are essentially in the realm of complex affine geometry. On the other hand, whenever a discrete subgroup Γ of $PSL(n+1, \mathbb{C})$ acts properly discontinuously on an invariant open set $\Omega \subset \mathbb{P}^n_{\mathbb{C}}$, the quotient space Ω/Γ is an orbifold equipped with a projective structure, and the study of holomorphic projective structures on complex manifolds and orbifolds is in itself a rich area of current research. This includes the theory of complex hyperbolic and complex affine manifolds. Thence the theory of complex Kleinian groups provides a means to study these important fields of mathematics in a unified way.

The study of complex Kleinian groups is indeed still in its childhood, and this monograph aims to contribute to the laying down of its foundations, studying basic concepts as for instance that of the limit set; constructions of discrete groups in higher dimensions; the uniformisation problem for two-dimensional complex orbifolds; classification problems: of the elements in $PSL(n+1, \mathbb{C})$ and their geometry and dynamics, of its discrete subgroups, of the complex structures one gets on quotients of open sets of $\mathbb{P}^n_{\mathbb{C}}$ which are invariant under the action of a discrete group; Teichmüller theory in higher dimensions; relations with twistor theory; etc.

Whenever one has a classical Kleinian group, one has a natural splitting of $\mathbb{P}^1_{\mathbb{C}}$ in two invariant subsets: one of these, say Ω, is where the action is discontinuous; this is also the equicontinuity set of the group, i.e., the points where the group forms a normal family. The other set Λ, its complement, is where the dynamics "concentrates". The set Ω plays a key role in complex geometry, as shown by the work of Ahlfors, Kra, Bers and many others. And the action on the limit set plays a key role for holomorphic dynamics, as shown by the work of D. Sullivan, W. Thurston, C. McMullen, M. Lyubich and many others.

One has a similar picture for conformal Kleinian groups in higher dimensions, and their study is the content of Chapter 1 of this monograph, which contains well-known material that we present in a way that supports the remaining chapters.

We refer to the literature, particularly to the recent articles of M. Kapovich listed in the bibliography, for a deeper and wider study of this important field of mathematics.

Chapter 1 provides a quick glance at the foundations of real hyperbolic geometry: its various models; its group of isometries; the relation between hyperbolic space and conformal geometry on the sphere at infinity; discrete groups of isometries; the limit set and discontinuity region; fundamental domains. The literature on these topics is vast, so we have made no attempt to give a comprehensive account of the subject. Instead, we give a number of examples and discuss some of the key ideas that help to get a feeling for this exciting subject. We also explain briefly some of the most celebrated theorems in the subject that we use in the sequel, such as Moore's ergodicity theorem, Mostow's rigidity, the Patterson-Sullivan measure, the ergodicity of action on the limit set, and Sullivan's theorem of nonexistence on invariant line-fields on the limit set.

In Chapter 2 we look at complex hyperbolic geometry, which is the complex analogue of real hyperbolic geometry. Its origin traces back to the work of É. Picard on differential equations in several complex variables. Later, G. Giraud made fundamental contributions to the subject through a series of papers, which are discussed in an appendix at the end of [67]. This is a fascinating branch of mathematics which has been having a fast development over the last few decades, thanks to the contributions of many authors as, notably, G. Mostow and P. Deligne, and more recently W. Goldman, J. Parker, N. Gusevskii, R. Schwartz, E. Falbel, M. Kapovich and several others. In this chapter we briefly describe the classical models for complex hyperbolic geometry, its group of holomorphic isometries, which is the projective Lorentz group $\mathrm{PU}(n,1)$, and methods for constructing discrete subgroups of it. We describe, following [67], the classification of the elements in $\mathrm{PU}(n,1)$ according to their geometry and dynamics, and for $n = 2$ also according to trace. We finish the chapter by giving the definition and basic properties of the limit set following [45].

In Chapter 3 we start our discussion on complex Kleinian groups. As mentioned earlier, these are, by definition, discrete subgroups of $\mathrm{PSL}(n+1, \mathbb{C})$ that act on $\mathbb{P}^n_{\mathbb{C}}$ with nonempty regions of discontinuity. This naturally includes the discrete groups of isometries of real and complex hyperbolic n-space, as well as all complex affine groups, and many more. In this general setting there is no well-defined notion of limit set when $n \geq 2$. We give an example in $\mathbb{P}^2_{\mathbb{C}}$, motivated by work of R. Kulkarni in the late 1970s, that illustrates the diversity of possibilities one has for defining this notion, unlike the situation in hyperbolic geometry where "the limit set" is a well-defined notion. There are several possible definitions of this concept, each with its own properties and characteristics. There is the Chen-Greenberg limit set for complex hyperbolic groups, i.e., subgroups of $\mathrm{PU}(n,1)$; there is the Kulkarni limit set; there is the complement of the region of discontinuity; the complements of the maximal regions where the action is properly discontinuous (in general there is no largest such region); and the complement of the region of equicontinuity. There are also other sets that can play the role of "the limit set" in different settings (e.g. the closure of the fixed points of loxodromic elements, whenever this makes sense). These notions are not always equivalent, as we explain in the text, and each of them has its own interest. Yet, in all cases one has a region Ω where the dynamics is "tame", and a set Λ where the dynamics concentrates.

When $n = 1$ a complex Kleinian group is nothing but a discrete group of automorphisms of $\mathbb{P}^1_\mathbb{C}$, which is the Riemann sphere. A natural next step is considering groups of automorphisms of $\mathbb{P}^2_\mathbb{C}$ and this is our focus of study in Chapters 4 to 8. The material in these chapters is mostly based on work done by Angel Cano and Juan-Pablo Navarrete, and more recently also in collaboration with Waldemar Barrera.

In Chapter 4 we look at the geometry and dynamics of the individual elements in $\mathrm{PSL}(3, \mathbb{C})$. In other words, we consider a single automorphism of $\mathbb{P}^2_\mathbb{C}$ and the cyclic group it generates. We know that, classically, the elements in $\mathrm{PSL}(2, \mathbb{C})$ are of three types: elliptic, parabolic and loxodromic (or hyperbolic). This classification is done in terms of their geometry and dynamics, and it can also be done algebraically, in terms of the trace of a lifting to $\mathrm{SL}(2, \mathbb{C})$. In fact there is a similar (geometric-dynamical) classification for the isometries of every Riemannian manifold of nonpositive curvature, and even more generally for isometries of $\mathrm{CAT}(0)$-spaces (see Remark 2.4.6). In Chapter 4 we show that the elements of $\mathrm{PSL}(3, \mathbb{C})$ can be also naturally classified into these three types, elliptic, parabolic and loxodromic, although $\mathbb{P}^2_\mathbb{C}$ has positive curvature and the action of $\mathrm{PSL}(3, \mathbb{C})$ is not by isometries. This classification is of course compatible with the classical one when the elements lie in $\mathrm{PU}(2, 1)$. The classification is done in terms of the geometry and dynamics, and also algebraically. It turns out that all the elliptic and parabolic elements are actually conjugate to elliptic and parabolic elements in $\mathrm{PU}(2, 1)$. The new phenomena appear when looking at the loxodromic elements. We remark that the geometric–dynamical classification of the elements in $\mathrm{PSL}(3, \mathbb{C})$ actually extends to higher dimensions, as shown in [39].

The next step in our study is considering in Chapter 5 subgroups of $\mathrm{PSL}(3, \mathbb{C})$ whose dynamics is governed by a subgroup of $\mathrm{PSL}(2, \mathbb{C})$. This is natural since the subgroups of $\mathrm{PSL}(2, \mathbb{C})$ are far better understood than the subgroups of $\mathrm{PSL}(3, \mathbb{C})$. The simplest way for doing so is by taking a discrete group Γ in $\mathrm{PSL}(2, \mathbb{C})$, lifting it to $\mathrm{SL}(2, \mathbb{C})$ and looking at its canonical inclusion in $\mathrm{SL}(3, \mathbb{C})$. This type of groups were called *suspensions* in [201] and this construction was generalised in [160], [41], getting interesting representations in $\mathrm{PSL}(3, \mathbb{C})$ of subgroups of $\mathrm{PSL}(2, \mathbb{C})$. Yet, there are many other situations in which one can get a lot of information about a given group in $\mathrm{PSL}(3, \mathbb{C})$ from a lower-dimensional one which "controls" its dynamics. This is the topic we study in Chapter 5. It is worth remarking that even if we start with a discrete group in $\mathrm{PSL}(3, \mathbb{C})$, it can happen that the corresponding control group in $\mathrm{PSL}(2, \mathbb{C})$ is nondiscrete. We thus make in this chapter a brief discussion of nondiscrete subgroups of $\mathrm{PSL}(2, \mathbb{C})$. Of course it makes no sense in this setting to speak of a "discontinuity region", since this is empty by definition. Here, it is the equicontinuity region which plays a significant role.

As we have said before, unlike the situation for subgroups of $PSL(2, \mathbb{R})$, in higher dimensions there is no unique notion of "the limit set" for complex Kleinian groups. There are instead several natural such notions, each with its own properties and characteristics, each providing a different kind of information about the geometry and dynamics of the group. Yet, in Chapter 6 we see that in dimension 2, the various natural definitions of limit set coincide generically, i.e., for complex Kleinian groups whose Kulkarni limit set has "enough" lines.

This is interesting also from the viewpoint of having a Sullivan dictionary between Kleinian groups and iteration theory in several complex variables. In fact we recall the important theorem in [60]), stating that in the space of all rational maps of degree d in $\mathbb{P}^n_{\mathbb{C}}$, for $n \geq 2$, those whose Fatou set is Kobayashi hyperbolic form an open dense set with the Zariski topology. Similarly, in this chapter we see that under certain "generic" conditions, the region of equicontinuity of a complex Kleinian group in $\mathbb{P}^2_{\mathbb{C}}$ coincides with the Kulkarni region of discontinuity, and it is the largest open invariant set where the group acts properly discontinuously. And this region is Kobayashi hyperbolic. Hence the results in this chapter, which are based on work by W. Barrera, A. Cano and J. P. Navarrete, give us a better understanding of the concept of "the limit set" in dimension 2, and they set down the first steps of a theory that points towards an analogous concept for Kleinian groups of the aforementioned theorem of Fornæss-Sibony.

In Chapter 7 we consider again complex hyperbolic Kleinian groups, i.e., discrete subgroups of $PU(n, 1)$, but we now look at their action on the whole projective space $\mathbb{P}^n_{\mathbb{C}}$, not only at the projective ball that serves as a model for complex hyperbolic space. We know from Chapter 3 that one has in this setting several possible definitions of the limit set. Here we compare the Kulkarni limit set and the complement of the region of equicontinuity, with the limit set in the sense of Chen-Greenberg. This is a subset of the sphere that bounds the projective ball that serves as a model for the complex hyperbolic space $\mathbb{H}^n_{\mathbb{C}}$. This allows us to get information about the action of the group on all of $\mathbb{P}^n_{\mathbb{C}}$ from its behaviour on the ball $\mathbb{H}^n_{\mathbb{C}}$.

In Chapter 8 we bring together the information obtained through Chapters 4, 5 and 7, to study discrete subgroups of $PSL(3, \mathbb{C})$ with a divisible set in $\mathbb{P}^2_{\mathbb{C}}$ in the sense of Y. Benoist. More precisely, we study subgroups of $PSL(3, \mathbb{C})$ acting on $\mathbb{P}^2_{\mathbb{C}}$ so that there is a nonempty open invariant set Ω where the group acts properly discontinuously and the quotient $M = \Omega/\Gamma$ is compact; we call such actions *quasi-cocompact*. This includes the cocompact case: When Γ has a largest region of discontinuity and the quotient is compact. The surface M is an orbifold naturally equipped with a projective structure. The material in this chapter is closely related to previous work by S. Kobayashi, T. Ochiai, Y. Inoue, B. Klingler and others, about compact complex surfaces with a projective structure. We give the complete classification of the divisible sets appearing in this way, the corresponding groups, the Kulkarni limit set and the topology of the quotient orbifold.

Chapters 9 and 10 focus on complex Kleinian groups in higher dimensions, and these are essentially based on the articles by José Seade and Alberto Verjovsky listed in the bibliography. The material in Chapter 9 is actually related to previous work by M. Nori aimed at construction of new compact complex manifolds. We also speak in that chapter about work by A. Cano.

Recall that the classical Schottky groups are subgroups of $PSL(2, \mathbb{C})$ obtained by considering disjoint families of circles in $\mathbb{P}^1_\mathbb{C} \cong \mathbb{S}^2$. These circles play the role of *mirrors* that split the sphere in two diffeomorphic halves which are interchanged by a conformal map, and these maps generate the Schottky group (see Chapter 1 for details). So the idea of constructing Schottky groups in higher dimensions is to construct *mirrors* in $\mathbb{P}^n_\mathbb{C}$ that split the space in two parts which are interchanged by a holomorphic authomorphism, and use these to construct discrete subgroups. This works fine on odd-dimensional projective spaces. One gets Schottky subgroups of $PSL(2n + 2, \mathbb{C})$ whose limit sets are solenoids with rich dynamics. Following [203], we determine the topology of the compact complex manifolds obtained as quotient $M_{\check{\Gamma}} := \Omega(\check{\Gamma})/\check{\Gamma}$ of the region of discontinuity divided by the action. We look at their Kuranishi space of versal deformations and prove that, for $n > 2$, every infinitesimal deformation of $M_{\check{\Gamma}}$ actually corresponds to an infinitesimal deformation of the group $\check{\Gamma}$ in the projective group $PSL(2n + 2, \mathbb{C})$. This is analogous to the classical Teichmüller theory for Riemann surfaces. Similar considerations were observed in [176] for $n = 1$, studying the so-called Pretzel Twistor spaces.

In even dimensions, one can show that there cannot be Schottky groups (see the text for a more precise statement). Yet, in these dimensions one does have mirrors, but unlike the odd-dimensional case, mirrors are now singular varieties, not smooth submanifolds, and all mirrors must intersect. Thus one gets *kissing-Schottky* groups as in the "Indra's Pearls" of [158], acting on $\mathbb{P}^2_\mathbb{C}$ and actually on all projective spaces. The examples of kissing-Schottky groups that we give in Chapter 9 are interesting because these groups are not elementary (see the text for the definition), nor affine, nor complex hyperbolic, in contrast with the groups that appear in Chapter 8.

Finally, Chapter 10 is based on [202]. Here we use twistor theory to construct complex Kleinian groups. Twistor theory is, no doubt, one of the jewels of mathematics in the 20th Century.

There are two different ways in which twistor theory can be considered. One is to see twistor theory as providing the geometrical setting for new and valuable mathematical methods in, for example, the treatment of Yang-Mills and other nonlinear equations. The other point of view, more ambitious, is the twistor programme for physics, in which it is held that, if the nature of the physical world is to be understood, then the usual description of space-time must be superseded by some form of twistor geometry. The mathematical foundations of this theory were developed by R. Penrose and a number of other great mathematicians such as M. Atiyah, N. Hitchin and others.

The "Penrose twistor programme" springs from the remarkable fact that there is a rich interplay between the conformal geometry of (even-dimensional) Riemannian manifolds and the complex geometry of their twistor spaces. In Chapter 10 we explain how this interplay can be also "pushed forward" to dynamics, providing interesting relations between conformal and holomorphic dynamics. One gets that every conformal Kleinian group can be regarded, canonically, as a complex Kleinian group via twistor theory. For example in dimension 4, one has a canonical embedding $\mathrm{Conf}_+(\mathbb{S}^4) \hookrightarrow \mathrm{PSL}(4, \mathbb{C})$ and the dynamics of conformal Kleinian groups in dimension 4 embeds in the dynamics of complex Kleinian groups in $\mathbb{P}^3_{\mathbb{C}}$. Furthermore, one can actually see things about conformal Kleinian groups $G \subset \mathrm{Conf}_+(\mathbb{S}^4)$ through their action on $\mathbb{P}^3_{\mathbb{C}}$ which are not visible through their action on $\mathbb{S}^4$. Similar statements hold in higher dimensions.

As mentioned before, the theory of complex Kleinian groups is still in its childhood, although its origin traces back to the work of Riemann, Poincaré, Klein, Picard, Giraud and many other great mathematicians. The knowledge we now have about classical Kleinian groups, as well as about complex hyperbolic groups, complex affine groups, geometric structures on complex manifolds, moduli spaces and Teichmüller theory, potential theory and iteration theory in several complex variables, inspire many questions and lines of further research in the topic of discrete subgroups of projective transformations, that are waiting to be explored.

Although the various chapters in this monograph add up to form a coherent unit, this monograph has been written so that each individual chapter can be read on its own. No doubt the theory of complex Kleinian groups will eventually become an important subfield of both, complex geometry and holomorphic dynamics, and we hope this monograph will help to lay down its foundations, and contribute to enhancing the interest in this fascinating topic.

Acknowledgments

This monograph is the outcome of several years of working in this beautiful area of mathematics. During this time we have profited from conversations with many friends and great mathematicians, to whom we are deeply indebted: Our heart-felt thanks to all of them. Although we are sure to leave out some important names, we want to mention here those we are able to remember.

First of all, we are much indebted to Alberto Verjovsky, for explaining to us many of the ideas and concepts included in this text, and for teaching us about the beauty of mathematics.

We are most grateful to professors Dennis Sullivan, John Erick Fornæss, Nessim Sibony, William Goldman, Nikolay Gusevskii, Misha Kapovich, Etienne Ghys, Misha Lyubich, Paulo Sad and John Parker. We have learned a lot from their works, their lectures, their conversations and their way of thinking. These have been to us like lights along the path. Our deepest thanks also to our colleagues Waldemar Barrera, Carlos Cabrera, Aubin Arroyo, Adolfo Guillot, Luis Loeza and Piotr Makienko. Each of them has contributed in one way or another to this monograph.

During this time we have also profited from the support of several institutions which have either hosted us or given us financial support, or both. Our deepest thanks to all of them, though we can only mention here a few of these. First of all, we are indebted to the Instituto de Matemáticas of the National University of Mexico (UNAM) and its Cuernavaca Unit, where two of us work, as well as to the Facultad de Matemáticas of the Universidad Autónoma de Yucatán (UADY), the home institution of the third one of us. Both of these institutions, and the people in them, have supported us in all ways.

We are most grateful to IMPA at Rio de Janeiro, Brazil, which always is like a home for us, offering us a most stimulating academic atmosphere; and to the Abdus Salam International Centre for Theoretical Physics at Trieste, Italy, where we have spent long periods of time, and where we profited from the School and Workshop "Discrete groups in Complex Geometry" in 2010, that left a mark on all of us.

During these years we have had several grants and support from the DGAPA of the National University of Mexico (UNAM), as well as from CONACYT, México, the CNRS, France, mostly through the Laboratoire International Associé Solomon Lefschetz (LAISLA).

Chapter 1

A Glance at the Classical Theory

Classical Kleinian groups are discrete subgroups of Möbius transformations which act on the Riemann sphere with a nonempty region of discontinuity. This includes Fuchsian groups, Schottky groups and many other interesting families.

Möbius transformations are all obtained as compositions of inversions on circles on the Riemann sphere $\mathbb{S}^2 \cong \mathbb{P}^1_{\mathbb{C}}$, and the group of all Möbius transformations is isomorphic to the group of orientation preserving isomorphisms of hyperbolic 3-space $\mathbb{H}^3_{\mathbb{R}}$.

When we go into "higher dimensions" there is a dichotomy: one may study discrete subgroups of isometries of hyperbolic n-space $\mathbb{H}^n_{\mathbb{R}}$; equivalently, these are also discrete groups of conformal automorphisms of the sphere at infinity, and they are all obtained by inversions on spheres of codimension 1. We refer to these as *conformal Kleinian groups* and their study is the goal of this chapter. Yet, by "higher dimensions" one may also mean discrete groups of holomorphic transformations of the complex projective space $\mathbb{P}^n_{\mathbb{C}}$ acting with a nonempty region of discontinuity. We call these *complex Kleinian groups*, following [201], [202], and their study is the substance of the rest of this monograph. Since we have a group isomorphism $\mathrm{PSL}(2, \mathbb{C}) \cong \mathrm{Conf}_+\mathbb{S}^2$, in the classical setting both concepts coincide: conformal and complex Kleinian groups are essentially the same thing (provided the conformal maps preserve also the orientation). In higher dimensions these concepts are different, but there are similarities and the theory we develop here often uses results and ideas from the conformal setting. Thence we begin this work by reviewing briefly the conformal Kleinian groups.

This is a fascinating theory with a vast literature. For an introduction to the subject we refer to the books of Maskit and Beardon. And for more updated accounts we refer to Misha Kapovich's book and the excellent notes [105] and [150].

1.1   Isometries of hyperbolic n-space. The conformal group

1.1.1   Poincaré models for hyperbolic space

We recall that a Riemannian metric g on a smooth manifold M means a choice of a positive definite quadratic form on each tangent space $T_x M$, varying smoothly over the points in M. Such a metric determines lengths of curves as usual, and so defines a metric on M by declaring that the distance between two points is the infimum of the lengths of curves connecting them.

A (local) isometry between any two such spaces is a map that preserves distances.

We now construct a model for hyperbolic n-space $\mathbb{H}_{\mathbb{R}}^n$. We learned this construction from Dennis Sullivan (see [216]) and it is based on work of Jakob Steiner (in 1825) on inversions in $\mathbb{R}^m$. We start by recalling that the classical inversion in $\mathbb{S}^1 \cong \widehat{\mathbb{R}} = \mathbb{R} \cup \{\infty\}$ is the map ι defined by $x \mapsto 1/x$, where we are assuming $1/0 = \infty$ and $1/\infty = 0$. This definition extends in the obvious way to intervals of arbitrary length and centre. Using this, we can define inversions in $\mathbb{S}^n \cong \widehat{\mathbb{R}} = \mathbb{R}^n \cup \{\infty\}$ on arbitrary $(n-1)$-spheres in $\mathbb{S}^n$ by considering all rays that emanate from the centre of the sphere and defining the inversion on each ray as we did in dimension 1. This includes inversions on $(n-1)$-spheres of maximal radius, which correspond to hyperplanes in $\mathbb{R}^n$. In that case the inversion is the usual Euclidean reflection.

Thus, for instance, if $C \subset \mathbb{R}^2$ is the circle of finite radius $r > 0$ and centre at the origin 0, then the inversion on C is defined by

$$\iota_r(x, y) = \frac{r^2}{\|(x, y)\|^2} (x, y).$$

It is clear that the inversion in $\mathbb{S}^n$ with respect to an $(n-1)$ sphere $\mathcal{S}$ has $\mathcal{S}$ as its fixed point set, and it interchanges the two halves of $\mathbb{S}^n$ separated by $\mathcal{S}$. In other words, $\mathcal{S}$ is like a *mirror* in $\mathbb{S}^n$, a concept that will play a key role later in Chapter 9.

The following theorem is due to Steiner:

Theorem 1.1.1. *Inversions in $\mathbb{S}^n$ are conformal maps that preserve spheres of all dimensions.*

Recall that a map between Riemannian manifolds is *conformal* if it preserves angles (measured in the usual way). So the theorem claims that inversions preserve angles and carry each d-sphere $S \subset \mathbb{S}^n$ into another d-sphere, for all $d \geq 0$.

Let $\text{Möb}(\mathbb{S}^n)$ be the group of diffeomorphisms of $\mathbb{S}^n \cong \widehat{\mathbb{R}} = \mathbb{R}^n \cup \{\infty\}$ generated by inversions on all $(n-1)$-spheres in $\mathbb{S}^n$, and let $\text{Möb}(\mathbb{B}^n)$ be the subgroup of $\text{Möb}(\mathbb{S}^n)$ consisting of maps that preserve the unit ball $\mathbb{B}^n$ in $\mathbb{R}^n$.

The proof of the following result is left as an exercise:

Theorem 1.1.2. *Let S_1 and S_2 be two $(n-1)$-spheres in $\mathbb{S}^n$, $n \geq 2$, and let ι_{S_1} be the inversion on the first of these spheres. Then $\iota_{S_1}(S_2) = S_2$ if and only if S_1 and S_2 intersect orthogonally (with respect to the usual round metric on $\mathbb{S}^n$). Hence the group $\text{Möb}(\mathbb{B}^n)$ is generated by inversions in $\mathbb{S}^n$ on $(n-1)$-spheres that meet orthogonally the boundary $\mathbb{S}^{n-1} = \partial\mathbb{B}^n$.*

Notice that if the $(n-1)$-sphere S_1 meets $\mathbb{S}^{n-1} = \partial\mathbb{B}^n$ orthogonally, then $\mathcal{C} := S_1 \cap \mathbb{S}^{n-1}$ is an $(n-2)$-sphere in $\mathbb{S}^{n-1}$ and the restriction to $\mathbb{S}^{n-1}$ of the inversion ι_{S_1} coincides with the inversion on $\mathbb{S}^{n-1}$ defined by the $(n-2)$-sphere $\mathcal{C}$. In other words one has a canonical group homomorphism $\text{Möb}(\mathbb{B}^n) \to \text{Möb}(\mathbb{S}^{n-1})$.

Conversely, given an $(n-2)$-sphere $\mathcal{C}$ in $\mathbb{S}^{n-1}$ there is a unique $(n-1)$-sphere S in $\mathbb{S}^n$ that meets $\mathbb{S}^{n-1}$ orthogonally at $\mathcal{C}$. The inversion

$$\iota_{\mathcal{C}} : \mathbb{S}^{n-1} \to \mathbb{S}^{n-1}$$

extends canonically to the inversion

$$\iota_S : \mathbb{B}^n \to \mathbb{B}^n\,,$$

thus giving a canonical group homomorphism $\text{Möb}(\mathbb{S}^{n-1}) \to \text{Möb}(\mathbb{B}^n)$, which is obviously the inverse morphism of the previous one. Thus one has:

Lemma 1.1.3. *There is a canonical group isomorphism $\text{Möb}(\mathbb{B}^n) \cong \text{Möb}(\mathbb{S}^{n-1})$, $\forall n \geq 2$.*

Remark 1.1.4. It is well-known that every conformal automorphism of a sphere $\mathbb{S}^m$, $m > 1$, is a composition of inversions. Thus one also has an isomorphism $\text{Möb}(\mathbb{B}^n) \cong \text{Conf}(\mathbb{S}^{n-1})$, $\forall n > 2$.

The following result shows that for $n \geq 3$, the property of being locally conformal is very strong and makes a remarkable difference with the holomorphic case in high dimensions, see [7], [8].

Theorem 1.1.5 (Liouville). *Let $n \geq 3$, and let $U, V \subset \mathbb{S}^n$ be open, connected and nonempty sets. Then every conformal injective function $f : U \to V$ is the restriction of an element in $\text{Möb}(\mathbb{S}^n)$.*

Definition 1.1.6. We call $\text{Möb}(\mathbb{B}^n)$ (and also $\text{Möb}(\mathbb{S}^n)$) *the general Möbius group* of the ball (or of the sphere).

The subgroup $\text{Möb}_+(\mathbb{B}^n)$ of $\text{Möb}(\mathbb{B}^n)$ of words of even length consists of the elements in $\text{Möb}(\mathbb{B}^n)$ that preserve the orientation. This is an index 2 subgroup of $\text{Möb}(\mathbb{B}^n)$. Similar considerations apply to $\text{Möb}(\mathbb{S}^n)$. We call $\text{Möb}_+(\mathbb{B}^n)$ and $\text{Möb}_+(\mathbb{S}^n)$ *Möbius groups* (of the ball and of the sphere, respectively).

We have:

Theorem 1.1.7. *The group $\text{Möb}(\mathbb{S}^n)$ of Möbius transformations is generated by the following transformations:*

(i) *Translations: $t(x) = x + a$, where $a \in \mathbb{R}^n$;*

(ii) *Dilatations (or homotecies): $t(x) = \lambda x$, where $\lambda \in (0, \infty)$;*

(iii) *Rotations: $t(x) = Ox$, where $O \in \mathrm{SO}(n)$;*

(iv) *The inversion: $t(x) = x/\|x\|^2$.*

In particular one has that $\mathrm{M\ddot{o}b}_+(\mathbb{B}^n)$ contains the orthogonal group $\mathrm{SO}(n)$ as the stabiliser (or isotropy) subgroup at the origin 0 of its action on the open ball $\mathbb{B}^n$. The stabiliser of 0 under the action of the full group $\mathrm{M\ddot{o}b}(\mathbb{B}^n)$ is $O(n)$. This implies that $\mathrm{M\ddot{o}b}_+(\mathbb{B}^n)$ acts transitively on the space of lines through the origin in $\mathbb{B}^n$. Moreover, $\mathrm{M\ddot{o}b}_+(\mathbb{B}^n)$ clearly acts also transitively on the intersection with $\mathbb{B}^n$ of each ray through the origin. Thus it follows that $\mathrm{M\ddot{o}b}_+(\mathbb{B}^n)$ acts transitively on $\mathbb{B}^n$.

Now consider the tangent space $T_0\mathbb{B}^n$ and fix the usual Riemannian metric on it, which is invariant under the action of $O(n)$. Given a point $x \in \mathbb{B}^n$, consider an element $\gamma \in \mathrm{M\ddot{o}b}(\mathbb{B}^n)$ with $\gamma(0) = x$. Let $D\gamma_0$ denote the derivative at 0 of the automorphism $\gamma : \mathbb{B}^n \to \mathbb{B}^n$. This defines an isomorphism of vector spaces $D\gamma_0 : T_0\mathbb{B}^n \to T_x\mathbb{B}^n$ and allows us to define a Riemannian metric on $T_x\mathbb{B}^n$. In this way we get a Riemannian metric at each tangent space of $\mathbb{B}^n$, which *a priori* might depend on the choices involved.

We claim that the above construction of a metric on the open ball is well defined, i.e., that the metric one gets on $T_x\mathbb{B}^n$ does not depend on the choice of the element $\gamma \in \mathrm{M\ddot{o}b}(\mathbb{B}^n)$ taking 0 into x. In fact, if $\eta \in \mathrm{M\ddot{o}b}(\mathbb{B}^n)$ is another element taking 0 into x, then $(\eta)^{-1} \circ \gamma$ leaves 0 invariant and is therefore an element in $O(n)$. Since the orthogonal group $O(n)$ preserves the metric at $T_0\mathbb{B}^n$, it follows that both maps, γ and η, induce the same metric on $T_x\mathbb{B}^n$. Hence this construction yields to a well-defined Riemannian metric on $\mathbb{B}^n$.

It is easy to see that this metric is complete and homogeneous with respect to points, directions and 2-planes. That is, given points $x, y \in \mathbb{B}^n$, lines ℓ_x, ℓ_y and 2-planes $\mathcal{P}_x, \mathcal{P}_y$, through these points, there is an element in $\mathrm{M\ddot{o}b}(\mathbb{B}^n)$ carrying x to y and ℓ_x to ℓ_y and an element taking x to y and $\mathcal{P}_x$ to $\mathcal{P}_y$. Hence this metric has constant (negative) sectional curvature.

Definition 1.1.8. The open unit ball $\mathbb{B}^n \subset \mathbb{R}^n$ equipped with the above metric serves as a model for the *hyperbolic n-space* $\mathbb{H}_{\mathbb{R}}^n$. The group $\mathrm{M\ddot{o}b}(\mathbb{B}^n)$ is its *group of isometries*, also denoted by $\mathrm{Iso}(\mathbb{H}_{\mathbb{R}}^n)$, and its index 2 subgroup $\mathrm{M\ddot{o}b}_+(\mathbb{B}^n)$ is the *group of orientation preserving isometries* of $\mathbb{H}_{\mathbb{R}}^n$, $\mathrm{Iso}_+(\mathbb{H}_{\mathbb{R}}^n)$.

Since $\mathrm{Iso}_+(\mathbb{H}_{\mathbb{R}}^n)$ acts transitively on $\mathbb{B}^n$ with isotropy $\mathrm{SO}(n)$, and the tangent bundle of $\mathbb{B}^n$ is obviously trivial, one has that, as a manifold, $\mathrm{Iso}_+(\mathbb{H}_{\mathbb{R}}^n)$ is diffeomorphic to $\mathrm{SO}(n) \times \mathbb{B}^n$. In particular $\mathrm{Iso}_+(\mathbb{H}_{\mathbb{R}}^2)$ is an open solid torus $\mathbb{S}^1 \times \mathbb{B}^2$.

This is the *open disc model* of Poincaré for hyperbolic geometry. We now describe the upper half-space model, also due to Poincaré, that we use in the sequel too. For this consider the closed ball $\overline{\mathbb{B}}^n \subset \mathbb{R}^n$, endow its interior with the hyperbolic metric as above, and think of $\mathbb{R}^n$ as the set of points in $\mathbb{R}^{n+1}$ with 0 in the last coordinate. Now project $\overline{\mathbb{B}}^n$ into the the upper half sphere,

$$\mathbb{S}^{n+1}_+ = \{(x_1, \ldots, x_{n+1}) \in \mathbb{R}^{n+1} \,|\, x_1^2 + \cdots + x_{n+1}^2 = 1 \text{ and } x_{n+1} \geq 0\}$$

via stereographic projection in $\mathbb{R}^{n+1}$ from the South pole $(0, \ldots, 0, -1) \in \mathbb{S}^{n+1}$. Now project $\mathbb{S}^{n+1}_+$ into the upper half space

$$\mathbb{R}^n_+ = \{(x_1, \ldots, x_{n+1}) \in \mathbb{R}^{n+1} \,|\, x_1 = 1 \text{ and } x_{n+1} \geq 0\}\,,$$

via stereographic projection from the point $(-1, 0, \ldots, 0)$. The composition of these two maps identifies the open ball $\mathbb{B}^n$ with the interior $\mathbb{H}^n_+$ of $\mathbb{R}^n_+$. Equipping $\mathbb{H}^n_+$ with the induced (hyperbolic) metric we get Poincaré's *upper half-space* model for hyperbolic n-space.

We finish this subsection with the following theorem, which summarises part of the previous discussion.

Theorem 1.1.9. *The Möbius group* Möb$(\mathbb{B}^n)$, *generated by inversions on* $(n-1)$-*spheres in* $\mathbb{S}^n \cong \mathbb{R}^n \cup \infty$ *which intersect transversally the boundary of the unit sphere* $\mathbb{S}^{n-1} = \partial\mathbb{B}^n$ *(with nonempty intersection), is isomorphic to the group* Iso$(\mathbb{H}^n)$ *of isometries of the n-dimensional hyperbolic space* $\mathbb{H}^n$. *This group is canonically isomorphic to* Conf$(\mathbb{S}^{n-1})$, *the group of conformal automorphisms of the* $(n-1)$-*sphere. As a manifold,* Iso$(\mathbb{H}^n)$ *is diffeomorphic to* $O(n) \times \mathbb{B}^n$, *so it is a noncompact Lie group of real dimension* $\frac{n(n-1)}{2} + n$. *The words in* Iso$(\mathbb{H}^n)$ *with even length form the index* 2 *subgroup* Iso$_+(\mathbb{H}^n)$ *of* Iso$(\mathbb{H}^n)$ *consisting of orientation preserving maps, which is canonically isomorphic to* Conf$_+(\mathbb{S}^{n-1})$.

In the sequel we denote the real hyperbolic space by $\mathbb{H}^n_\mathbb{R}$, to distinguish it from the complex hyperbolic space $\mathbb{H}^n_\mathbb{C}$ (of real dimension $2n$) that we will consider in later chapters. Also, we denote by $\mathbb{S}^{n-1}_\infty$ the sphere at infinity, that is, the boundary of $\mathbb{H}^n_\mathbb{R}$ in $\mathbb{S}^n$. We set $\overline{\mathbb{H}}^n_\mathbb{R} := \mathbb{H}^n_\mathbb{R} \cup \mathbb{S}^{n-1}_\infty$.

Remark 1.1.10 (Models for real hyperbolic geometry). There are several other classical models for real hyperbolic geometry: the projective ball model, the hyperboloid model, the upper-half sphere model and the Siegel domain model. The *upper-half sphere model* was briefly mentioned above and serves, among other things, to pass geometrically from the disc model to the upper-half space model and back. We refer to Thurston's book for descriptions of several other models for the hyperbolic n-space. These are briefly discussed below.

The hyperboloid model, also called *Lorentz or Minkowski model,* is very much related to the models we use in the sequel to study complex hyperbolic geometry (and so are the projective ball and the Siegel domain models that we describe below). For this we look at the upper hyperboloid $\mathcal{P}$ of the two-sheeted hyperboloid defined by the quadratic function

$$x_1^2 + \cdots + x_n^2 - x_{n+1}^2 = -1\,.$$

Its group of isometries is now $O(n,1)_o$, the subgroup of the Lorentz group $O(n,1)$ consisting of transformations that preserve $\mathcal{P}$. This model is generally credited to Poincaré too, though it seems that K. Weierstrass (and probably others) used it before. Its geodesics are the intersections of $\mathcal{P}$ with linear 2-planes in $\mathbb{R}^{n+1}$ passing through the origin; every linear space passing through the origin meets $\mathcal{P}$ in a totally geodesic subspace.

The projective ball model is also called the Klein, or Beltrami-Klein, model. For this we look at the disc $\mathcal{D}$ in $\mathbb{R}^{n+1}$ defined by

$$\mathcal{D} := \{(x_1, \cdots, x_{n+1}) \,|\, x_1^2 + \cdots + x_n^2 < 1 \text{ and } x_{n+1} = 1 \}.$$

That is, we look at the points in $\mathbb{R}^{n+1}$ where the quadratic form $Q(x) = x_1^2 + \cdots + x_n^2 - x_{n+1}^2$ is negative and $x_{n+1} = 1$. Notice that stereographic projection from the origin determines a bijection between $\mathcal{D}$ and $\mathcal{P}$.

Finally, *the Siegel domain (or paraboloid) model* for $\mathbb{H}_{\mathbb{R}}^n$ is obtained by looking at the points $(x_1, \ldots, x_n)$ in $\mathbb{R}^n$ that satisfy $2x_n > x_1^2 + \cdots + x_{n-1}^2$. This is bounded by a paraboloid, and it is equivalent to the upper half-space model. The *Cayley transform* provides an equivalence between this domain and the unit ball in $\mathbb{R}^n$. This is done in Chapter 2 for complex hyperbolic geometry.

1.1.2   Möbius groups in dimensions 2 and 3

The two- and three-dimensional cases are classical and can be regarded simultaneously. Consider the open 3-ball $\mathbb{B}^3$ and its boundary $\partial\mathbb{B}^3$, which is the 2-sphere, that we regard as being the Riemann sphere $\mathbb{S}^2$, i.e., the usual 2-sphere equipped with a complex structure, making it biholomorphic to the extended complex plane $\widehat{\mathbb{C}} = \mathbb{C} \cup \infty$, also called the *Cauchy* plane.

It is explained in many text books that in this dimension, an orientation preserving diffeomorphism of $\mathbb{S}^2$ is conformal if and only if it is holomorphic. This is essentially a consequence of the Cauchy-Riemann equations. Moreover, every holomorphic automorphism of the Riemann sphere is a Möbius transformation $z \mapsto \frac{az+b}{cz+d}$, where a, b, c, d are complex numbers such that $ad - bc = 1$.

Let us look now at the group $\mathrm{SL}(2, \mathbb{C})$ of 2×2 complex matrices with determinant 1. This group acts linearly on $\mathbb{C}^2$, so it acts on the complex projective line $\mathbb{P}_{\mathbb{C}}^1$ which is biholomorphic to the Riemann sphere $\mathbb{S}^2 \cong \mathbb{C} \cup \infty := \widehat{\mathbb{C}}$. The induced action of $\mathrm{SL}(2, \mathbb{C})$ on $\mathbb{P}_{\mathbb{C}}^1$ is via the Möbius transformations:

$$z \mapsto \frac{az + b}{cz + d} \, .$$

Thus one has a natural projection

$$\mathrm{SL}(2, \mathbb{C}) \longrightarrow \mathrm{Conf}_+(\mathbb{S}^2) \cong \mathrm{Iso}_+(\mathbb{H}_{\mathbb{R}}^3),$$

given by $\begin{pmatrix} a & b \\ c & d \end{pmatrix} \mapsto \dfrac{az + b}{cz + d}$. This is in fact a homomorphism of groups: the product of two matrices in $\mathrm{SL}(2, \mathbb{C})$ maps to the composition of the corresponding Möbius transformations.

It is clear that the above projection is surjective. Furthermore, two matrices in $\mathrm{SL}(2, \mathbb{C})$ define the same Möbius transformation if and only if they differ by multiplication by ± 1. Hence the group $\mathrm{PSL}(2, \mathbb{C}) \cong \mathrm{SL}(2, \mathbb{C})/\{\pm I\}$ can be identified with the group of all Möbius transformations $\{\frac{az+b}{cz+d}\}$, which is isomorphic to the group of orientation preserving isometries of the hyperbolic 3-space. This coincides with the group of holomorphic automorphisms of the Riemann sphere; it also coincides with $\mathrm{Conf}_+(\mathbb{S}^2)$, the group of orientation preserving conformal automorphisms on $\mathbb{S}^2$.

Summarizing:

Theorem 1.1.11. *One has the following isomorphisms of groups:*

$$\mathrm{Iso}_+(\mathbb{H}^3_{\mathbb{R}}) \cong \underset{+}{\mathrm{M\ddot{o}b}}(\mathbb{B}^3) \cong \mathrm{Conf}_+(\mathbb{S}^2)$$

$$\cong \left\{ \frac{az+b}{cz+d} \; ; \; a, b, c, d, \in \mathbb{C} \, , \, ad - bc = 1 \right\} \cong \mathrm{PSL}(2, \mathbb{C}) \, .$$

Now recall that a Möbius transformation $\frac{az+b}{cz+d}$ with $ad - bc = 1$ preserves the upper half plane $\mathcal{H} \subset \mathbb{C}$ if and only if a, b, c, d are real numbers. These correspond to compositions of inversions in $\widehat{\mathbb{C}} = \mathbb{C} \cup \infty$ on circles (or lines) orthogonal to the x-axis. Hence these are isometries of $\mathbb{H}^2_{\mathbb{R}}$ and one has:

Theorem 1.1.12.

$$\mathrm{Iso}_+(\mathbb{H}^2_{\mathbb{R}}) \cong \underset{+}{\mathrm{M\ddot{o}b}}(\mathbb{B}^2) \cong \left\{ \frac{az+b}{cz+d} \; ; \; a, b, c, d, \in \mathbb{R} \, , \, ad - bc = 1 \right\} \cong \mathrm{PSL}(2, \mathbb{R}) \, .$$

1.1.3   Geometric classification of the elements in $\mathrm{Iso}_+(\mathbb{H}^n_{\mathbb{R}})$

We now classify the elements of $\mathrm{Iso}_+(\mathbb{H}^n_{\mathbb{R}})$ in terms of their fixed points. We start with the case $n = 2$ which is classical and there is a vast literature about this topic (see for instance [144], [19], [184]). By the theorem above, an isometry of the hyperbolic plane can be regarded as a Möbius transformation T given by $z \mapsto \frac{az+b}{cz+d}$ with $a, b, c, d, \in \mathbb{R}$ and $ad - bc = 1$. The fixed points of T are the points where $T(z) = z$. These are the solutions of the equation

$$z = \frac{(a - d) \pm \sqrt{(d - a)^2 + 4bc}}{2c} \, .$$

Since the coefficients a, b, c, d are all real numbers we have the following three possibilities:

(i) $(d - a)^2 + 4bc < 0$;

(ii) $(d - a)^2 + 4bc = 0$;

(iii) $(d-a)^2 + 4bc > 0$.

Assuming, as we do, $ad - bc = 1$, we have:

$$(d-a)^2 + 4bc = (a+d)^2 - 4,$$

and $a + d$ is the trace of the matrix $\begin{pmatrix} a & b \\ c & d \end{pmatrix}$, so we call $\mathrm{Tr}(T) := a + d$ *the trace* of T. Then the three cases above can be written as::

(i) $0 \leq \mathrm{Tr}^2(T) < 4$. The map T is called *elliptic*;

(ii) $\mathrm{Tr}^2(T) = 4$. The map T is called *parabolic*

(iii) $\mathrm{Tr}^2(T) > 4$. The map T is called *hyperbolic*.

In the first case the map T has one fixed point in $\mathbb{H}^2_{\mathbb{R}}$, regarded as the upper half-plane $\mathcal{H} = \{\Im z > 0\}$; the other fixed point is the complex conjugate of the previous one, so it is in the lower half-plane. In the second case T has only one fixed point (of multiplicity 2) and this is contained in the x-axis (union ∞), which is the "boundary" of the hyperbolic plane, the sphere at infinity. In the third case T has two distinct fixed points, both contained in the sphere at infinity.

If T is elliptic, then one can conjugate it by an automorphism of the Riemann sphere to make it have its fixed points at 0 and ∞, and T becomes a rotation around the origin, $T(z) = e^{i\theta} z$.

If T is parabolic, then it is conjugate in $\mathrm{PSL}(2, \mathbb{C})$ to a map of the form $S(z) = z + k$, with $k \in \mathbb{R}$ constant. This map is a translation and has ∞ as its fixed point.

If T is hyperbolic, then it is conjugate in $\mathrm{PSL}(2, \mathbb{C})$ to a map of the form $S(z) = \lambda^2 z$, with λ real and $\neq \pm 1$. This map has 0 and ∞ as fixed points and all other points move along straight lines through the origin. This description is good in some sense, but it is not satisfactory because the map S does not preserve $\mathcal{H}$, which is our model for $\mathbb{H}^2_{\mathbb{R}}$. To describe its dynamics in $\mathcal{H}$ it is better to consider its fixed points x_1, x_2, and assume for simplicity that both are finite and contained in the real-axis. These two points determine a unique geodesic in $\mathbb{H}^2_{\mathbb{R}}$, namely the unique half-circle in $\mathcal{H}$ with end-points x_1, x_2 and meeting orthogonally the x-axis. This geodesic is invariant under T. Moreover, given any other point $x \in \mathcal{H}$, there is a unique circle passing through x_1, x_2 and x. These circles fill out the whole space $\mathbb{C}$ and they are invariant under T, so they are unions of orbits. When the fixed points are taken to be 0 and ∞, these circles become the straight lines through the origin, or the meridians through the North and South poles if we think of T as acting on the Riemann sphere.

If we consider now an isometry T of $\mathbb{H}^3_{\mathbb{R}}$ and we think of it as a Möbius transformation with (possibly) complex coefficients, then we have again three possibilities:

(i) The map T has two distinct fixed points which are both complex conjugate numbers. In this case T is said to be *elliptic*, as before. Again, T is conjugate in $\mathrm{PSL}(2, \mathbb{C})$ to a rotation.

(ii) The map T has only one fixed point which is real. In this case T is said to be *parabolic* and it is conjugate in $\mathrm{PSL}(2, \mathbb{C})$ to a translation.

(iii) The map T has two distinct fixed points which are both real numbers. In this case, as before, T is conjugate in $\mathrm{PSL}(2, \mathbb{C})$ to a map of the form $z \mapsto \lambda^2 z$, but this time λ can be a complex number with $|\lambda| \neq 1$. In this case T is said to be *loxodromic*. Now T leaves invariant the geodesic in $\mathbb{H}^3$ that has end-points at the fixed points of T and the dynamics of all other points is a translation along that geodesic, together with a rotation around it. The number λ is called the *multiplier* of T. When this number is real the map is said to be *hyperbolic*, and in that case there is no rotation, only translation along the geodesic.

In order to give a similar classification in higher dimensions it is convenient to look at these transformations "from the inside" of the hyperbolic space $\mathbb{H}_{\mathbb{R}}^n$ (see [105] for a deeper and more complete description of this classification). Let T be an isometry of $\mathbb{H}_{\mathbb{R}}^n$ and pick up a point $p \in \mathbb{H}_{\mathbb{R}}^n$ such that the points p, $T(p)$ and $T^2(p)$ are not in a Euclidean straight line. Let $\mathcal{L}$ be the line that bisects the angle that they form, and look at the lines $T^{-1}(\mathcal{L})$, $\mathcal{L}$ and $T(\mathcal{L})$. There are three possibilities:

(i) These three lines intersect in $\mathbb{H}_{\mathbb{R}}^n$.

(ii) These three lines intersect at the $(n-1)$-sphere at infinity of $\mathbb{H}_{\mathbb{R}}^n$.

(iii) These three lines do not intersect, neither in $\mathbb{H}_{\mathbb{R}}^n$ nor at the sphere at infinity.

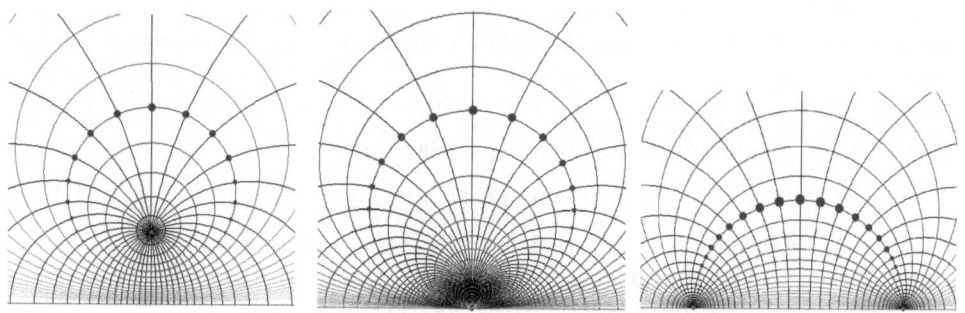

Figure 1.1: The three types of isometries

In the first case T has a fixed point at the meeting point of the three lines. The map T is said to be *elliptic*. These maps form an open set in $\mathrm{Iso}(\mathbb{H}_{\mathbb{R}}^n)$.

In the second situation the three lines are *parallel* in hyperbolic space and one has a fixed point at infinity. The map is a translation and it is said to be *parabolic*; this can be regarded as a limit case between the other two.

The last case is when the lines are *ultra-parallel*, i.e., they do not meet in $\overline{\mathbb{H}}_{\mathbb{R}}^n$ (see Thurston's book for more on the topic). Now T leaves invariant the geodesic γ that minimises the length between the lines L and $T(L)$. In this case T is a translation along γ and a rotation around it. The end-points of γ are fixed points of T. These maps are called *loxodromic* (or just *hyperbolic*) and they also form an open set in $\text{Iso}(\mathbb{H}_{\mathbb{R}}^n)$.

Remark 1.1.13. In Chapter 2 we will see that in the case of the complex hyperbolic space, one has a similar classification of its isometries into elliptic, parabolic and loxodromic (or hyperbolic), depending on their dynamics. Actually the classification of the isometries into these three categories holds in the more general setting of the isometries of spaces of nonpositive curvature, and even more generally (for instance for CAT(0)-spaces, see Remark 2.4.6). We refer to [13], [32] for clear accounts on this subject.

1.1.4   Isometric spheres

An isometric sphere for a Möbius transformation g is an sphere $S \subset \mathbb{S}^n \setminus \{\infty\}$ where g behaves like an isometry with respect to the euclidian metric. Such a geometric object helps to the understanding of the dynamics of the transformation, as for instance for the construction of fundamental domains and to proving important convergence properties when we iterate the map.

More precisely, let us think of the n-sphere as being $\mathbb{R}^n \cup \{\infty\}$ and assume g satisfies $g(\infty) \neq \infty$, which we can always do up to conjugation. Set $p = g^{-1}(\infty)$ and $q = g(\infty)$. If $p \neq q$ we let τ be the reflection through the hyperplane which contains the point $(p+q)/2$ and is orthogonal to the line which passes through p and q. Otherwise, if $p = q$, we define $\tau = \text{Id}$. For each $R > 0$, let S_R be the sphere with centre p and radius R, let σ_R be the inversion with respect to S_R, and consider the map

$$b = g^{-1} \tau \sigma_R.$$

Then $b(\infty) = \infty$ and $b(p) = p$. Thus b is of the form

$$b(x) = \lambda_R O_R(x - p) + p, \tag{1.1.13}$$

for some $\lambda_R > 0$ and $O_R \in O(n)$. Hence $g(S_R)$ is a sphere with radius $R\lambda_R$ and centre q. So g carries the spheres with centre at p into spheres with centre at q.

Now take $R = 1$, then from (1.1.13) one gets

$$g(\lambda_1 O_1(x - p) + p) = \tau \sigma_1.$$

Thence the image under g of the sphere with centre p and radius r is the sphere with centre q and radius λ_1/r, so we arrive at the following proposition-definition (we refer to [19] for a proof of this proposition).

Definition-Proposition 1.1.14. The unique sphere centred at p whose image under g is a sphere with the same radius is called the *isometric sphere*. It will be denoted by I_g.

We recall that a transformation g is said to be an involution if g^2 is the identity.

Proposition 1.1.15. *Let g be a Möbius transformation different from the identity and such that ∞ is not fixed by g. Then g is of the form*

$$g = \tau \sigma O,$$

where O is a rotation with centre at $g^{-1}(\infty)$, σ is the reflection on the isometric sphere, τ is the identity if g is an involution or else it is the reflection on the plane orthogonal to the line which joins $g(\infty)$ and $g^{-1}(\infty)$ and contains the point $(g(\infty) + g^{-1}(\infty))/2$.

The following result enables us to measure the "distortion" of sets under the action of Möbius transformations, see [19] for a proof.

Theorem 1.1.16. *Let g be a Möbius transformation such that $g(\infty) \neq \infty$ and let R_g be the radius of the corresponding isometric sphere. If E is a closed set such that $g^{-1}(\infty) \notin E$, we have:*

$$\mathrm{diam}_{\mathrm{Eucl}}(g(E)) \leq 2R_g^2/\rho,$$

where ρ denotes the Euclidean distance between $g^{-1}(\infty)$ and E. Moreover, if $\infty \in E$ we also have

$$\mathrm{diam}_{\mathrm{Eucl}}(g(E)) \geq R_g^2/\rho.$$

As a consequence of this theorem one gets:

Theorem 1.1.17. *Let (g_m) be a sequence of Möbius transformations acting on $\mathbb{S}^n$ such that there is an $r > 0$ so that if we set $U = \{x \in \mathbb{R}^n : |x| > r\}$, then the sequence satisfies that $g_k^{-1}(g_m(U)) \cap U = \emptyset$ for each $k \neq m$. Let r_m denote the radius of the isometric sphere of g_m. Then*

$$\lim_{m \to \infty} r_m = 0.$$

Proof. Let δ_m be the distance between U and $g_m^{-1}(\infty)$. Thus

$$\mathrm{diam}(g_m(U)) \geq r_m^2/\delta_m \geq r_m^2/r.$$

Since $g_m(U)$ is an Euclidean ball, we deduce that its radius should be at least $r_m^2/2r$. In particular,

$$\mathrm{Vol}_{\mathrm{Eucl}}(g_m(U)) \geq (1/2^m r^m) r_m^{2m}.$$

We now observe that the balls $g_m(U)$ form a bounded set and they are pairwise disjoint. This implies that

$$\infty > \sum_{m>1} \mathrm{Vol}_{\mathrm{Eucl}}(g_m(U)) \geq \sum_{m>1} (1/2^m r^m) r_m^{2m},$$

and the result follows. $\qquad\square$

The above results will be used in the sequel for studying convergence properties of the orbits of points under the action of a discrete group.

1.2   Discrete subgroups

In this section we look at discrete subgroups of $\mathrm{Iso}(\mathbb{H}^n_{\mathbb{R}})$, i.e., subgroups of $\mathrm{Iso}(\mathbb{H}^n_{\mathbb{R}})$ which are discrete as topological subspaces. This is equivalent to saying that the identity element has a neighbourhood in $\mathrm{Iso}(\mathbb{H}^n_{\mathbb{R}})$ which has no element of the subgroup other than the identity. Every such subgroup is necessarily countable.

We give some basic definitions and several important examples of such groups. We start the section with a discussion on discontinuous actions.

1.2.1   Properly Discontinuous Actions

In this subsection G is assumed to be a group acting on a smooth manifold M by diffeomorphisms. Recall that the *stabiliser* of a point $x \in M$, also called the *isotropy*, is the subgroup $G_x \subset G$ defined by

$$G_x = \{g \in G \,|\, g(x) = x\}\,;$$

The *orbit* of x under the action of G is the set:

$$Gx = \{y \in M \,|\, y = g(x) \text{ for some } g \in G\}\,.$$

Definition 1.2.1. The action of G is *discontinuous at* $x \in M$ if there is a neighbourhood U of x such that the set

$$\{g \in G \,|\, gU \cap U \neq \varnothing\}$$

is finite. The set of points in M at which G acts discontinuously is called the *region of discontinuity*. This set is also called *the regular set* of the action. The action is *discontinuous on* M if it is discontinuous at every point in M.

The following proposition provides an equivalent definition of discontinuous actions.

Proposition 1.2.2. *The group G acts discontinuously on M if and only if for each $x \in M$ the isotropy group is finite and for every compact subset K of M if follows that*

$$\mathrm{card}(K \cap Gx) < \infty\,.$$

We have the following well-known result.

Proposition 1.2.3. *If the G-action on M is discontinuous, then the G-orbits have no accumulation points in M. That is, if (g_m) is a sequence of distinct elements of G and $x \in M$, then the sequence $(g_m(x))$ has no limit points. Conversely, if G satisfies this condition, then G acts discontinuously on M.*

One gets the following corollary, which is specially relevant for this monograph since it applies to all the cases we envisage:

Corollary 1.2.4. *Let* $\mathrm{Diff}\,(M)$ *denote the group of diffeomorphisms of* M *endowed with the compact-open topology. If* G *acts discontinuously on* M *by diffeomorphisms, then* G *is a discrete subset of* $\mathrm{Diff}\,(M)$.

Notice that for the case when $f : M \to M$ is a diffeomorphism of a smooth manifold, G. D. Birkhoff introduced in 1927 the notion of wandering and nonwandering points of f (see [26] or [209, p. 749]). By definition, $x \in M$ is called a *wandering point* of f when there is a neighbourhood U of x such that $\cup_{|m|>0} f^m(U) \cap U = \emptyset$, where for $m > 0$, f^m denotes the mth iterate $f \circ \cdots \circ f$, while for $m < 1$, f^m denotes the mth iterate of f^{-1}. A point is called *nonwandering* if it is not a wandering point. Of course we can think of the family of diffeomorphisms $\{f^m\}_{m\in\mathbb{Z}}$ as defining an action of the integers $\mathbb{Z}$ on M, with $f^0 \equiv \mathrm{Id}$. In this case the wandering points are precisely the points in M where the action is discontinuous.

Definition 1.2.5. Let G be as before and consider a subgroup $H \subset G$. We say that a set $Y \subset M$ is *precisely invariant* under H if for all $h \in H$ and for all $g \in G \setminus H$ one has

$$h(Y) = Y \quad \text{and} \quad g(Y) \cap Y = \emptyset.$$

Proposition 1.2.6. *The group* G *acts discontinuously on* M *if and only if for all* $x \in M$ *one has that the stabiliser* G_x *is finite and there exists a neighbourhood* U *of* x *that is precisely invariant under* G_x.

Proof. It is clear that if for all $x \in M$ one has that the stabiliser G_x is finite and there exists a neighbourhood U of x that is precisely invariant under G_x, then the action is discontinuous. Conversely, let us assume that G acts discontinuously on M and let $x \in M$. Then there is a neighbourhood U of x such that $H_x = \{g \in G \mid gU \cap U \neq \emptyset\}$ is finite. Trivially the stabliser G_x is finite and contained in H_x. Now, since M is a Hausdorff space, for each $g \in H_x \setminus G_x$ let U_g and U_{gx} be open neighbourhoods of x and gx respectively, such that $U_g \cap U_{gx} = \emptyset$. Set

$$W = \bigcap_{g \in H_x \setminus G_x} (U \cap U_g) \cap g^{-1}(U_{gx} \cap gU).$$

Then W is an open neighbourhood of x which satisfies

$$\{g \in G \mid gW \cap W \neq \emptyset\} = G_x.$$

Finally, set $\widetilde{W} = \bigcap_{g \in G_x} gW$. Clearly $\widetilde{W}$ is an open neighbourhood of x which is precisely invariant under G_x. □

Remark 1.2.7. Some authors use the name *properly discontinuous actions* for group actions satisfying that for all $x \in X$ one has that the stabiliser G_x is finite and there exists a neighbourhood U of x that is precisely invariant under G_x. It is clear that such an action is necessarily discontinuous. However, it is easy to construct examples of discontinuous actions which fail to satisfy that all points have a neighbourhood which is precisely invariant under the stabiliser. For this statement to be true we need to impose certain conditions on the topology of X. We remark too that in some places in the literature, the term *properly discontinuous actions* actually means a discontinuous action which is also free, i.e., all the stabilisers are trivial (see for instance [114] and [211]). Yet, in the modern literature the term *"properly discontinuous actions"* has a different meaning, that we explain in the definition below. This is what we will understand in this monograph by a properly discontinuous action.

Definition 1.2.8. Let G act on the manifold M by diffeomorphisms. The action is said to be *properly discontinuous* if for each nonempty compact set $K \subset M$ the set

$$\{g \in G \,|\, gK \cap K \neq \varnothing\},$$

is finite.

It is clear that every properly discontinuous action is *a fortiori* discontinuous. In Example 1.2.11 below we show that the converse statement is false generally speaking. Notice also that in the literature, some authors say that an action satisfies *the Sperner's condition* if it satisfies the condition stated in Definition 1.2.8 (see for instance [114]).

The following propositions provide equivalent ways of defining properly discontinuous actions:

Proposition 1.2.9. *The group G acts properly discontinuously on M if and only if for every pair of compact subsets K_1, K_2 of M, there are only a finite number of elements $g \in G$ such that $g(K_1) \cap K_2 \neq \emptyset$.*

Proposition 1.2.10. *Let G act properly discontinuously on M. Then the orbits of the action on compact sets have no accumulation points. That is, if (g_m) is a sequence of distinct elements of G and $K \subset M$ is a nonempty compact set, then the sequence $(g_m(K))$ has no limit points. Conversely, if G satisfies this condition, then G acts properly discontinuously on M.*

It is clear that the second proposition implies the previous one. We refer to [114] for the proof of Proposition 1.2.10.

The following example shows that discontinuous actions are not necessarily properly discontinuous:

Example 1.2.11. Let G be the cyclic group induced by the transformation $g : \mathbb{C}^2 \to \mathbb{C}^2$ given by $g(z, w) = (\frac{1}{2}z, 2w)$. Clearly G acts discontinuously on $\mathbb{C}^2 \setminus \{0\}$, but if we let S be the set

$$S = \{(z, w) \in \mathbb{C}^2 \,\big|\, |z| = |w| = 1\},$$

then the set of cluster points of its orbit is $\{(z, w) \in \mathbb{C}^2 \,|\, z = 0\} \cup \{(z, w) \in \mathbb{C}^2 \,|\, w = 0\}$, the union of the two coordinate axis.

Notice that if G acts discontinuously and freely on a manifold M, then the quotient map $\pi : M \to M/G$ is a covering map and the group of automorphisms of the covering is G itself, $\mathrm{Aut}(M \to M/G) = G$. Yet, the example above shows that the quotient M/G may not be a Hausdorff space, even if the action is free and discontinuous. We notice that in this same example, the axis is the set of accumulation points of the orbits of compact sets in $\mathbb{C}^2 \setminus \{(0,0)\}$. If we remove the axis, we get a a properly discontinuous action on their complement, and in that case the quotient is indeed Hausdorff. This is a general fact for properly discontinuous actions (see [114], [132]).

The proposition below says that in the case of conformal automorphisms of the n-sphere, every discontinuous action is *a fortiori* properly discontinuous. That is, in this setting both concepts are equivalent (we refer to [185] for a proof).

Proposition 1.2.12. *Let M be an n-sphere and G a discrete group of conformal maps of M. Then the regular set of G is the largest open set of M where the action of G is properly discontinuous.*

In the next chapter we will see that similar statements hold for discrete subgroups of isometries of complex hyperbolic space. Similar, but weaker, statements hold also for discrete subgroups of $\mathrm{PSL}(n, \mathbb{C})$ provided one considers the appropriate region of discontinuity (see Chapter 3).

1.2.2 The limit set and the discontinuity region.

We consider again a subgroup G of $\mathrm{Iso}(\mathbb{H}_{\mathbb{R}}^{n+1})$, and we think of it as acting on $\overline{\mathbb{H}}_{\mathbb{R}}^{n+1} := \mathbb{H}_{\mathbb{R}}^{n+1} \cup \mathbb{S}_{\infty}^{n}$. In this case Definition 1.2.1 becomes:

Definition 1.2.13. The *region of discontinuity* of G is the set $\Omega = \Omega(G)$ of all points in $\overline{\mathbb{H}}_{\mathbb{R}}^{n+1}$ which have a neighbourhood that intersects only finitely many copies of its G-orbit.

The following result is in some sense analogous to Montel's theorem. This is the *convergence property* in M. Kapovich's language (see for instance [105] or [106, p. 495]), and it enables us to prove basic properties about actions of groups on the hyperbolic space. We will see later that the same property holds in complex hyperbolic geometry and this is why the Chen-Greenberg limit set of complex hyperbolic discrete groups shares most of the properties one has for the limit set of discrete groups in real hyperbolic geometry.

Lemma 1.2.14. *Let (γ_m) be a sequence of distinct elements of a discrete group $G \subset \mathrm{M\ddot{o}b}(\mathbb{S}^n)$. Then either it contains a convergent subsequence, or it converges to a constant map away from a point in $\mathbb{S}^n$. That is, there exist a subsequence, still denoted by (γ_m), and points $x, y \in \mathbb{S}^n$ such that:*

(i) *γ_m converges uniformly to the constant function y on compact sets of $\mathbb{S}^n - \{x\}$.*

(ii) γ_m^{-1} *converges uniformly to the constant function* x *on compact sets of* $\mathbb{S}^n -$
$\{y\}$.

To prove this result we use the following lemma:

Lemma 1.2.15. *Think of hyperbolic space* $\mathbb{H}_\mathbb{R}^{n+1}$ *as being the unit* $(n + 1)$-*ball equipped with the hyperbolic metric; its boundary is* $\mathbb{S}^n$. *Let* (γ_m) *be a sequence of distinct elements of a discrete group* $G \subset \mathrm{Iso}(\mathbb{H}_\mathbb{R}^{n+1})$. *Then the set of accumulation points of the orbits* $(\gamma_m)(x)$ *is contained in* $\mathbb{S}^n$.

Proof. Let us assume that the lemma is false and there is a subsequence of (γ_m), still denoted by (γ_m), and $y \in \mathbb{H}^{n+1}$ such that $\gamma_m(\infty) \xrightarrow[n\to\infty]{} y$.

We can equip $\mathbb{H}_-^{n+1} = \{x \in \mathbb{R}^{n+1} \mid |x| > 1\} \cup \{\infty\}$ with a metric of constant curvature -1 for which the group of preserving orientation isometries is given by $\mathrm{M\ddot{o}b}(\mathbb{H}_\mathbb{R}^{n+1})$; denote such metric by d. Now let $z \in \mathbb{H}_-^{n+1}$, then

$$d(\gamma_m(z), y) \leq d(\gamma_m(z), \gamma_m(\infty)) + d(y, \gamma_m(\infty)) = d(z, x) + d(y, \gamma_m(x)).$$

Thus the set $\{\gamma_m(z) : m \in \mathbf{N}\}$ is relatively compact. Since $(\gamma_m) \subset \mathrm{Iso}(\mathbb{H}_-^{n+1})$ the Arzelà-Ascoli theorem yields that there is a subsequence of (γ_m), still denoted by (γ_m), and $\gamma : \mathbb{H}_-^{n+1} \to \mathbb{H}_-^{n+1}$ such that $\gamma_m \xrightarrow[n\to\infty]{} \gamma$ in the compact-open topology. Clearly γ is an isometry and therefore G is nondiscrete. Which is a contradiction. $\square$

Proof of Lemma 1.2.14. By Lemma 1.1.3 we may assume that $G \subset \mathrm{Iso}(\mathbb{H}_\mathbb{R}^{n+1})$. Now consider the centres of the isometric spheres I_m of γ_m and I_{-m} of γ_m^{-1}, i.e., $\gamma^{-1}(\infty)$ and $\gamma_m(\infty)$. By Lemma 1.2.15 we can assume these sequences are convergent and if we say that that $g_m^1(\infty)$ converges to y and $g_m(\infty)$ converges to x, we must have that $x, y \in \partial(\mathbb{H}_\mathbb{R}^{n+1})$. Let $z \in \mathbb{R}^{n+1} \setminus \{y\}$ and W be a neighbourhood of z such that $d_{Euc}(W, y) > 0$. By Theorem 1.1.17 and Lemma 1.2.15, the radius of the isometric spheres tends to zero. Thus we can assume that W is in the outside of all the isometric spheres. Finally, by Proposition 1.1.15 we know that g_m sends the outside of I_m to the inside of I_{-m}, so the result follows. $\square$

Theorem 1.2.16. *Let* G *be a subgroup of* $\mathrm{Iso}(\mathbb{H}_\mathbb{R}^{n+1})$. *The following three conditions are equivalent:*

(i) *The subgroup* $G \subset \mathrm{Iso}(\mathbb{H}_\mathbb{R}^{n+1})$ *is discrete.*

(ii) *The region of discontinuity of* G *in* $\mathbb{H}_\mathbb{R}^{n+1}$ *is all of* $\mathbb{H}_\mathbb{R}^{n+1}$.

(iii) *The region of discontinuity of* G *in* $\mathbb{H}_\mathbb{R}^{n+1}$ *is nonempty.*

Proof. It is clear that (ii) $\implies$ (iii) $\implies$ (i). Let us prove that (i) implies (ii). Let K be a compact set and assume that $K(G) = \{\gamma \in G : \gamma K \cap K \neq \emptyset\}$ is countable. Then by Lemma 1.2.14 there is a sequence $(\gamma_m) \subset K(G)$ and points $x, y \in \partial\mathbb{H}_\mathbb{R}^{n+1}$ such that γ_m converges uniformly to y on compact sets of $\overline{\mathbb{H}}_\mathbb{R}^{n+1} - \{x\}$. Let U be a neighbourhood of y disjoint from K. Then there is a natural number n_o such that $\gamma_m(K) \subset U$ for $m > n_o$. In particular we deduce $\gamma_m(K) \cap K = \emptyset$ for all $m > n_0$, which is a contradiction, and the result follows. $\qquad\square$

Notice that by continuity, it is clear that if the region of discontinuity of G in $\mathbb{S}^n$ is nonempty, then the region of discontinuity of G in $\mathbb{H}_{\mathbb{R}}^{n+1}$ is nonempty and therefore G is discrete.

Definition 1.2.17. Let G be a discrete subgroup of $\mathrm{Iso}(\mathbb{H}^{n+1})$. The *limit set* of G, denoted by $\Lambda(G)$ or simply Λ, is the set of accumulation points in $\overline{\mathbb{H}}_{\mathbb{R}}^{n+1}$ of orbits of points in $\mathbb{H}_{\mathbb{R}}^{n+1}$.

One has:

Theorem 1.2.18. *Let G be as above. Then the limit set is contained in the sphere at infinity $\mathbb{S}_{\infty}^n = \partial \mathbb{H}_{\mathbb{R}}^{n+1}$ and is independent of the choice of orbit.*

Proof. Let $x, y \in \mathbb{H}_{\mathbb{R}}^{n+1}$ and p a cluster point of Gy. Then there exists a sequence $(g_m) \subset G$ such that $g_m(y)$ converges to p. By Lemma 1.2.14 it follows that q also is a cluster point of $(g_m(x))$, which ends the proof. $\qquad\square$

Theorem 1.2.19. *Let G be a discrete subgroup of $\mathrm{Iso}(\mathbb{H}_{\mathbb{R}}^{n+1})$. The limit set of G is the complement of the region of discontinuity in $\mathbb{S}_{\infty}^n$.*

Proof. First we show that Λ lies in the complement of the region of discontinuity. Let $y \in \Lambda(G)$, then there is a point $p \in \mathbb{H}_{\mathbb{R}}^{n+1}$ and a sequence (γ_m) such that $\gamma_m(p) \to x$. From Lemma 1.2.14 we conclude that there exists $x \in \partial \mathbb{H}_{\mathbb{R}}^{n+1}$ such that we can assume that γ_m converges uniformly to the constant y on compact sets of $\overline{\mathbb{H}}_{\mathbb{R}}^{n+1} - \{x\}$. Let U be any neighbourhood of y. Then there is a natural number m_0 for which $\gamma_m(y) \in U$ for $m \geq m_0$.

Now let $q \in \partial \mathbb{H}_{\mathbb{R}}^{n+1}$ be a point in the discontinuity region, and assume that $q \in \Lambda(G)$. By the previous argument we deduce that q does not belong to the discontinuity region, which is a contradiction. Hence the discontinuity region is contained in the complement of $\Lambda(G)$. In others words, the complement of the discontinuity region is contained in $\Lambda(G)$. $\qquad\square$

It is clear from its definition that the limit set $\Lambda(G)$ is a closed G-invariant set, and it is empty if and only if G is finite (since every sequence in a compact set contains convergent subsequences).

Definition 1.2.20. Let G be a group acting on a manifold X. The *equicontinuity region* of G, denoted $\mathrm{Eq}\,(G)$, is the set of points $z \in X$ for which there is an open neighbourhood U of z such that $G\,|_U$ is a normal family.

Recall that a collection of transformations is a *normal family* if and only if every sequence of distinct elements has a subsequence which converges uniformly on compact sets.

One has:

Theorem 1.2.21. *Let G be a discrete subgroup of* $\mathrm{Iso}(\mathbb{H}_\mathbb{R}^{n+1})$. *Then the equicontinuity region of* G *coincides with the discontinuity region* $\mathbb{S}^n \setminus \Lambda(G)$.

Proof. Observe that by Lemma 1.2.14, it is enough to show that $\mathrm{Eq}\,(G) \subset \Omega(G)$. Let $x \in \mathrm{Eq}\,(G)$ and assume that $x \in \Lambda(G)$, thus by Lemma 1.2.14 there is a sequence (γ_m) and a point y such that γ_m converges uniformly to the constant function y on compact sets of $\overline{\overline{\mathbb{H}}}_\mathbb{R}^{n+1} - \{x\}$. Since $x \in \mathrm{Eq}\,(G)$, it follows that γ_m converges uniformly to the constant function y on $\overline{\overline{\mathbb{H}}}_\mathbb{R}^{n+1}$. Let $q \in \mathbb{H}_\mathbb{R}^{n+1}$ and U be a neighbourhood of y such that $U \cap \mathbb{H}_\mathbb{R}^{n+1} \subset \mathbb{H}_\mathbb{R}^{n+1} - \{q\}$. The uniform convergence implies that there is a natural number n_0 such that $\gamma_m(\overline{\overline{\mathbb{H}}}_\mathbb{R}^{n+1}) \subset U \cap \mathbb{H}_\mathbb{R}^{n+1} \subset \mathbb{H}_\mathbb{R}^{n+1} - \{q\}$ for each $m > m_0$. This is a contradiction since each γ_m is a homeomorphism. $\qquad\square$

We have:

Theorem 1.2.22. *Let G be discrete group such that its limit set has more than two points, then it has infinitely many points.*

Proof. Assume that $\Lambda(G)$ is finite with at least three points. Thus

$$\widetilde{G} = \bigcap_{x \in \Lambda(G)} \mathrm{Isot}(x, G)$$

is a normal subgroup of G with finite index. Moreover, since each element in $\mathrm{Iso}(\mathbb{H}_\mathbb{R}^{n+1})$ has at most two fixed points in $\partial\mathbb{H}_\mathbb{C}^{n+1}$ we conclude that $\widetilde{G}$ is trivial and therefore G is finite, which is a contradiction. $\qquad\square$

Definition 1.2.23. The group G is *elementary* if its limit set has at most two points.

Theorem 1.2.24. *If G is not an elementary group, then its action on the limit set is minimal. That is, the closure of every orbit in $\Lambda(G)$ is all of $\Lambda(G)$.*

Proof. Let $x, y \in \Lambda(G)$, then there is a sequence $(g_m) \subset G$ and a point $p \in \mathbb{H}_\mathbb{R}^{n+1}$ such that $g_m(p)$ converges to y. By Lemma 1.2.14 there is a point $q \in \partial\mathbb{H}_\mathbb{R}^{n+1}$, such that we can assume that g_m converges uniformly to y on compact sets of $\overline{\overline{\mathbb{H}}}_\mathbb{R}^{n+1}$. Now, it is well know (see [19]) that there is a transformation $g \in G$ such that $g(x) \neq x$. thus we can assume that $x \neq q$ and therefore we conclude that $g_m(x)$ converges to y. $\qquad\square$

Corollary 1.2.25. *If G a nonelementary Kleinian group, then $\Lambda(G)$ is a nowhere dense perfect set.*

In other words, if G is nonelementary, then $\Lambda(G)$ has empty interior and every orbit in the limit set is dense in $\Lambda(G)$.

Remark 1.2.26. It is noticed in [202] that if the limit set of a nonelementary conformal group acting on $\mathbb{S}^n$ is a compact smooth k-manifold N, for some $0 < k \leq n$, then N is a round sphere $\mathbb{S}^k$. The proof, by Livio Flaminio, is a direct consequence, via stereographic projection of $\mathbb{S}^n$ into the tangent plane of $\mathbb{S}^n$ at a hyperbolic fixed point of the group, of the following fact: if M is a closed k-submanifold of $\mathbb{R}^n$ which is invariant under a homothetic transformation, then M is a k-dimensional subspace of $\mathbb{R}^n$.

1.2.3 Fundamental domains

Given a discrete subgroup G of $\mathrm{Iso}(\mathbb{H}_\mathbb{R}^{n+1})$ one has:

Definition 1.2.27. A *fundamental domain* for the action of G on $\mathbb{H}_\mathbb{R}^{n+1}$ is an open set $F \subset \mathbb{H}_\mathbb{R}^{n+1}$ satisfying the following two conditions:

(i) If $\bar{F}$ denotes the (topological) closure of F in $\mathbb{H}_\mathbb{R}^{n+1}$, then for arbitrary distinct elements g_1, g_2 in G one has that $g_1(F)$ and $g_2(\bar{F})$ do not meet, i.e., $g_1(F) \cap g_2(\bar{F}) = \emptyset$.

(ii) The union $\cup_{g \in G}\, g(\bar{F})$ of all the sets $g(\bar{F})$ for all $g \in G$, is the whole space $\mathbb{H}_\mathbb{R}^{n+1}$.

In other words, the various images of $\bar{F}$ by the elements of G cover the whole space and they are pairwise disjoint except for points in the boundary of $\bar{F}$.

Fundamental domains appear frequently in various contexts in geometry and dynamics, not only in hyperbolic geometry, and the definition is the same. They are important for various reasons, but basically for the fact that we can know essentially everything about the action of the group on the region of discontinuity just by looking at the fundamental domain and the way the group relates the points in its boundary. This should become clear in the sequel.

Observe that a fundamental domain for a group G is by no means unique. In particular, if F is a fundamental domain, then every translate $g(F)$ of it by an element in G is also a fundamental domain. In fact there can be entirely distinct fundamental domains for a certain group action.

There are several methods for constructing fundamental domains for discrete group actions, and we refer to [19] for more on this subject. Here we give a method for constructing a fundamental domain for $G \subset \mathrm{Iso}(\mathbb{H}_\mathbb{R}^{n+1})$ which is rather simple. The same method works in general whenever we have a topological group acting on a locally compact metric space where the metric satisfies that, given any two points x_1, x_2, there is a third "middle" point x_0 of same distance to both points x_1, x_2. In the case we envisage here the sides of the fundamental domain F we get are pieces of hyperbolic $n - 1$-planes, so F is said to be *polyhedral*, and it is called a *Dirichlet fundamental domain*, or a *Dirichlet region*, for G.

The construction is easy: we select a point $x_o \in \mathbb{H}_\mathbb{R}^{n+1}$ which is not fixed by any element of G, and consider the set F of those points in $\mathbb{H}_\mathbb{R}^{n+1}$ whose hyperbolic distance d_h to x_o is strictly less than its distance to every other point in the G-orbit of x_o. That is,

$$F = \{x \in \mathbb{H}_\mathbb{R}^{n+1} \mid d_h(x, x_0) < d_h(x, g(x_0)) \text{ for all } g \in G\}.$$

One can show that F is indeed a fundamental domain for G.

This region F is convex in the hyperbolic sense, being the intersection of half spaces. Its boundary is a countable collection of $(n-1)$-dimensional faces which are G-related in pairs, and the face-pairing transforms are a set of generators for G.

We now introduce the important concept of geometrically finite groups. We refer to [105] for a deeper discussion of the subject. For this, we recall that if H is a closed subset of the sphere at infinity $\mathbb{S}^n_\infty := \overline{\mathbb{H}}^{n+1} \setminus \mathbb{H}^{n+1}$, then its *convex hull*, $\mathrm{Hull}(H)$, is the smallest convex subset of $\mathbb{H}^{n+1}$ whose closure in $\overline{\mathbb{H}}^{n+1}$ meets $\mathbb{S}^n_\infty$ exactly at H. When H is the limit set $\Lambda(G)$ of a discrete subgroup G of $\mathrm{Iso}(\mathbb{H}^{n+1})$, and the cardinality of $\Lambda(G)$ is not 1, then $\mathrm{Hull}(\Lambda(G))$ is an invariant set. Now, for all $\epsilon > 0$ we may consider the ϵ-neighbourhood $\mathrm{Hull}_\epsilon(\Lambda(G))$ of the convex hull; this is also an invariant set.

Definition 1.2.28. A discrete subgroup G of $\mathrm{Iso}(\mathbb{H}^{n+1})$ is *geometrically finite* if:

- G is finitely generated; and

- The quotient $\mathrm{Hull}_\epsilon(\Lambda(G))/G$ has finite hyperbolic volume, for some $\epsilon > 0$.

One has the following theorem due to Marden and Thurston in dimension 2. The proof in general is given in [30]:

Theorem 1.2.29. *A discrete subgroup $G \subset \mathrm{Iso}(\mathbb{H}^{n+1}_{\mathbb{R}})$ is geometrically finite if it has a convex fundamental domain (not necessarily a Dirichlet domain) with finitely many faces.*

Perhaps because of this important result, often in the literature one finds that geometrically finite groups are defined as those having a convex fundamental domain with finitely many faces. Yet, there are examples in [7] of geometrically finite subgroups of $\mathrm{Iso}(\mathbb{H}^4)$ that do not admit convex fundamental domains with finitely many faces.

It is clear that if a Dirichlet region has finitely many faces, then the group is finitely generated. Furthermore, in dimensions 2 and 3, if a group is geometrically finite, then every fundamental domain for it has finitely many faces and the group is finitely generated (see [140]). In dimension 2 the converse is true: every finitely generated group is geometrically finite. However this is false in dimension 3, by [77]. Moreover, in higher dimensions there are discrete groups for which some fundamental domain is a convex polyhedron with finitely many faces, but there is some other fundamental domain which is a convex polyhedron with infinitely many faces. We will not deal with these situations in this monograph; we refer to the literature for more on the subject (see for instance [6], [7]).

1.2.4 Fuchsian groups

Our first examples of discrete subgroups of isometries of hyperbolic space are the Fuchsian groups:

Definition 1.2.30. A (classical) *Fuchsian group* is a discrete subgroup of PSL(2, ℝ).

From the previous section we know that a Fuchsian group can be equivalently defined to be a discrete subgroup of orientation preserving isometries of the hyperbolic plane $\mathbb{H}^2_\mathbb{R}$.

Example 1.2.31 (Surface groups brummel). Let S be a Riemann surface of genus $g > 1$. By the Riemann-Kobe uniformisation theorem, S is of the form $\mathbb{H}^2_\mathbb{R}/G$ where G is a Fuchsian group which acts freely on $\mathbb{H}^2_\mathbb{R}$, so it has no elliptic elements. The group G is isomorphic to the fundamental group $\pi_1(S)$, acting on $\mathbb{H}^2_\mathbb{R}$ by deck transformations.

Let us give other examples of Fuchsian groups. We think of PSL(2, ℝ) as the group of orientation preserving isometries on the 2-disc $\mathbb{D}$ equipped with the hyperbolic metric.

Example 1.2.32 (Triangle groups). Given integers $p, q, r \geq 2$ such that

$$1/p + 1/q + 1/r < 1 \,,$$

let $T = T_{p,q,r}$ be a triangle in $\mathbb{D}$ bounded by geodesics, with angles π/p, π/q and π/r. Recall these geodesics are segments of circles in $\widehat{\mathbb{R}}^2$ orthogonal to the boundary of $\mathbb{D}$, so we have isometries of $\mathbb{H}^2_\mathbb{R}$ defined by the inversions on these three circles, the "sides" of T.

Let G^* be the group of isometries of $\mathbb{H}^2_\mathbb{R}$ generated by the inversions on the three sides of T. Then it is easy to see that G^* is a discrete subgroup of isometries of $\mathbb{H}^2_\mathbb{R}$ and T is a fundamental domain for this action. However G^* is not a subgroup of PSL(2, ℝ); to get a Fuchsian group one must take the index 2 subgroup G of G^* of orientation preserving maps, i.e., words of even length. In this case, if g_1 is the inversion on one of the sides of T, then the double triangle $F = T \cup g_1(T)$ is a fundamental domain for G.

The action of G has three orbits of fixed points in $\mathbb{H}^2_\mathbb{R}$, which correspond to the three vertices of T. The corresponding isotropy subgroups are cyclic of orders p, q, r respectively. The quotient space $\mathbb{H}^2_\mathbb{R}/G$, which equals F/G, is the 2-sphere with three marked points; it has an orbifold structure. (See Chapter 8.)

Of course one could start with a geodesic polygon P with r sides in $\mathbb{H}^2_\mathbb{R}$ having rational angles, and get a corresponding Fuchsian group with fundamental domain a double polygon. The corresponding quotient space is always the 2-sphere with r marked points.

Notice that in all these examples the corresponding limit set is the whole circle at infinity.

Example 1.2.33 (Ideal Triangle groups). Now take three points in the boundary $\mathbb{S}^1$ of the disc $\mathbb{D}$ and think of it as the sphere at infinity of the hyperbolic plane. Select three distinct (arbitrary) points in this circle $\mathbb{S}^1$ and join them by geodesics in $\mathbb{H}^2_\mathbb{R}$. We get an *ideal triangle* in $\mathbb{H}^2_\mathbb{R}$, that is a triangle $T \subset \mathbb{H}^2_\mathbb{R}$ with vertices at infinity; each pair of sides converging to a vertex are therefore parallel. Now make

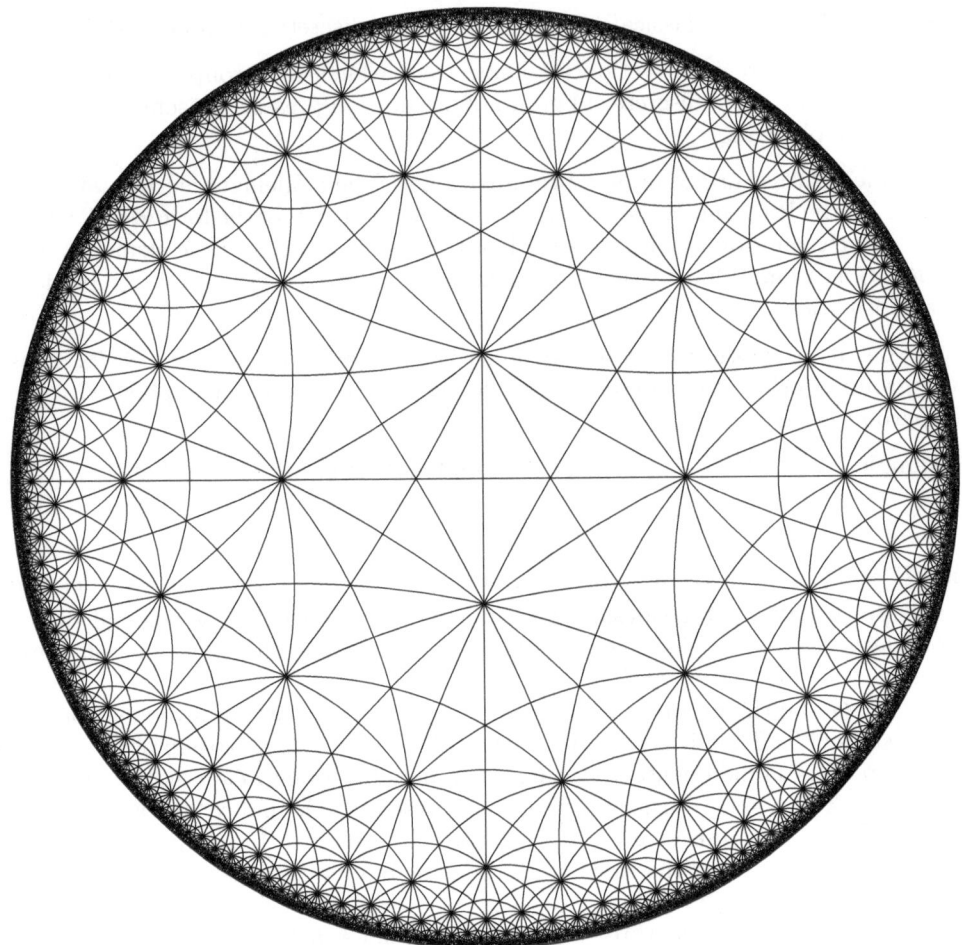

Figure 1.2: The triangle group $\langle 2, 3, 7 \rangle$

the same construction as in the previous example to get a corresponding triangle group G which is Fuchsian, with fundamental domain a double ideal triangle. The quotient space $\mathbb{H}^2_{\mathbb{R}}/G$ is now the Riemann sphere with three punctures. Notice also that just as in the previous example, the limit set is the whole circle at infinity. However, the closure of F is not compact in $\mathbb{H}^2_{\mathbb{R}}$, as it was in the previous example; yet, F has finite hyperbolic area.

We remark that one can think of this example as a limiting case of the previous one, in which we are gradually moving away the circles that define the edges of T, keeping them orthogonal to the boundary $\mathbb{S}^1$, until the circles meet tangentially at three points in $\mathbb{S}^1$. More generally, given the circle $\mathbb{S}^1$, consider an

Figure 1.3: An ideal triangle group

arbitrary family of circles orthogonal to $\mathbb{S}^1$ and such that each circle is tangent to the two adjacent circles, forming a "necklace of pearls". Then the inversions in these circles leave invariant the disc $\mathbb{D}$ and give rise to a Fuchsian group with fundamental domain the union of the polyhedron in $\mathbb{H}_{\mathbb{R}}^2$ that they bound union a copy of it by one of the inversions.

Example 1.2.34. Now continue to "deform" the previous example and separate the circles, so that they are still orthogonal to $\mathbb{S}^1$ but they are pairwise disjoint. Now the fundamental domain of the full group of inversions is the "polygon" bounded by the circles; it hits the circle $\mathbb{S}^1$ at infinity in a fundamental domain for its action

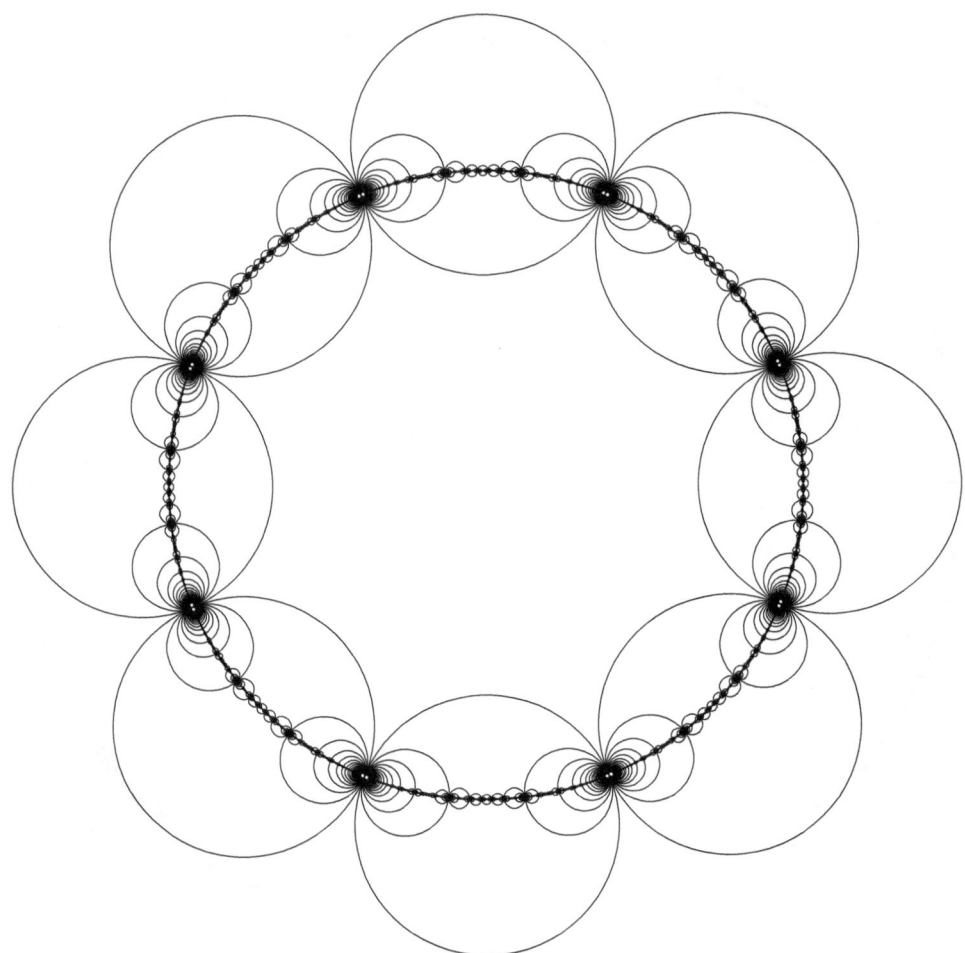

Figure 1.4: A necklace of pearls

on the region of discontinuity in this circle. The limit set is now a Cantor set in the circle. The double polygon is a fundamental domain for the corresponding Fuchsian group, and its limit set coincides with that of the full group of inversions. In this case the region of discontinuity in the circle is nonempty and the fundamental domain in $\mathbb{H}^2_{\mathbb{R}}$ has infinite hyperbolic area.

In general, given a Fuchsian group G and a fundamental domain F one has the following three possibilities, as illustrated by the previous examples:

Case I. The closure of the fundamental domain is compact in $\mathbb{H}^2_{\mathbb{R}}$. The group is said to be *cocompact*.

Case II. The fundamental domain has finite area but its closure hits the circle at infinity at isolated points. The group is said to be *cofinite*.

Groups of the two types described in cases I and II are called *Fuchsian groups of the first kind*. They all have in common that the limit set is the whole circle $\mathbb{S}^1$.

Case III. The fundamental domain has infinite area and its closure hits the sphere at infinity in a fundamental domain for its action there.

This type of groups are called *Fuchsian groups of the second kind*. Their limit set is a Cantor set contained in the circle $\mathbb{S}^1$.

This motivates the following more general definition of a Fuchsian group.

Definition 1.2.35. A *Fuchsian group* is a discrete subgroup of isometries of the hyperbolic n-space, whose limit set is contained in a (round) $(n-2)$-sphere, which is necessarily contained in the sphere at infinity. The group is said to be *quasi-Fuchsian* if its limit set is contained in a quasi-sphere of dimension $n-2$, (i.e., a topological manifold which is a sphere up to a quasi-conformal homeomorphism).

A "round" sphere means a sphere which bounds a totally geodesic $(n-1)$-disc in $\mathbb{H}^n$. We may also think of a Fuchsian group as a discrete group of conformal automorphisms of the $(n-1)$-sphere (the sphere at infinity) such that its limit set is contained in an equator.

1.2.5 Kleinian and Schottky groups

Another classical example of discrete subgroups of hyperbolic isometries is provided by the Schottky groups that we now discuss (see [143] for details). As a historical fact we mention that the classical Schottky groups were introduced by Friederich Schottky in the late 1880s (see [192]), before Klein and Poincaré began the general theory of Kleinian groups. The name Schottky group was coined by Poincaré.

A *Schottky group* is a group generated by Möbius transformations $A_1, \ldots, A_g$, $g \geq 1$, with the following geometric restriction: there is a collection of $2g$ pairwise disjoint regions in $\widehat{\mathbb{C}}$, say $R_1, S_1, \ldots, R_g, S_g$, bounded by Jordan curves, so that $A_j(R_j) = \widehat{\mathbb{C}} - \overline{S_j}$ for all $j = 1, \ldots, g$. The generators $A_1, \ldots, A_g$ are called a Schottky set of generators; the domains $R_1, S_1, \ldots, R_g, S_g$ are a fundamental set of domains, and the integer g is *the genus* of the Schottky group. A Schottky group is *classical* if all the Jordan curves corresponding to some set of generators can be chosen to be circles.

All Schottky groups are finitely generated free groups such that all nontrivial elements are loxodromic. Conversely Maskit showed that any finitely generated free Kleininan group such that all nontrivial elements are loxodromic is a Schottky group.

If we denote by G a Schottky group of genus g, then for $g > 2$ one has that $\Lambda(G)$ is a Cantor set, and the quotient $\Omega(G)/G$ is a closed Riemann surface of

genus g. The converse holds and it is known as the Koebe retrosection theorem
(see [124]):

Theorem 1.2.36 (Retrosection theorem). *Every closed Riemann surface of genus
greater than 1 can be uniformised by a suitable Schottky group.*

Let us mention that Schottky groups may also be characterised as follows:

Theorem 1.2.37 (Maskit, 1967). *A group $G \subset \mathrm{PSL}(2, \mathbb{C})$ is a Schottky group if
and only if G is a finitely generated, purely loxodromic Kleinian group which is
isomorphic to a free group.*

The notion of a Schottky group has also been extended in the following geometric way, which is the starting point for Chapter 9 of this monograph. Consider
an arbitrary family of pairwise disjoint 2-discs $D_1, \ldots, D_r$ in the 2-sphere with
boundaries the circles $C_1, \ldots, C_r$. Let $\iota_1, \ldots, \iota_r$ be the inversions on these r cir-
cles, and let G be the subgroup of $\mathrm{Iso}(\mathbb{H}^3_\mathbb{R}) \cong \mathrm{M\ddot{o}b}(\mathbb{B}^3) \cong \mathrm{Conf}(\mathbb{S}^2)$ generated by
these maps. Then G is called a Schottky group. Its index 2 subgroup of words of
even length is a classical Schottky group in the previous sense.

Notice that G has a nonempty region of discontinuity and the complement
in $\mathbb{S}^2$ of the union $D_1 \cup \cdots \cup D_r$ is a fundamental domain for G.

Definition 1.2.38. A discrete subgroup of $\mathrm{Iso}(\mathbb{H}^{n+1}) \cong \mathrm{Conf}(\mathbb{S}^n)$ is *Kleinian* if it
acts on $\mathbb{S}^n$ (the sphere at infinity) with a nonempty region of discontinuity. We
refer to these as *Conformal Kleinian groups*.

So, the previous discussion shows that Fuchsian and Schottky groups are
special types of Kleinian groups.

Remark 1.2.39. Nowadays the term "Kleinian group" is being often used for an
arbitrary discrete subgroup of hyperbolic motions, regardless of whether or not
the region of discontinuity is empty.

Continuing with the previous Example 1.2.33, choose circles $C_1, \ldots, C_r$ in $\widehat{\mathbb{C}}$
so that there is a common circle C orthogonal to all of them and, moreover, they
form a "necklace of pearls" as at the end of that example. The limit set is the
whole circle C. Now slightly perturb the circles $C_1, \ldots, C_r$, keeping them round
but dropping the condition that they have a common orthogonal circle.

We consider two settings:

(i) We perturb the circles slightly, so that each circle overlaps with its neighbors. Then one has (this is not obvious) that the limit set becomes a fractal curve
of Hausdorff dimension between 1 and 2, and choosing appropriate deformations
one can cover the whole range of Hausdorff dimension between 1 and 2. This
beautiful example is a special case of a more general result in [31].

A limiting case would be when the circles are tangent to its two neighbors
(as initially) but there is no common orthogonal circle, in this case we say that
the group G is a *Kissing-Schottky group* (cf. [158]).

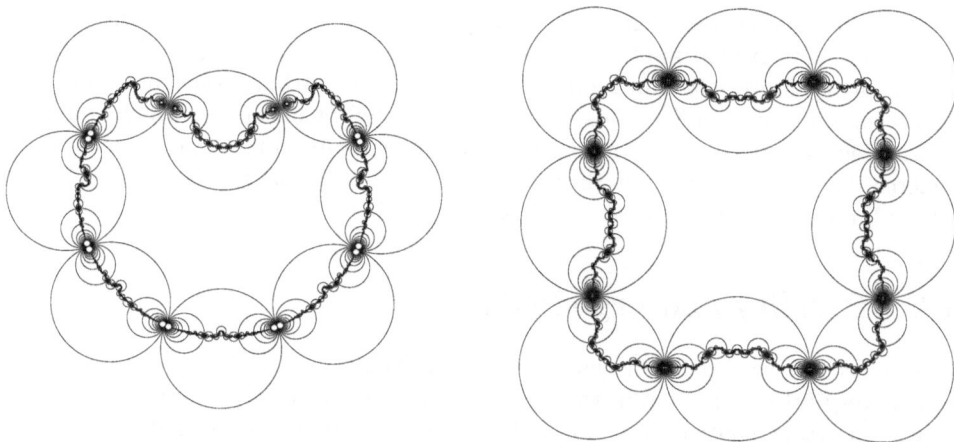

Figure 1.5: Deformations of a Fuchsian group: The limit set is a quasi-circle

(ii) We perturb the circles so that they become pairwise disjoint. Then the limit set is a Cantor set. We get a Schottky group as above. If we preserve the condition that there is a common orthogonal circle which is orthogonal to all circles $C_1, \ldots, C_r$, then the group is Fuchsian and we are in the situation envisaged in Example 1.2.34.

These are all examples of Kleinian groups.

Recall there is a canonical inclusion $\mathrm{Conf}(\mathbb{S}^n) \hookrightarrow \mathrm{Conf}(\mathbb{S}^{n+1})$ and every subgroup G of $\mathrm{Conf}(\mathbb{S}^n)$ can be regarded as acting on $\mathbb{S}^{n+1}$ leaving invariant an equator, which contains its limit set. Hence one has:

Proposition 1.2.40. *Every discrete subgroup of* $\mathrm{Conf}(\mathbb{S}^n)$ *is Kleinian when regarded as a subgroup of* $\mathrm{Conf}(\mathbb{S}^{n+1})$.

This justifies the modern use of the term "Kleinian" to denote simply a discrete subgroup of $\mathrm{Conf}(\mathbb{S}^n)$, with no condition on its limit set.

Remark 1.2.41. The previous examples obtained by reflections on circles in $\mathbb{S}^2$ that form a "necklace of pearls" give as limit set a fractal circle in the 2-sphere. Many other fractal curves can be obtained in this way, showing that the limit sets of Schottky (and kissing-Schottky) groups provide a dynamical method to obtain fractal sets of remarkable beauty and complexity.

A natural question is whether a topological n-sphere ($n \geq 1$) which is not a round sphere can be the limit set of a higher-dimensional geometrically finite Kleinian group. In this case one can show that the sphere is necessarily fractal (possibly unknotted) and self-similar. Examples of wild knots in $\mathbb{S}^3$ which are limit sets of geometrically finite Kleinian groups have been obtained in [144], [105], [90] and [79]. In [91] the author gives an example of a wild 2-sphere in $\mathbb{S}^4$ which is the limit set of a geometrically finite Kleinian group. In [28] the authors construct an infinite number of wild knots $\mathbb{S}^n \hookrightarrow \mathbb{S}^{n+2}$ for $n = 1, \ldots, 5$, which are limit sets of geometrically finite Kleinian groups. Also, in [92] the authors construct an action of a Kleinian group on $\mathbb{S}^3$, whose limit set Λ is a wild Cantor set, which is a kind of Antoine's necklace. This construction can be generalised to all odd-dimensional spheres, and it is shown that there exist discrete, real analytic actions on $\mathbb{S}^{2n+1}$ whose limit set is a Cantor set wildly embedded in $\mathbb{S}^{2n+1}$.

1.3   Rigidity and ergodicity

In this section we briefly discuss several rigidity and ergodicity theorems which play a very important role in the theory of actions of discrete groups. There are other important contributions along the lines of "rigidity and ergodicity", as for instance: the ergodicity of the geodesic flow on compact manifolds of constant negative curvature, proved by Eberhard Hopf; Ragunathan's conjecture, proved by Marina Ratner; Birkhoff's Ergodic theorem, and several others that we will not discuss here, restricting our discussion to the results that we use in this monograph.

1.3.1   Moore's Ergodicity Theorem

Important research on the theory of surfaces of constant negative curvature was carried out around the turn of the 19th century by F. Klein, H. Poincaré and others, in connection with complex function theory. This gave rise to the theory of Kleinian and Fuchsian groups. If Γ is a discrete subgroup of $\mathrm{PSL}(2, \mathbb{R})$ with no torsion elements, then the quotient $M := \mathbb{H}^2_\mathbb{R}/\Gamma$ is a hyperbolic 2-dimensional manifold, which is naturally equipped with the hyperbolic metric. Thus one has the geodesics in M, and a well-defined geodesic flow on the unit tangent bundle $T_1(M)$ of M.

    The metric on M, being $\mathrm{PSL}(2, \mathbb{R})$-invariant, defines an invariant measure on M; so one has an invariant measure also on $T_1(M)$, and one can look at the geodesic flow from a measure-theoretical point of view. This was done by E. Hopf in 1939, proving that for compact hyperbolic surfaces, the geodesic flow is ergodic. This theorem was generalised by Hopf himself and by many other authors, in several directions. One can mention for instance the remarkable works of M. Ratner, S. G. Dani, G. A. Margulis, M. Raghunathan, S. V. Fomin, I. M. Gelfand, D. V. Anosov, Y. G. Sinai and many others (see for instance Zimmer's book for more on the subject).

    Moore's Ergodicity Theorem (1966) can be thought of as belonging to this class of results. It is well known that if G is a Lie group and $H \subset G$ is a closed subgroup, then there is a unique invariant measure class on the quotient manifold G/H (where two measures are said to be in the same measure class if they have the same zero-measure sets). Moore's theorem answers the following natural question: *If G is a semisimple Lie group and H, Γ are closed subgroups, when is the action* *of H on G/Γ ergodic?* Recall that a (connected) Lie group is said to be semisimple if it has no nontrivial abelian connected normal closed subgroup.

Moore's proof makes use of the power of representation theory of Lie groups in the study of certain classes of flows on manifolds. The starting point is the so-called "Mautner phenomenon" for unitary representations. This is a general assertion of the form that "if $\gamma(t)$ is a one-parameter subgroup of a Lie group G, ρ is a unitary representation of G on a Hilbert space $\mathfrak{H}$ and v is a vector in $\mathfrak{H}$ which is fixed by $\gamma(t)$, i.e., $\rho(\gamma(t))v = v$ for all t, then v must also be fixed by a generally much larger subgroup H of G". The existence of pairs (G, H) satisfying the Mautner phenomenon was first exhibited in [204] and [62], and then studied and clarified in [145].

Moore refined the notion of a Mautner phenomenon in various ways, one of these leading to the ergodicity theorem. He proved that if G is a connected, semisimple Lie group with no compact factors, and H is a noncompact subgroup, then the pair (G, H) satisfies that for any unitary representation ρ of G and a vector x in the representation space of ρ, the equation $\rho(h)x = x$ for all $h \in H$ implies that $\rho(g)x = x$ for all $g \in G$. Then he showed that given (G, H) as above, if Γ is a closed subgroup of G such that G/Γ has finite invariant measure, then H acts ergodically on G/Γ.

More generally, Moore's theorem can be stated as follows (we refer to [233, 2.2.15] for a proof of this theorem):

Theorem 1.3.1 (Moore's Ergodicity Theorem). *Let $G = \prod G_i$ be a finite product where each G_i is a connected noncompact simple Lie group with finite centre. Suppose S is an irreducible ergodic G-space with a finite invariant measure. If $H \subset G$ is a closed noncompact subgroup, then H is ergodic on S.*

For instance, if $\Gamma \subset \mathrm{SL}(2, \mathbb{R})$ is a discrete subgroup which is cofinite (i.e. $\mathbb{H}^2_{\mathbb{R}}/\Gamma$ has finite area), then Γ acts ergodically on $\mathbb{S}^1 \cong \mathrm{SL}(2, \mathbb{R})/P$ where P is the noncompact subgroup of upper triangulable matrices. In this case $\mathbb{S}^1$ is the limit set of Γ and we obtain that the action of Γ on its limit set is ergodic. This is a special case of a more general theorem, that we briefly explain below, due to S. J. Patterson in dimension 2 and to D. P. Sullivan in higher dimensions.

1.3.2 Mostow's rigidity theorem

It has been known since Riemann that compact Riemann surfaces of genus > 0 have a rich deformation theory. Indeed, the conformal equivalence classes of compact Riemann surfaces of genus $g > 1$ make up a complex manifold of dimension $3g - 3$. This implies that the fundamental group Γ of one such surface, considered as a subgroup of $\mathrm{PSL}(2, \mathbb{R})$ via the uniformisation theorem, admits a continuous family of deformations which are not conjugate to Γ.

In 1960 Selberg made the remarkable discovery that certain higher-dimensional symmetric compact manifolds of negative curvature behave very differently in the sense that their fundamental groups, treated as discrete subgroups of the

Lie groups of isometries of their universal covering, cannot be deformed nontrivially. This led eventually to Mostow's Rigidity Theorem for compact hyperbolic manifolds [153], and more generally for simply connected Riemannian symmetric spaces of negative curvature having no factors of rank 1 [154]:

Theorem 1.3.2 (Mostow Rigidity Theorem). *Let Γ_1 and Γ_2 be discrete subgroups of $\mathrm{Iso}(\mathbb{H}_{\mathbb{R}}^n)$ with compact quotients $M_i := \mathrm{Iso}(\mathbb{H}_{\mathbb{R}}^n)/\Gamma_i$, $i = 1, 2$, with $n \geq 3$. If Γ_1 is isomorphic to Γ_2, then Γ_1 and Γ_2 are conjugate in $\mathrm{Iso}(\mathbb{H}_{\mathbb{R}}^n)$.*

In other words, the mere fact that Γ_1 and Γ_2 are isomorphic as abstract groups implies that they are actually conjugate in $\mathrm{Iso}(\mathbb{H}_{\mathbb{R}}^n)$, which is remarkable. We refer to Section 5 in [105] for a discussion on the relations amongst several types of equivalences for discrete subgroups of $\mathrm{Iso}(\mathbb{H}_{\mathbb{R}}^n)$ as well as for a wider view of Mostow's rigidity theorem and the known ways for proving it.

This theorem can be reformulated geometrically by saying that every isomorphism $\phi : \pi_1(M_1) \to \pi_1(M_2)$ between the fundamental groups of compact hyperbolic manifolds (or orbifolds) of dimension 3 or more, can be realised as being induced by an isometry $h : M_1 \to M_2$. In particular M_1 and M_2 are diffeomorphic. Furthermore, the hypothesis of the manifolds being compact can be relaxed, just demanding that they have finite-volume. We refer to [150] for a short and clear proof of Mostow's rigidity theorem, which is along the same lines of Mostow's original proof. Here we sketch the main ideas.

The manifolds M_1, M_2, being of the form $\mathrm{Iso}(\mathbb{H}_{\mathbb{R}}^n)/\Gamma_i$, are Eilenberg-MacLane spaces of type $K(\Gamma_i, 1)$. Therefore the isomorphism ϕ between their fundamental groups can be realised by a homotopy equivalence $\Phi : M_1 \to M_2$. One then has a lifting $\widetilde{\Phi}$ of Φ to the universal covering:

$$\widetilde{\Phi} : \mathbb{H}_{\mathbb{R}}^n \to \mathbb{H}_{\mathbb{R}}^n .$$

Then one can show [150, Lemma 6.15] that $\widetilde{\Phi}$ is a quasi-isometric isomorphism of $\mathbb{H}_{\mathbb{R}}^n$. Next we observe that every quasi-isometric isomorphism f of $\mathbb{H}_{\mathbb{R}}^n$ extends to a self-map F of the sphere at infinity $\mathbb{S}_{\infty}^{n-1}$, the extension being defined as follows: given $x \in \mathbb{S}_{\infty}^{n-1}$, take a geodesic ray γ landing at x; consider the curve $f \circ \gamma$, this is a "quasi-geodesic". Let δ be a geodesic ray that "shadows" the quasi-geodesic $f \circ \gamma$ (i.e., δ is bounded distance away from $f \circ \gamma$). Then define $F(x)$ as the endpoint of δ. The next step is to show [150, theorems 6.12 and 6.14] that since f is a quasi-isometry, then F is a quasi-conformal homeomorphism of $\mathbb{S}_{\infty}^{n-1}$, and it conjugates the action of Γ_1 to that of Γ_2. Moreover, the map F is differentiable almost everywhere, by fundamental results on quasi-conformal mappings. In fact McMullen points out that all these arguments work equally well for $n = 2$: it is in the final step that we need the hypothesis $n > 2$ in order to prove, using the ergodicity of the geodesic flow, that F is actually conformal and therefore corresponds to an isometry of $\mathbb{H}_{\mathbb{R}}^n$.

The following remarkable theorem from [228] extends Mostow's theorem for noncompact manifolds. The proof is based on work on quasi-conformal homeomorphisms of $\mathbb{S}^{n-1}$ due to Mostow, Tukia and others.

Theorem 1.3.3 (Tukia's Rigidity Theorem). *Let $\mathbb{H}_{\mathbb{R}}^n$ denote the n-dimensional hyperbolic space. Suppose that $n > 2$ and let Γ_1 be a discrete subgroup of $\mathrm{Iso}_+\mathbb{H}_{\mathbb{R}}^n$. Let f be a quasi-conformal homeomorphism of $\mathbb{H}_{\mathbb{R}}^n$ inducing a conjugation φ between Γ_1 and another subgroup $\Gamma_2 \subset \mathrm{Iso}_+\mathbb{H}_{\mathbb{R}}^n$. Suppose that for every $k > 0$ there exists a compact set $D \subset \mathbb{H}_{\mathbb{R}}^n$ such that f is k-quasi-conformal outside the orbit of D. Then the extension of f to the boundary $\mathbb{S}^{n-1}$ of $\mathbb{H}_{\mathbb{R}}^n$ is a conformal map, and φ is a conjugation by a Möbius transformation.*

Mostow's Rigidity Theorem is a deep and fundamental theorem in the theory of hyperbolic geometry and Kleinian groups, opening a gate for striking rigidity theorems, as sketched above. We may consider, more generally, discrete groups Γ of semisimple Lie groups G. Recall that the rank of G is the maximal dimension of an abelian subgroup which can be diagonalised (over $\mathbb{R}$ or $\mathbb{C}$, as the case may be). A discrete subgroup $\Gamma \subset G$ is called a *lattice* if the quotient G/Γ has finite volume; the lattice is cocompact (or uniform) if G/Γ is actually compact. For instance $\mathrm{SL}(n, \mathbb{R})$ has rank $n - 1$ over $\mathbb{R}$. Mostow's rigidity theorem was extended in [154] for uniform lattices in semisimple groups of rank > 1 with trivial centre and no compact factors, and then in [141] for the nonuniform case. This is known as the *Mostow-Margulis Rigidity Theorem*. The theorem also holds for groups of rank 1 excluding $\mathrm{PSL}(2, \mathbb{R})$, by [154] in the uniform case and by [179] for general lattices. In [150] there is a version of Mostow rigidity that works for hyperbolic surfaces.

We refer to Raghunathan's book "Discrete subgroups of Lie groups" for a clear and comprehensive account on the topic with material known by the early 1970's. This is also discussed in R. J. Zimmer's book "Ergodic theory and semisimple groups", published in 1984, which includes Margulis' work on rigidity and arithmeticity of lattices in semisimple groups. Of course there is also Margulis' book, published in 1991, which gives a coherent presentation of the theory of lattices, their structure and their classification. *Margulis superrigidity theorem* is a striking result stating that all lattices in any simple Lie group of $\mathbb{R}$-rank ≥ 2 are superrigid. This means that given such a Lie group G, a lattice Γ in it, and a homomorphism $\phi\Gamma \to \mathrm{GL}(n, \mathbb{R})$, for any n, then ϕ actually extends (virtually) to a homomorphism $\widetilde{\phi} \to \mathrm{GL}(n, \mathbb{R})$ so that its image is contained in the Zariski closure of $\phi(\Gamma)$. As usual, "virtually" means that for this statement to hold we may possibly have to restrict our attention to a subgroup of finite index in Γ. Superrigidity is so strong that Margulis theorem implies that all lattices in such a group G are arithmetic.

1.3.3 On the Patterson-Sullivan measure

At the end of the 1970s, S. J. Patterson in [172] gave a geometric construction of a family $(\sigma_x)_{x \in \mathbb{H}^{n+1}}$ of measures associated to every discrete subgroup $\Gamma \subset \mathrm{Iso}\,\mathbb{H}_{\mathbb{R}}^2$, with support in the limit set, which are quasi-invariant under the action of Γ and δ_Γ-conformal. D. P. Sullivan in [213] extended Patterson's construction to higher dimensions. In some sense, the Patterson-Sullivan measures give the proportion of elements of the orbit Γx that go to a given zone in $\mathbb{S}^n$. This construction laid down the foundations for a measure theoretical study of the limit sets arising from discrete groups.

From our knowledge of ergodic theory, we know that quasi-invariant measures, rather than the invariant ones, are the correct type of measures to look for in the study of the dynamics of discrete groups (c.f. Subsection 1.3.1). The remarkable construction of the Patterson-Sullivan measures, that we briefly describe below, fits along this line of ideas.

We consider a discrete group $\Gamma \subset \mathrm{Iso}(\mathbb{H}^{n+1})$ acting on the hyperbolic space, that we think of as being the unit ball $\mathbb{B}^{n+1}$ endowed with the hyperbolic metric. We are interested in the distribution of the orbits in $\mathbb{H}^{n+1}$ and the way in which they approach the sphere at infinity. For each $x \in \mathbb{B}^{n+1}$, we denote by $B(x, R)$ the ball of radius R centred at x. We use the orbital counting function defined by:

$$N(R, x) = \mathrm{card}(\{B(x, R) \cap \Gamma x\}).$$

One has that there exists a constant A depending on Γ, such that

$$N(R, x) < A \cdot e^{rn}.$$

This orbital counting function plays an important role in the sequel, and it is very much related to the so-called critical exponent of the group, and to the Hausdorff dimension of the limit set. The critical exponent is a concept, a number, associated to the rate at which the orbits tend to the sphere at infinity $\mathbb{S}^n_\infty$, which is the unit sphere in $\mathbb{R}^{n+1}$. In fact, notice that given any point $x \in \mathbb{B}^{n+1}$, a way to study how its orbit tends to $\mathbb{S}^n_\infty$ is by looking at the numbers $1 - |\gamma(x)|$ for $\gamma \in \Gamma$, or in other words, by looking at the distance to the centre 0 of points in the orbit . If $d_h(x, y)$ denotes the hyperbolic distance, then it is not hard to see that one has

$$d_h(0, x) = \log\left(\frac{1 + |x|}{1 - |x|}\right). \tag{1.3.4}$$

Together with the triangle inequality, this implies that for all $x, y \in \mathbb{B}^{n+1}$, one has

$$1 - |\gamma(x)| \leq 2\, e^{d_h(x,y)}\, (1 - |\gamma(y)|),$$

which implies that the ratios

$$\frac{1 - |\gamma(x)|}{1 - |\gamma(y)|}, \quad \text{with } \gamma \in \Gamma,$$

are comparable in the sense that they lie between finite limits. Hence, a good way to study the rate at which the orbits tend to ∞ is to consider the convergence of the series

$$\sum_{\gamma \in \Gamma} (1 - |\gamma(0)|)^\alpha, \tag{1.3.5}$$

for various $\alpha > 0$. Since the hyperbolic distance to 0 is given by equation (1.3.4), it is an exercise to see that the convergence or divergence of the series in equation (1.3.5) coincides with the convergence or divergence of the Poincaré series

$$\sum_{\gamma \in \Gamma} e^{-\alpha\, d_h(0, \gamma(0))} . \tag{1.3.6}$$

And this latter series is often easier to handle than the previous one. In fact, using the orbital counting function it is easy to see that the series (1.3.6) always converges for $\alpha > n$. So it makes sense to define:

Definition 1.3.7. The *critical exponent* of the group $\Gamma \subset \mathrm{Iso}(\mathbb{H}^{n+1})$ is the number defined by:

$$\delta_\Gamma = \inf \left\{ \alpha \in \mathbb{R} \ \Big| \ \sum_{\gamma \in \Gamma} e^{-\alpha\, d_h(0, \gamma(0))} < \infty \right\} .$$

The critical exponent depends only on Γ and it can be expressed in terms of the orbital counting function as:

$$\delta_\Gamma = \limsup_{\substack{R \\ R \to \infty}} \frac{1}{R} \log N(R, 0).$$

This is a nonnegative number which is at most n. The Poincaré series in equation (1.3.6) converges for $\alpha > \delta_\Gamma$ and diverges for $\alpha < \delta_\Gamma$. For $\alpha = \delta_\Gamma$, the series may converge or not. The group Γ is said to be of *convergence type* if the above series converges when α is the critical exponent. Otherwise the group is said to be of *divergence type*.

One has (see for instance [57] or [163, 1.6.2 and 1.6.3]):

Theorem 1.3.8. *If the group* Γ *is such that its limit set is not the whole sphere* $\mathbb{S}^n_\infty$, *then the series* (1.3.6) *converges at the exponent* n. *If the limit set is all of* $\mathbb{S}^n_\infty$ *and* Γ *is cofinite, then its critical exponent is* $\delta_\Gamma = n$ *and the group is of divergence type.*

For $x, y \in \mathbb{B}^{n+1}$ and $s > \delta_\Gamma$, set

$$g_s(x, y) = \sum_{\gamma \in \Gamma} e^{-s\, d_h(x, \gamma(y))} .$$

The idea now is to construct a measure for x, y, s by placing a Dirac point mass of weight $\frac{e^{-s\, d_h(x, \gamma(y))}}{g_s(x, y)}$ at each point in the orbit of y. One then aims to get a measure in the limit as $s \to \delta_\gamma^+$. It turns out that if the group is of divergence type, this procedure works fine. However, if the group is of convergence type, one does not get anything interesting.

To overcome this difficulty, the point masses are multiplied by a factor $h(d_h(x, \gamma(y)))$ which does not change the critical exponent of the series, but ensures divergence at that exponent. Then one has the corresponding Poincaré series

$$P(s, x) = \sum_{\gamma \in \Gamma} h(d_h(x, \gamma(x))) e^{-sd_h(x, \gamma(x))} ,$$

and one may argue as for groups of divergence type. The function $h : \mathbb{R}_+ \to \mathbb{R}$ must be continuous, nondecreasing and with the following properties:

(i) The corresponding Poincaré series converges for $s > \delta_\Gamma$ and diverges for $s \le \delta_\Gamma$.

(ii) For every $\epsilon > 0$ there exists r_0 such that if $r > r_0$, $t > 1$, then $h(t + r) \le e^{\epsilon t} h(r)$.

(iii) $h(r + t) \le h(r)h(t)$.

Now one has:

Definition 1.3.9. Let Γ be a discrete subgroup of Möbius transformations acting on the unit ball $\mathbb{B}^{n+1}$ and let $x \in \mathbb{B}^{n+1}$. Define the *Patterson-Sullivan measure* μ_x associated to Γ at x as the weak limit of the series,

$$\frac{1}{P(s, x)} \sum_{\gamma \in \Gamma} h(d_h(x, \gamma(x))) \, e^{-sd(x, \gamma(x))} \, D(\gamma x) , \qquad (1.3.9)$$

as s tends by the above to the critical exponent δ_Γ, where $D(\gamma x)$ is the Dirac point mass of weight 1 at γx.

Theorem 1.3.10. *Let Γ be a nonelementary group of Möbius transformations of $\mathbb{H}_\mathbb{R}^{n+1}$, $n \ge 1$. Then for each $x \in \mathbb{H}_\mathbb{R}^{n+1}$ there exists a Patterson-Sullivan measure μ_x associated to Γ.*

In fact, if the group is convex cocompact, then its critical exponent equals the Hausdorff dimension of the limit set ([163, Theorem 4.6.4]).

The following result describes basic properties of the Patterson-Sullivan measures (we refer to [218] or [163] for the proof of these statements).

Theorem 1.3.11. *Let Γ be a nonelementary group of Möbius transformations of $\mathbb{H}_\mathbb{R}^{n+1}$. For every $x \in \mathbb{H}_\mathbb{R}^{n+1}$, let μ_x be a Patterson-Sullivan measure associated to Γ. Then:*

(i) *The measure μ_x is concentrated in the limit set $\Lambda(\Gamma)$.*

(ii) *For each $\gamma \in \Gamma$ one has: $\gamma^* \mu_x = \mu_{\gamma^{-1}(x)}$.*

(iii) *For each $\gamma \in \Gamma$ one has: $\gamma^* \mu_x = |\gamma'_x|^{\delta_\Gamma} \mu_x$.*

(iv) *For every $x, y \in \mathbb{H}_{\mathbb{R}}^{n+1}$ the measures μ_x, μ_y are absolutely continuous with respect to each other and the Radon-Nikodym derivative satisfies*

$$\frac{d\mu_x}{d\mu_y}(z) = \left(\frac{P(x, z)}{P(y, z)}\right)^{\delta_\Gamma}$$

where $P(x, z) = \frac{1 - |x|^2}{|x - z|^2}$ is the Poisson kernel.

(v) *If Γ is nonelementary and geometrically finite, then the measure μ_x has no atomic part.*

(vi) *For each Borel set A and each $\gamma \in \Gamma$ it follows that:*

$$\mu(\gamma(A)) = \int_A |\gamma'|^{\delta_\Gamma} \, d\mu, \qquad (1.3.12)$$

where $|\gamma'|$ denotes the unique positive number making $\frac{\gamma'}{|\gamma'|}$ an orthogonal matrix.

Any quasi-invariant, finite, positive Borel measure with support on the limit set and satisfying items (ii) and (iv), with δ_Γ replaced by an α, is said to be a *conformal measure (or density)* of dimension α for Γ. And every conformal density satisfies equation (1.3.12).

Conformal measures were introduced by Patterson in dimension 2, and then by Sullivan in general. These appear as powerful tools in several major applications. One of these is the investigation of the geodesic flow of the quotient manifold. These appear also in relation with some rigidity problems connected to the barycentre method; see for instance the characterization of symmetric spaces due to [138] and [25]. Conformal measures appear as well in the study of several geometric and number-theoretic problems related to fractal sets in holomorphic dynamics, see for instance [149].

Now we look at the action of the group $\Gamma \subset \text{Iso}(\mathbb{H}_{\mathbb{R}}^{n+1})$ on the sphere $\mathbb{S}_\infty^n$. We know from the previous discussion that if Γ is nonelementary, then one has the conformal densities μ_x of dimension δ_Γ, the critical exponent of the group. These define conformal measures on the sphere, with support on the limit set of the group. One has the following theorem from [218]:

Theorem 1.3.13. *Assume a discrete group $\Gamma \subset \text{Iso}(\mathbb{H}_{\mathbb{R}}^{n+1})$ admits a Γ-invariant conformal density σ of dimension $\alpha > 0$. Then the action of Γ on $\mathbb{S}_\infty^n$ is ergodic with respect to the measure class defined by σ if and only if the only Γ-invariant conformal densities of dimension α are constant multiples of σ.*

Theorem 1.3.14. *Let $\Gamma \subset \text{Iso}(\mathbb{H}_{\mathbb{R}}^{n+1})$ be a geometriclly finite group and σ a conformal invariant density of dimension $\alpha > 0$, then Γ is ergodic on $\mathbb{S}^n \times \mathbb{S}^n$ with respect to $\sigma_x \times \sigma_x$.*

In analogy with the theorem of E. Hopf about ergodicity of the geodesic flow with respect to a certain measure, associated to the Lebesgue measure, D. Sullivan discovered that the Patterson-Sullivan measures determine an invariant measure for the geodesic flow on the unit tangent bundle of the manifold $\mathbb{H}^{n+1}/\Gamma$. More precisely:

Let $\mathcal{U}$ be the unitary tangent bundle of $\mathbb{H}^{n+1}$, that is,

$$\mathcal{U} = \{(x,\eta) : x \in \mathbb{H}^{n+1}, \eta \in T_x(\mathbb{H}^{n+1}) \text{ is unitary}\}.$$

This space can be described as follows: given any $(x,\eta) \in \mathcal{U}$, there is exactly one geodesic ν passing through x whose velocity vector at x is η. Set ν_+ and ν_- for the beginning and the end points of this geodesic and let z be the euclidean mid point of the geodesic ν. Notice that the geodesics in $\mathbb{H}^{n+1}$ correspond bijectively with the pairs of distinct point in the sphere at infinity $\mathbb{S}^n = \mathbb{S}^n_\infty$. Thus we can define a homeomorphism

$$\phi : \mathcal{U} \to ((\mathbb{S}^n \times \mathbb{S}^n) \setminus \{(y,y) \in \mathbb{S}^n \times \mathbb{S}^n\}) \times \mathbb{R},$$

by

$$\phi(z,x) = \begin{cases} (\nu_-,\nu_+,d(x,z)) & \text{if } x \in \widetilde{z,\nu_+}, \\ (\nu_-,\nu_+,-d(x,z)) & \text{if } x \in \widetilde{z,\nu_-} \end{cases}$$

where $\widetilde{x,\nu_\pm}$ is the connected component of $\nu \setminus \{z\}$ containing $\nu_\pm$.

From now on the spaces $\mathcal{U}$ and $\phi(\mathcal{U})$ both will be represented by $\mathcal{U}$.

Now we can define the flow g_t on $\mathcal{U}$ which serves as basis for the geodesic flow

$$g_t(\nu_-,\nu_+,s) = (\nu_-,\nu_+,t+s).$$

Observe that the group of Möbius transformations leaving invariant $\mathbb{H}^{n+1}_\mathbb{R}$ acts naturally on $\mathcal{U}$ by

$$\gamma(x,\eta) = \left(\gamma(x), \frac{\gamma'(x)}{|\gamma'(x)|}\eta\right).$$

Using this equation it is easy to show that:

(i) The action of $\text{Möb}(\mathbb{H}^{n+1}_\mathbb{R})$ commutes with the flow g_t.

(ii) Whenever $\Gamma \in \text{Möb}(\mathbb{H}^{n+1}_\mathbb{R})$ is a discrete group, the action of Γ on $\mathcal{U}$ is properly discontinuous.

(iii) Given a Patterson-Sullivan measure μ_x for Γ we have a Γ-invariant measure m_σ on $\mathcal{U}$ defined by the following differential (see [163, Chapter 8] for details):

$$dm_\sigma(\nu_-,\nu_+,s) = \frac{d\mu_x(\eta_-)d\mu_x(\eta_+)ds}{|\nu_+ - \nu - |^{2\delta_\gamma}}.$$

Now consider the quotient $\mathcal{U}/\Gamma$. Since g_t commutes with the action of the group Möb($\mathbb{H}_{\mathbb{R}}^{n+1}$), it follows that g_t induces a flow g'_t on the manifold $\mathcal{U}/\Gamma$, which is called *the geodesic flow* and m_σ induces a measure m'_σ on $\mathcal{U}/\Gamma$ which is g'_t-invariant.

Recall that the geodesic flow on $\mathcal{U}$ is said to be conservative if for every set $A \subset \mathcal{U}$ with $\mu_x(A) > 0$, we have $mu_x(A \cap g_t(A)) > 0$ for an infinite sequence (t_n).

Let us now recall the definition of the conical limit set of Γ:

Definition 1.3.15. A point $z \in \mathbb{S}^n$ in the limit set Λ of Γ is said to be a conical limit point if for each $x \in \mathbb{H}_{\mathbb{R}}^{n+1}$, regarded as the unit ball, there is a sequence $(\gamma_m) \subset \Gamma$ such that the sequence

$$\frac{|z - \gamma_m(x)|}{1 - |\gamma_m(x)|},$$

remains bounded.

Intuitively, conical points are the points in the limit set which are approximated by orbits within a cone. More precisely, a point $z \in \mathbb{S}^n$ is a conical limit point for Γ if there is a geodesic σ in $\mathbb{H}_{\mathbb{R}}^{n+1}$ ending at z and such that for every $x \in \mathbb{H}_{\mathbb{R}}^{n+1}$, there are infinitely many Γ-images of x within a bounded hyperbolic distance from σ.

For instance, it is clear that every fixed point of a loxodromic element in Γ is a conical limit point. On the contrary, a fixed point of a parabolic element in Γ is a limit point which cannot be conical, since every parabolic element is conjugate to a translation. In fact one has that for geometrically finite groups, the whole limit set Λ consists fully of conical limit points and fixed points of parabolic elements.

The following theorem can be stated for conformal densities in general (see [163]), however for the sake of simplicity we state it only for the Patterson-Sullivan measures.

Theorem 1.3.16. *Let Γ be a Möbius group preserving $\mathbb{H}_{\mathbb{R}}^{n+1}$, μ_x a Patterson-Sullivan measure for Γ of exponent δ_Γ, and g'_t the geodesic flow on $\mathcal{U}/\Gamma$. Let m'_{μ_x} be the invariant measure on $\mathcal{U}$ constructed above. Then the following statements are equivalent:*

(i) *The geodesic flow g'_t is m'_{μ_x}-conservative.*

(ii) *The geodesic flow g'_t is m'_{μ_x}-ergodic.*

(iii)
$$\sum (1 - |\gamma(0)|)^{\delta_\Gamma} = \infty.$$

(iv) *Γ is $(\sigma \times \sigma)$-ergodic on $\mathbb{S}^n \times \mathbb{S}^n$.*

(v) *The conical limit set has full μ_x-measure.*

Remark 1.3.17. The construction of conformal densities (and Patterson-Sullivan measures) generalises to the case of complex hyperbolic groups, that we study in the sequel. This generalises also to quaternionic hyperbolic groups and actually for every nonelementary discrete group of isometries of a simply connected complete Riemannian manifold with negative curvature, see for instance [180, 181].

1.3.4   Sullivan's theorem on nonexistence of invariant lines

D. P. Sullivan in [215] proved the following remarkable theorem:

Theorem 1.3.18. *Let Γ be a finitely generated discrete group of isometries of the hyperbolic space $\mathbb{H}^3$, so that Γ is also a group of conformal transformations of the sphere $\mathbb{S}^2$. Then there is no Γ-invariant measurable vector field on the conservative part of $\mathbb{S}^2$ for the action with respect to Lebesgue measure. Hence there is no measurable Γ-invariant field of tangent lines supported on the limit set.*

A measurable version of the Riemann mapping theorem due to Ahlfors and Bers gives a one-to-one correspondence between measurable vector fields and quasi-conformal homeomorphisms, up to composition with a conformal homeomorphism. Hence Sullivan's theorem says that on the conservative part of $\mathbb{S}^2$ there are no quasi-conformal deformations of Γ. Hence this theorem of Sullivan is in the same spirit as Mostow's rigidity theorem.

This theorem has important applications in geometry and dynamics, and we refer to [148] and [166] for more on the topic. In the sequel we use (in Chapter 10) a generalization of Sullivan's theorem proved in [202], which is based on a deep theorem by L. Flaminio and R. J. Spatzier that we explain below.

It is worth saying that it is not known whether or not the equivalent of Theorem 1.3.18 holds for rational maps on the Riemann sphere. This would have important applications to iteration theory and there are several people working on that problem.

Recall that a subset of $\mathrm{Iso}(\mathbb{H}^n_{\mathbb{R}})$, regarded as an algebraic variety, is said to be *Zariski closed* if it is the set of solutions (zeroes) of a collection of polynomial equations on $\mathrm{Iso}(\mathbb{H}^n_{\mathbb{R}})$. Given a subset V of $\mathrm{Iso}(\mathbb{H}^n_{\mathbb{R}})$, its Zariski closure is the (unique) smallest algebraic subset of $\mathrm{Iso}(\mathbb{H}^n_{\mathbb{R}})$ that contains V. It is not hard to see that if V is a group, then its Zariski closure is also a group.

Definition 1.3.19. A discrete subgroup Γ of $\mathrm{Iso}_+(\mathbb{H}^n_{\mathbb{R}})$ is said to be *Zariski-dense* in $\mathrm{Iso}_+(\mathbb{H}^n_{\mathbb{R}})$ if its Zariski closure is all of $\mathrm{Iso}_+(\mathbb{H}^n_{\mathbb{R}})$.

We know from the previous section that if Γ is a geometrically-finite discrete subgroup of $\mathrm{Iso}_+(\mathbb{H}^n_{\mathbb{R}})$, then there is a unique Patterson-Sullivan measure, up to constant multiples. We denote this by μ. One has:

Theorem 1.3.20 ([59]). *Let Γ be a geometrically-finite discrete subgroup of $\mathrm{Iso}_+(\mathbb{H}^n_{\mathbb{R}})$ which is Zariski-dense. Let μ be the Patterson-Sullivan measure on the limit set Λ of Γ. Then every Γ-invariant measurable distribution on $\Lambda \subset \mathbb{S}^{n-1}$ by subspaces* *of dimension d is μ-almost everywhere trivial, i.e., either $d = n - 1$ or $d = 0$. Furthermore, the restriction to Λ of the bundle of orthonormal n-frames on $\mathbb{S}^n$ is locally $U_x \times SO(n)$, with U_x an open set in Λ, so we can endow it with the measure $\widetilde{\mu}$ which is (locally) the product of μ with the invariant measure measure on $SO(n)$. Then Γ acts on this bundle, via the differential, and this action is ergodic with respect to $\widetilde{\mu}$.*

This theorem is used in [59] to study rigidity properties of the horospherical foliations of geometrically finite hyperbolic manifolds, generalizing Marina Ratner's theorem.

One also has the following theorem of [202], which improves Theorem 1.3.20. This result will be used in the sequel, in Chapter 10.

Theorem 1.3.21. *Let Γ be a geometrically-finite and Zariski-dense discrete subgroup of $\mathrm{Iso}(\mathbb{H}^{n+1}_{\mathbb{R}})$. Let μ be the Patterson-Sullivan measure on the limit set Λ. Let $0 < p < n$. Let $\Gamma_{\Lambda,p}$ be the the restriction to Λ of the Grassmannian fibre bundle, $G_{n,p}(\mathbb{S}^n)$, of p-dimensional subspaces of $T\mathbb{S}^n$. Then Γ acts, via the differential, minimally on $G_{\Lambda,p}$. Furthermore, Γ acts ergodically on $G_{\Lambda,p}$ with respect to the measure $\widetilde{\mu}$ which is (locally) the product of μ with the homogeneous measure on $G_{n,p} = SO(n)/(SO(p) \times SO(n - p))$.*

The proof of ergodicity in Theorem 1.3.21 follows easily from Theorem 1.3.20 above. In fact, notice that locally one has a product structure: given any point $x \in \Lambda$, there is an open neighbourhood U_x of x in $\mathbb{S}^n$ such that, $G_{\Lambda,p}|_{U_x} \cong (\Lambda \cap U_x) \times G_{n,p} \cong (\Lambda \cap U_x) \times (SO(n)/(SO(p) \times SO(n - p)))$. Furthermore, the action of Γ on $G_{\Lambda,p}$ sends a fibre isometrically onto a fibre, since Γ preserves angles. The action of Γ on $G_{\Lambda,p}$ is a factor of the action of Γ on $\mathcal{F}_{\Lambda,n}$, the restriction to Λ of the bundle of orthonormal n-frames on $\mathbb{S}^n$, which locally is $U_x \times SO(n)$. This action is ergodic with respect to $\widetilde{\mu}$ by [59]. Hence the action on $G_{\Lambda,p}$ is also ergodic.

In order to prove the minimality of the action stated in the theorem above, the next step, once we have the ergodicity, is to show that the action of Γ on $G_{\Lambda,p}$ carries fibres isometrically into fibres. This is an exercise and we refer to [202] for details. Then the minimality of the action follows from the ergodicity together with the following general theorem (2.3.1 in [202]): (Only the first statement below is needed for Theorem 1.3.21, but the second statement will be used in Chapter 10.)

Theorem 1.3.22. *Let X and Y be compact metric spaces, and let G be a compact group which acts minimally on X. Let $\pi : E \to X$ be a locally trivial fibre bundle with fibre Y, such that G acts on E as a skew-product, $g(x, y) = (g(x), F_{(g,x)}(y))$, where $F_{(g,x)} : Y \to Y$ is an isometry for each $(g, x) \in G \times X$. Let H be a minimal subset of E for the action of G. Then:*

(i) *The restriction of π to H, $\pi|_H : H \to X$, is a locally trivial fibre bundle, whose fibres are homogeneous spaces on which G acts transitively.*

(ii) *If the space Y is the 2-sphere, then either the minimal set H is all of E, or else the bundle $\pi : E \to X$ admits a section.*

This theorem implies that if the action of Γ on $G_{\Lambda,p}$ is not minimal, then there exists a minimal set $F \subset G_{\Lambda,p}$ which is a fibre bundle over Λ and whose fibre F_x, at $x \in \Lambda$, is a proper submanifold of the fibre $\{x\} \times G_{n,p}$. All submanifolds F_x are isometric. Consider a small tubular neighbourhood $\mathcal{U}_x$ of F_x in the fibre $\{x\} \times G_{n,p}$ consisting of points at distance less than $\epsilon > 0$ of F_x. Then, for ϵ small, the union $\mathcal{U} := \cup_{x \in \Lambda} \mathcal{U}_x$ is a measurable set of positive measure which is Γ-invariant and whose measure varies with ϵ. This contradicts ergodicity, so the action is minimal. $\square$

Chapter 2

Complex Hyperbolic Geometry

In complex hyperbolic geometry we consider an open set biholomorphic to an open ball in $\mathbb{C}^n$, and we equip it with a particular metric that makes it have constant negative holomorphic curvature. This is analogous to but different from the real hyperbolic space. In the complex case, the sectional curvature is constant on complex lines, but it changes when we consider real 2-planes which are not complex lines.

Complex hyperbolic geometry contains in it the real version: while every complex $\mathbb{C}^k$-*plane* in $\mathbb{H}_{\mathbb{C}}^n$ is biholomorphically isometric to $\mathbb{H}_{\mathbb{C}}^k$, every totally real k-plane in $\mathbb{H}_{\mathbb{C}}^n$ is isometric to the real hyperbolic space $\mathbb{H}_{\mathbb{R}}^k$. Moreover, every complex line in $\mathbb{H}_{\mathbb{C}}^n$ is biholomorphically isometric to $\mathbb{H}_{\mathbb{R}}^2$.

There are three classical models for complex hyperbolic space $\mathbb{H}_{\mathbb{C}}^n$: the unit ball model in $\mathbb{C}^n$, the projective ball model in $\mathbb{P}_{\mathbb{C}}^n$ and the Siegel domain model. In this monograph we normally use the projective ball model, and the symbol $\mathbb{H}_{\mathbb{C}}^n$ will be reserved for that. These models for complex hyperbolic n-space are briefly discussed in Sections 2 and 3, where we also explain and define some basic notions which are used in the sequel, such as chains, bisectors and spinal spheres, horospherical coordinates and the Heisenberg geometry at infinity. We refer to [67], [71], [168], [200], [169] for rich accounts on complex hyperbolic geometry.

In Section 4 we discuss the geometry, dynamics and algebraic properties of the elements in the projective Lorentz group $PU(n,1)$, which is the group of holomorphic isometries of $\mathbb{H}_{\mathbb{C}}^n$. We give in detail Goldman's proof of the classification according to trace, since this paves the ground for the discussion in Chapter 4.

In Section 5 we discuss methods and sources for constructing complex hyperbolic Kleinian groups, i.e., discrete subgroups of $PU(n,1)$. For further reading on this we refer to the excellent survey articles [105], [168] and the bibliography in them. As explained in [168], the methods for constructing complex hyperbolic lattices can be roughly classified into four types: i) Arithmetic constructions; ii) reflection groups and construction of appropriate fundamental domains; iii) algebraic constructions, via the Yau-Miyaoka uniformisation theorem; and iv) as monodromy groups of certain hypergeometric functions. In fact, a fundamental problem in complex hyperbolic geometry is whether or not there are nonarithmetic lattices in dimensions ≥ 3 (see for instance [50]).

Finally, Section 6 is based on [45]. Here we discuss the definition of limit set for complex hyperbolic Kleinian groups and give some of its properties. This is analogous to the corresponding definition in real hyperbolic geometry. The similarities in both settings spring from the fact that in both situations one has the convergence property (see [105, §3.2]): every sequence of isometries either contains a convergent subsequence or contains a subsequence which converges to a constant map away from a point on the sphere at infinity. This property is not satisfied by discrete subgroups of $\mathrm{PSL}(n + 1, \mathbb{C})$ in general, and this is why even the concept of limit set becomes intriguing in that general setting. This will be explored in the following chapters.

2.1   Some basic facts on Projective geometry

We recall that the complex projective space $\mathbb{P}^n_{\mathbb{C}}$ is defined as

$$\mathbb{P}^n_{\mathbb{C}} = (\mathbb{C}^{n+1} - \{0\})/\sim,$$

where "$\sim$" denotes the equivalence relation given by $x \sim y$ if and only if $x = \lambda y$ for some nonzero complex scalar λ. This is a compact connected complex n-dimensional manifold, diffeomorphic to the orbit space $\mathbb{S}^{2n+1}/\mathrm{U}(1)$, where $\mathrm{U}(1)$ is acting coordinate-wise on the unit sphere in $\mathbb{C}^{n+1}$.

We notice that the usual Riemannian metric on $\mathbb{S}^{2n+1}$ is invariant under the action of $\mathrm{U}(1)$ and therefore descends to a Riemannian metric on $\mathbb{P}^n_{\mathbb{C}}$, which is known as the *Fubini-Study* metric.

If $[\ ]_n : \mathbb{C}^{n+1} - \{0\} \to \mathbb{P}^n_{\mathbb{C}}$ represents the quotient map, then a nonempty set $H \subset \mathbb{P}^n_{\mathbb{C}}$ is said to be a projective subspace of dimension k if there is a $\mathbb{C}$-linear subspace $\widetilde{H} \subset \mathbb{C}^{n+1}$ of dimension $k+1$ such that $[\widetilde{H}]_n = H$. If no confusion arises, we will denote the map $[\ ]_n$ just by $[\ ]$. Given a subset P in $\mathbb{P}^n_{\mathbb{C}}$, we define

$$\langle P \rangle = \bigcap \{l \subset \mathbb{P}^n_{\mathbb{C}} \mid l \text{ is a projective subspace and } P \subset l\}.$$

Then $\langle P \rangle$ is a projective subspace of $\mathbb{P}^n_{\mathbb{C}}$, see for instance [125]. In particular, given $p, q \in \mathbb{P}^n_{\mathbb{C}}$ distinct points, $\langle \{p, q\} \rangle$ is the unique proper complex projective subspace passing through p and q. Such a subspace will be called a complex (projective) line and denoted by $\overleftrightarrow{p, q}$; this is the image under $[\ ]_n$ of a two-dimensional linear subspace of $\mathbb{C}^{n+1}$. Observe that if ℓ_1, ℓ_2 are different complex lines in $\mathbb{P}^2_{\mathbb{C}}$, then $\ell_1 \cap \ell_2$ consists of exactly one point.

If $e_1, \ldots, e_{n+1}$ denotes the elements of the standard basis in $\mathbb{C}^{n+1}$, we will use the same symbols to denote their images under $[\ ]_n$.

It is clear that every linear automorphism of $\mathbb{C}^{n+1}$ defines a holomorphic automorphism of $\mathbb{P}^n_{\mathbb{C}}$, and it is well known (see for instance [37]) that every automorphism of $\mathbb{P}^n_{\mathbb{C}}$ arises in this way. Thus one has

Theorem 2.1.1. *The group of projective automorphisms is:*

$$\mathrm{PSL}(n+1,\mathbb{C}) := \mathrm{GL}\,(n+1,\mathbb{C})/(\mathbb{C}^*)^{n+1} \cong \mathrm{SL}(n+1,\mathbb{C})/\mathbb{Z}_{n+1}\,,$$

where $(\mathbb{C}^*)^{n+1}$ is being regarded as the subgroup of diagonal matrices with a single nonzero eigenvalue, and we consider the action of $\mathbb{Z}_{n+1}$ (viewed as the roots of the unity) on $\mathrm{SL}(n+1,\mathbb{C})$ given by the usual scalar multiplication.

This result is in fact a special case of a more general, well-known theorem, stating that every holomorphic endomorphism $f : \mathbb{P}_\mathbb{C}^n \to \mathbb{P}_\mathbb{C}^n$ is induced by a polynomial self-map $F = (F_0,\dots,F_n)$ of $\mathbb{C}^{n+1}$ such that $F^{-1}(0) = \{0\}$ and the components F_i are all homogeneous polynomials of the same degree. The case we envisage here is when these polynomials are actually linear.

We denote by $[[\;]]_{n+1} : \mathrm{SL}(n+1,\mathbb{C}) \to \mathrm{PSL}(n+1,\mathbb{C})$ the quotient map. Given $\gamma \in \mathrm{PSL}(n+1,\mathbb{C})$ we say that $\tilde{\gamma} \in \mathrm{GL}\,(n+1,\mathbb{C})$ is a lift of γ if there is an scalar $r \in \mathbb{C}^*$ such that $r\tilde{\gamma} \in \mathrm{SL}(n,\mathbb{C})$ and $[[r\tilde{\gamma}]]_{n+1} = \gamma$.

Notice that $\mathrm{PSL}(n+1,\mathbb{C})$ acts transitively, effectively and by biholomorphisms on $\mathbb{P}_\mathbb{C}^n$, taking projective subspaces into projective subspaces.

There are two classical ways of decomposing the projective space that will play a significant role in the sequel; each provides a rich source of discrete subgroups of $\mathrm{PSL}(n+1,\mathbb{C})$. The first is by thinking of $\mathbb{P}_\mathbb{C}^n$ as being the union of $\mathbb{C}^n$ and the "hyperplane at infinity":

$$\mathbb{P}_\mathbb{C}^n = \mathbb{C}^n \cup \mathbb{P}_\mathbb{C}^{n-1}\,.$$

A way for doing so is by writing

$$\mathbb{C}^{n+1} = \mathbb{C}^n \times \mathbb{C} = \{(Z, z_{n+1})\,|\, Z = (z_1,\dots,z_n) \in \mathbb{C}^n \text{ and } z_{n+1} \in \mathbb{C}\}.$$

Then every point in the hyperplane $\{(Z,1)\}$ determines a unique line through the origin in $\mathbb{C}^{n+1}$, i.e., a point in $\mathbb{P}_\mathbb{C}^n$; and every point in $\mathbb{P}_\mathbb{C}^n$ is obtained in this way except for those corresponding to lines (or "directions") in the hyperplane $\{(Z,0)\}$, which form the "hyperplane at infinity" $\mathbb{P}_\mathbb{C}^{n-1}$.

It is clear that every affine map of $\mathbb{C}^{n+1}$ leaves invariant the hyperplane at infinity $\mathbb{P}_\mathbb{C}^{n-1}$. Furthermore, every such map carries lines in $\mathbb{C}^{n+1}$ into lines in $\mathbb{C}^{n+1}$, so the map naturally extends to the hyperplane at infinity. This gives a natural inclusion of the affine group

$$\mathrm{Aff}\,(\mathbb{C}^n) \cong \mathrm{GL}\,(n,\mathbb{C}) \ltimes \mathbb{C}^n\,,$$

in the projective group $\mathrm{PSL}(n+1,\mathbb{C})$. Hence every discrete subgroup of $\mathrm{Aff}\,(\mathbb{C}^n)$ is a discrete subgroup of $\mathrm{PSL}(n+1,\mathbb{C})$.

The second classical way of decomposing the projective space that plays a significant role in this monograph leads to complex hyperbolic geometry, which we study in the following section. For this we think of $\mathbb{C}^{n+1}$ as being a union

$V_- \cup V_0 \cup V_+$, where each of these sets consists of the points $(Z, z_{n+1}) \in \mathbb{C}^{n+1}$ satisfying that $\|Z\|^2$ is respectively smaller, equal to or larger than $|z_{n+1}|$.

We will see in the following section that the projectivisation of V_- is an open $(2n)$-ball $\mathbb{B}$ in $\mathbb{P}_{\mathbb{C}}^n$, bounded by $[V_0]$, which is a sphere. This ball $\mathbb{B}$ serves as model for complex hyperbolic geometry. Its full group of holomorphic isometries is $\mathrm{PU}(n,1)$, the subgroup of $\mathrm{PSL}(n+1, \mathbb{C})$ of projective automorphisms that preserve $\mathbb{B}$. This gives a second natural source of discrete subgroups of $\mathrm{PSL}(n+1, \mathbb{C})$, those coming from complex hyperbolic geometry.

We finish this section with some results about subgroups of $\mathrm{PSL}(n + 1, \mathbb{C})$ that will be used later in the text.

Proposition 2.1.2. *Let $\Gamma \subset \mathrm{PSL}(n + 1, \mathbb{C})$ be a discrete group. Then Γ is finite if and only if every element in Γ has finite order.*

This proposition follows from the theorem below (see [182, Theorem 8.29]) and its corollary; see [160], [41] for details.

Theorem 2.1.3 (Jordan). *For any $n \in \mathbb{N}$ there is an integer $S(n)$ with the following property: If $G \subset \mathrm{GL}(n, \mathbb{C})$ is a finite subgroup, then G admits an abelian normal subgroup N such that $\mathrm{card}(G) \leq S(n)\mathrm{card}(N)$.*

Corollary 2.1.4. *Let G be a countable subgroup of $\mathrm{GL}(3, \mathbb{C})$, then there is an infinite commutative subgroup N of G.*

Proof of Proposition 2.1.2. If every element in G has finite order, then by Selberg's lemma (see for instance [186]) it follows that G has an infinite set of generators, say $\{\gamma_m\}_{m \in \mathbb{N}}$. Define
$$A_m = \langle \gamma_1, \ldots, \gamma_m \rangle,$$
then by Selberg's lemma A_m is finite and by Theorem 2.1.3 there is a normal commutative subgroup $N(A_m)$ of A_m such that
$$\mathrm{card}(A_m) \leq S(3)\mathrm{card}(N(A_m)).$$

Assume, without loss of generality, that $\mathrm{card}(A_m) = k_0\mathrm{card}(N(A_m))$ for some k_0 and every m. Set
$$n_m = \max\{o(g) : g \in N(A_m)\},$$
where $o(g)$ represents the order of g, and consider the following cases:

Case 1. The sequence $(n_m)_{m \in \mathbb{N}}$ is unbounded.

In this case we can assume that
$$k_0 n_j < n_{j+1} \text{ for all } j.$$

Now for each m, consider $\gamma_m \in N(A_m)$ such that $o(\gamma_m) = n_m$. Thus $\gamma_m^{k_0} \in \bigcap_{m \leq j} N(A_j)$ and $\gamma_m^{k_0} \neq \gamma_j^{k_0}$. Hence $\langle \gamma_m^{k_0} : m \in \mathbb{N} \rangle$ is an infinite commutative subgroup of G.

Case 2. The sequence $(n_m)_{m \in \mathbb{N}}$ is bounded.

We may assume that $n_m = c_0$ for every index m. Let us construct the following sequence:

Step 1. Assume that $\operatorname{card}(N(A_1)) > k_0 c_o^3$. For each $m > 1$, consider the map

$$\phi_{1,m} : N(A_1) \longrightarrow A_m/N(A_m)$$

given by $l \mapsto N(A_m)l$. Since $\operatorname{card}(N(A_1)) > \operatorname{card}(A_m/N(A_m)) = k_0$, we deduce that $\phi_{1,m}$ is not injective. Then there is an element $w_1 \neq id$ and a subsequence $(B_n)_{n \in \mathbb{N}} \subset (A_n)_{n \in \mathbb{N}}$ such that $\operatorname{card}(N(B_1)) > k_0 c_o^4$ and $w_1 \in \bigcap_{m \in \mathbb{N}} N(B_m)$.

Step 2. For every $m \in \mathbb{N}$ consider the map $\phi_{2,m} : N(B_1)/\langle w_1 \rangle \longrightarrow B_m/N(B_m)$ given by $\langle w_1 \rangle l \mapsto N(B_m)l$. As in step 1 we can deduce that there is an element w_2 and a subsequence $(C_n)_{n \in \mathbb{N}} \subset (B_n)_{n \in \mathbb{N}}$ such that $\operatorname{card}(N(C_1)) > k_0 c_o^5$ and $w_2 \in \bigcap_{m \in \mathbb{N}} N(C_m) - \langle w_1 \rangle$.

Step 3. For $m \in \mathbb{N}$ consider the map $\phi_{3,m} : N(C_1)/\langle w_1, w_2 \rangle \longrightarrow C_m/N(C_m)$ given by $\langle w_1, w_2 \rangle l \mapsto N(C_m)l$. As in step 2 we deduce that there is an element w_3 and a subsequence $(D_n)_{n \in \mathbb{N}} \subset (C_n)_{n \in \mathbb{N}}$ such that $\operatorname{card}(N(A_1)) > k_0 c_o^6$ and $w_3 \in \bigcap_{m \in \mathbb{N}} N(D_m) - \langle w_1, w_2 \rangle$.

Continuing this process ad infinitum we deduce that $\langle w_m : m \in \mathbb{N} \rangle$ is an infinite commutative subgroup. $\qquad \square$

2.2 Complex hyperbolic geometry. The ball model

There are three classical models for complex hyperbolic n-space, namely:

(i) the unit ball model in $\mathbb{C}^n$;

(ii) the projective ball model in $\mathbb{P}_{\mathbb{C}}^n$; and

(iii) the Siegel domain model in $\mathbb{C}^n$.

Let us discuss first the ball models for complex hyperbolic space. For this, let $\mathbb{C}^{n,1}$ denote the vector space $\mathbb{C}^{n+1}$ equipped with the Hermitian form $\langle , \rangle$ given by

$$\langle u, v \rangle = u_1 \bar{v}_1 + \cdots + u_n \bar{v}_n - u_{n+1} \bar{v}_{n+1},$$

where $u = (u_1, u_2, \ldots, u_{n+1})$ and $v = (v_1, v_2, \ldots, v_{n+1})$. This form corresponds to the Hermitian matrix

$$H = \begin{pmatrix} 1 & 0 & 0 \\ 0 & 1 & 0 \\ 0 & 0 & -1 \end{pmatrix}.$$

One obviously has $\langle u, v \rangle = uHv^*$, where v^* is the Hermitian adjoint of v, i.e., it is the column vector with entries $\bar{v}_i$, $i = 1, \ldots, n+1$. Notice H has n positive eigenvalues and a negative one, so it has signature $(n, 1)$.

As before, we think of $\mathbb{C}^{n+1} \approx \mathbb{C}^{n,1}$ as being the union $V_- \cup V_0 \cup V_+$ of negative, null and positive vectors z, depending respectively (in the obvious way) on the sign of $\langle z, z \rangle = |z_1|^2 + \cdots + |z_n|^2 - |z_{n+1}|^2$. It is clear that each of the three sets $V_- \cup V_0 \cup V_+$ is a union of complex lines; that is, if a vector v is in V_-, then every complex multiple of v is a negative vector, and similarly for V_0 and V_+. The set V_0 is often called *the cone of light*, or the space of null vectors for the quadratic form $Q(z) = \langle z, z \rangle$.

We now look at the intersection of V_0 and V_- with the hyperplane in $\mathbb{C}^{n,1}$ defined by $z_{n+1} = 1$. For V_0 we get the $(2n-1)$-sphere

$$\mathbb{S} := \{ (z_1, \ldots, z_{n+1}) \in \mathbb{C}^{n+1} \,|\, |z_1|^2 + \cdots + |z_n|^2 = 1 \}.$$

For V_- we get the ball $\mathbb{B}$ bounded by $\mathbb{S}$:

$$\mathbb{B} := \{ (z_1, \ldots, z_{n+1}) \in \mathbb{C}^{n+1} \,|\, |z_1|^2 + \cdots + |z_n|^2 < 1 \}.$$

This ball, equipped with the complex hyperbolic metric serves as a model for complex hyperbolic geometry, this is the unit ball model of complex hyperbolic space. We refer to [169] for details on this model and for beautiful explanations about the way in which it relates to the Siegel domain model that we explain in the following section.

To get the complex hyperbolic space we must endow $\mathbb{B}$ with the appropriate metric. We do so in a similar way to how we did it in Chapter 1 for the real hyperbolic space. Consider the group $\mathrm{U}(n,1)$ of elements in $\mathrm{GL}\,(n+1, \mathbb{C})$ that preserve the above Hermitian form. That is, we consider matrices A satisfying $A^*HA = H$, where A^* is the Hermitian transpose of A (that is, each column vector v with entries $v_0, v_1, \ldots, v_n$, is replaced by its transpose v^*, the row vector $(\bar{v}_0, \bar{v}_1, \ldots, \bar{v}_n)$). It is easy to see that $\mathrm{U}(n,1)$ acts transitively on $\mathbb{B}$ with isotropy $\mathrm{U}(n)$ (see [67, Lemma 3.1.3]). In fact $\mathrm{U}(n,1)$ acts transitively on the space of negative lines in $\mathbb{C}^{n,1}$.

Let $\underline{0} = (0, \ldots, 0, 1)$ denote the centre of the ball $\mathbb{B}$, consider the space $T_{\underline{0}}\mathbb{B} \cong \mathbb{C}^n$ tangent to $\mathbb{B}$ at $\underline{0}$, and put on it the usual Hermitian metric on $\mathbb{C}^n$. Now we use the action of $\mathrm{U}(n,1)$ to spread the metric to all tangent spaces $T_x \mathbb{H}_{\mathbb{C}}^n$, using that the action is transitive and the isotropy is $\mathrm{U}(n)$, which preserves the usual metric on $\mathbb{C}^n$. We thus get a Hermitian metric on $\mathbb{B}$, which is clearly homogeneous. This is the complex hyperbolic metric, and the ball $\mathbb{B}$, equipped with this metric, serves as a model for complex hyperbolic n-space $\mathbb{H}_{\mathbb{C}}^n$. This is the *unit ball model for complex hyperbolic space*. The boundary $\partial \mathbb{H}_{\mathbb{C}}^n$ is called *the sphere at infinity* (it is called *the absolute* in [67]).

Notice that this way of constructing a model for the complex hyperbolic space $\mathbb{H}_{\mathbb{C}}^n$ is entirely analogous to the method used in Chapter 1 to construct the real hyperbolic space $\mathbb{H}_{\mathbb{R}}^n$. Yet, there is one significant difference. In the real case the action of $\mathrm{Iso}(\mathbb{H}_{\mathbb{R}}^n)$ on the unit ball has isotropy $O(n)$ and this group acts transitively on the spaces of lines and 2-planes through a given point. Thence $\mathbb{H}_{\mathbb{R}}^n$ has constant sectional curvature. In the complex case, the corresponding isotropy group is $U(n)$, which acts transitively on the space of complex lines through a given point, but it does not act transitively on the space of real 2-planes: a totally real plane cannot be taken into a complex line by an element in $U(n)$. Therefore the sectional curvature of $\mathbb{H}^n_{\mathbb{C}}$ is not constant, though it has constant holomorphic curvature.

Observe that for $n = 1$ one gets the complex hyperbolic line $\mathbb{H}^1_{\mathbb{C}}$. This corresponds to the unit ball

$$\{(z_1, z_2) \in \mathbb{C}^2 \,|\, |z_1| < 1 \text{ and } z_2 = 1\}.$$

Notice that $U(1)$ is isomorphic to $SO(2)$, hence $\mathbb{H}^1_{\mathbb{C}}$ is biholomorphically isometric to the open ball model of the real hyperbolic space $\mathbb{H}^2_{\mathbb{R}}$. Moreover, since $U(n, 1)$ acts transitively on the space of negative lines in $\mathbb{C}^{n,1}$, every such line can be taken into the line spanned by the vector $\{(0, \ldots, 0, 1)\} \subset \mathbb{C}^{n,1}$ and the above considerations essentially show that the induced metric on the unit ball in this complex line corresponds to the usual real hyperbolic metric on the ball model for $\mathbb{H}^2_{\mathbb{R}}$. That is: *every complex line that meets $\mathbb{H}^n_{\mathbb{C}}$ determines an embedded copy of $\mathbb{H}^1_{\mathbb{C}} \cong \mathbb{H}^2_{\mathbb{R}}$* (see [67, §1.4] or [169, §5.2] for clear accounts on $\mathbb{H}^1_{\mathbb{C}}$).

It is now easy to construct the projective ball model for complex hyperbolic space, which is the model we actually use in the sequel. For this we notice that if a complex line through the origin $0 \in \mathbb{C}^{n+1}$ is null, then it meets the above sphere $\mathbb{S}$ at exactly one point. Hence the projectivisation $(V_0 \setminus \{0\})/\mathbb{C}^*$ of V_0 is diffeomorphic to the $(2n - 1)$-sphere $\mathbb{S}$. Similar considerations apply for the negative lines, so the projectivisation $[V_-]$ is the open $2n$-ball $[\mathbb{B}]$ bounded by the sphere $[S] = [V_-]$.

The ball $[\mathbb{B}]$ in $\mathbb{P}^n_{\mathbb{C}}$ can be equipped with the metric coming from the complex hyperbolic metric in $\mathbb{B}$, and we get the *projective ball model for complex hyperbolic space*. From now on, unless it is stated otherwise, the symbol $\mathbb{H}^n_{\mathbb{C}}$ will denote this model for complex hyperbolic space. The corresponding Hermitian metric is the *Bergmann metric*, up to multiplication by a constant. It is clear from the above construction that the projective Lorentz group $PU(n, 1)$ acts on $\mathbb{H}^n_{\mathbb{C}}$ as its the group of holomorphic isometries.

In [67, 3.1.7] there is an algebraic expression for the distance function in complex hyperbolic space which is useful, among other things, for making computations. For this, recall first that in Euclidean space the distance function is determined by the usual inner product, and this is closely related with the angle between vectors x, y:

$$\cos(\angle(x, y)) = \frac{|x \cdot y|}{\|x\| \|y\|} = \sqrt{\hat{\delta}(x, y)},$$

where $\hat{\delta}(x, y) = \frac{(x \cdot y)(y \cdot x)}{(x \cdot x)(y \cdot y)}$. Similarly, the Bergmann metric can be expressed (up to multiplication by a constant) as follows: given points $[x], [y] \in \mathbb{H}^n_{\mathbb{C}}$, their complex hyperbolic distance ρ is:

$$\rho([x], [y]) = 2 \cosh^{-1}(\sqrt{\delta(x, y)});$$

$$\delta(x, y) = \frac{\langle x, y \rangle \langle y, x \rangle}{\langle x, x \rangle \langle y, y \rangle},$$

where $\langle , \rangle$ is now the Hermitian product given by the quadratic form Q used to construct $\mathbb{H}^n_{\mathbb{C}}$.

2.2.1 Totally geodesic subspaces

Given a point $z \in \mathbb{H}^n_{\mathbb{C}}$ one wishes to know what the geodesics through z, and more generally the totally geodesic subspaces of $\mathbb{H}^n_{\mathbb{C}}$, look like. Here is a first answer. Consider a complex projective line $\mathcal{L}$ in $\mathbb{P}^n_{\mathbb{C}}$ passing by z. Then $\mathcal{L} \cap \mathbb{H}^n_{\mathbb{C}}$ is a holomorphic submanifold of $\mathbb{H}^n_{\mathbb{C}}$ and we already know (see the previous section or [67, Theorem 3.1.10]) that $\mathcal{L} \cap \mathbb{H}^n_{\mathbb{C}}$ is isometric to $\mathbb{H}^1_{\mathbb{C}}$; the latter can be regarded as $\mathbb{H}^2_{\mathbb{R}}$ equipped with Poincare's ball model for real hyperbolic geometry. This "2-plane" $\mathcal{L} \cap \mathbb{H}^n_{\mathbb{C}}$ is totally geodesic, i.e., every geodesic in $\mathbb{H}^n_{\mathbb{C}}$ joining two points in $\mathcal{L} \cap \mathbb{H}^n_{\mathbb{C}}$ is actually contained in $\mathcal{L} \cap \mathbb{H}^n_{\mathbb{C}} \cong \mathbb{H}^1_{\mathbb{C}}$. This type of surfaces in $\mathbb{H}^n_{\mathbb{C}}$ are called *complex geodesics* (they are also called *complex slices*). They have constant negative curvature for the Bergman metric, and we assume this metric has been scaled so that these slices have constant sectional curvature -1 (see [67]). The intersection of the projective line $\mathcal{L}$ with the boundary $\partial \mathbb{H}^n_{\mathbb{C}}$ is a circle $\mathbb{S}^1$. This kind of circles in the sphere at infinity, which bound a complex slice, are called *chains*.

Notice that two distinct points z_1, z_2 in $\mathbb{H}^n_{\mathbb{C}}$ determine a unique line in $\mathbb{P}^n_{\mathbb{C}}$, so there is a unique complex geodesic $\mathcal{L}$ passing through them. There is also a unique real geodesic in $\mathcal{L} \cong \mathbb{H}^2_{\mathbb{R}}$ passing through these points. Of course this statement can be easily adapted to include the case when either one, or both, of these points is in the boundary $\partial \mathbb{H}^n_{\mathbb{C}}$.

Each real geodesic in $\mathbb{H}^n_{\mathbb{C}}$ is determined by its end points in the sphere $\partial \mathbb{H}^n_{\mathbb{C}}$. For each point $q \in \partial \mathbb{H}^n_{\mathbb{C}}$, the real geodesics in $\mathbb{H}^n_{\mathbb{C}}$ which end at q are parametrised by the points in $\mathbb{R}^{2n-1} \approx \partial \mathbb{H}^n_{\mathbb{C}} \backslash q$ and they form a *parabolic pencil* (see [67, Sections 7.27, 7.28]). Each of these real geodesics σ is contained in a complex geodesic Σ asymptotic to q, and the set of all complex geodesics end at q has the natural structure of an $(n-1)$-dimensional complex affine plane.

In fact, since each complex geodesic Σ corresponds to the intersection of the ball $\mathbb{H}^n_{\mathbb{C}}$ with a complex projective line $\mathcal{L}$, one has that Σ is asymptotic to all points in $\mathcal{L} \cap \partial \mathbb{H}^n_{\mathbb{C}}$, which form a circle $\mathbb{S}^1$. If the complex geodesic Σ ends at $q \in \partial \mathbb{H}^n_{\mathbb{C}}$, it follows that the real geodesics in Σ asymptotic to q are parametrised by $\mathbb{R} \approx (\mathcal{L} \cap \partial \mathbb{H}^n_{\mathbb{C}}) \setminus \{q\}$.

More generally, if $\mathcal{P}$ is a complex projective subspace of $\mathbb{P}^n_{\mathbb{C}}$ of dimension k that passes through the point $z \in \mathbb{H}^n_{\mathbb{C}}$, then $\mathcal{P} \cap \mathbb{H}^n_{\mathbb{C}}$ is obviously a complex holomorphic submanifold of $\mathbb{H}^n_{\mathbb{C}}$. Then Theorem 3.1.10 in [67] tells us that $\mathcal{P} \cap \mathbb{H}^n_{\mathbb{C}}$ is actually a totally geodesic subspace of $\mathbb{H}^n_{\mathbb{C}}$ which is biholomorphically isometric to $\mathbb{H}^k_{\mathbb{C}}$. Such a holomorphic submanifold of $\mathbb{H}^n_{\mathbb{C}}$ is called a $\mathbb{C}^k$-*plane*; so a $\mathbb{C}^k$-*plane* is a complex geodesic. The boundary of a $\mathbb{C}^k$-*plane* is a sphere of real dimension

$2k-1$ called a $\mathbb{C}^k$-*chain*. This is the set of points where the corresponding projective plane meets the sphere at infinity $\partial\mathbb{H}_{\mathbb{C}}^n$. A $\mathbb{C}^1$-*chain* is simply a chain.

Goldman shows in his book that behind $\mathbb{C}^k$-planes, there is only another type of totally geodesic subspaces in $\mathbb{H}_{\mathbb{C}}^n$: The totally real projective subspaces:

Definition 2.2.1. Let $\widetilde{\mathcal{R}}^{k+1}$ be a linear real subspace of $\mathbb{C}^{n,1}$ of real dimension $k+1$ which contains negative vectors. We say that $\widetilde{\mathcal{R}}^{k+1}$ is *totally real* with respect to the Hermitian form Q if $J(\widetilde{\mathcal{R}}^{k+1})$ is Q-orthogonal to $\widetilde{\mathcal{R}}^{k+1}$, where J denotes complex multiplication by i. A *totally real subspace of* $\mathbb{H}_{\mathbb{C}}^n$ means the intersection with $\mathbb{H}_{\mathbb{C}}^n$ of the projectivisation $\mathcal{R}^k := [\widetilde{\mathcal{R}}^{k+1}]$ of a totally real projective subspace $\widetilde{\mathcal{R}}^{k+1}$ of $\mathbb{C}^{n,1}$. Such a plane in $\mathbb{H}_{\mathbb{C}}^n$ is called an $\mathbb{R}^k$-*plane*. (Of course this can only happen if $k \leq n$.) An $\mathbb{R}^2$-plane is called *a real slice*.

It is easy to see that $\mathrm{PU}(n,1)$ acts transitively on the set of all $\mathbb{R}^k$-planes in $\mathbb{H}_{\mathbb{C}}^n$, for every k. One has (see [67, Section 3.1]):

Theorem 2.2.2. *Every totally geodesic submanifold of* $\mathbb{H}_{\mathbb{C}}^n$ *is either a* $\mathbb{C}^k$-*plane or an* $\mathbb{R}^k$-*plane. In particular* $\mathbb{H}_{\mathbb{C}}^n$ *has no totally geodesic real submanifolds of codimension* 1 *(for* $n > 1$*). Furthermore:*

(i) *Every* $\mathbb{C}^k$-*plane, with its induced metric, is biholomorphically isometric to* $\mathbb{H}_{\mathbb{C}}^k$*. Every complex line in* $\mathbb{H}_{\mathbb{C}}^n$ *is biholomorphically isometric to Poincaré's ball model of* $\mathbb{H}_{\mathbb{R}}^2$*.*

(ii) *Ever* $\mathbb{R}^k$-*plane, with its induced metric, is isometric to the real hyperbolic space* $\mathbb{H}_{\mathbb{R}}^k$ *equipped with the Beltrami-Klein model for hyperbolic geometry.*

(iii) *In particular,* $\mathbb{H}_{\mathbb{C}}^n$ *has two types of real* 2*-planes which are both totally geodesic: complex slices, and real slices. Furthermore, it is at these two types of* 2*-planes where* $\mathbb{H}_{\mathbb{C}}^n$ *attains its bounds regarding sectional curvature: the sectional curvature in* $\mathbb{H}_{\mathbb{C}}^n$ *varies in the interval* $[-1, -\frac{1}{4}]$ *with the upper bound corresponding to the curvature of real slices and the lower one being attained at complex slices.*

2.2.2 Bisectors and spines

Recall that in real hyperbolic geometry the group of isometries is generated by inversions on spheres of codimension 1 that meet orthogonally the sphere at infinity. This type of spheres are totally geodesic. These spheres also determine the sides of the Dirichlet fundamental domains. In complex hyperbolic space, there are no totally real submanifolds of codimension 1, and a reasonable substitute are the bisectors (or equidistant hypersurfaces), that we now define. These were introduced by Giraud (see [67]) and also used by Moser, who called them spinal surfaces, to construct the first examples of nonarithmetic lattices in $\mathrm{PU}(n,1)$.

Given points z_1, z_2 in $\mathbb{H}_{\mathbb{C}}^n$, we denote by $\rho(z_1, z_2)$ their complex hyperbolic distance.

Definition 2.2.3. Let z_1, z_2 be two distinct points in $\mathbb{H}^n_{\mathbb{C}}$.

(i) The *bisector of z_1 and z_2* is the set

$$\mathfrak{E}\{z_1, z_2\} = \{z \in \mathbb{H}^n_{\mathbb{C}} \,|\, \rho(z, z_1) = \rho(z, z_2)\}.$$

(ii) The boundary of a bisector is called *a spinal sphere* in $\partial\mathbb{H}^n_{\mathbb{C}}$.

(iii) The complex geodesic $\Sigma = \Sigma(z_1, z_2)$ in $\mathbb{H}^n_{\mathbb{C}}$ spanned by z_1, z_2 is called the *complex spine* (or simply the $\mathbb{C}$-*spine*) of the bisector $\mathfrak{E}\{z_1, z_2\}$ with respect to z_1 and z_2.

(iv) The *spine* $\sigma = \sigma(z_1, z_2)$ of $\mathfrak{E}\{z_1, z_2\}$ with respect to z_1 and z_2 is the intersection of the bisector with the complex spine of z_1 and z_2:

$$\sigma := \mathfrak{E}\{z_1, z_2\} \cap \Sigma = \{z \in \Sigma \,|\, \rho(z, z_1) = \rho(z, z_2)\}.$$

Notice that by the above discussion, the complex spine Σ is the intersection with $\mathbb{H}^n_{\mathbb{C}}$ of projectivisation of a complex linear 2-space in $\mathbb{C}^{n,1}$. Thence one has a (holomorphic) orthogonal projection $\pi_{\Sigma} : \mathbb{H}^n_{\mathbb{C}} \to \Sigma$ induced by the orthogonal projection in $\mathbb{C}^{n,1}$ (with respect to the corresponding quadratic form). Then one has the following theorem, which is essentially the *Slice Decomposition Theorem* of Giraud and Mostow (see Theorem 5.1, its corollary 5.1.3 and lemma 5.1.4 in [67]). Recall that a bisector is a real hypersurface in the complex manifold $\mathbb{H}^n_{\mathbb{C}}$ and therefore comes equipped with a natural CR-structure and a Levi-form.

Theorem 2.2.4. *The bisector $\mathfrak{E}$ is a real analytic hypersurface in $\mathbb{H}^n_{\mathbb{C}}$ diffeomorphic to $\mathbb{R}^{2n-1}$, which fibres analytically over the spine σ with projection being the restriction to $\mathfrak{E}$ of the orthogonal projection $\pi_{\Sigma} : \mathbb{H}^n_{\mathbb{C}} \to \Sigma$:*

$$\mathfrak{E} = \pi_{\Sigma}^{-1}(\sigma) = \bigcup_{s \in \sigma} \pi_{\Sigma}^{-1}((s)).$$

Furthermore, each bisector $\mathfrak{E}$ is Levi-flat and the slices $\pi_{\Sigma}^{-1}((s)) \subset \mathfrak{E}$ are its maximal holomorphic submanifolds. Hence $\mathfrak{E}$, the spine, the complex spine and the slices are independent of the choice of points z_1, z_2 used to define them.

Of course, spinal spheres are diffeomorphic to $\mathbb{S}^{2n-2}$. We remark too [67, Theorem 5.1.6] that the above association $\mathfrak{E} \rightsquigarrow \sigma$ defines a bijective correspondence between bisectors and geodesics in $\mathbb{H}^n_{\mathbb{C}}$: every real geodesic σ is contained in a unique complex geodesic Σ; then one has an orthogonal projection $\pi_{\Sigma} : \mathbb{H}^n_{\mathbb{C}} \to \Sigma$ and the bisector is $\mathfrak{E} = \pi_{\Sigma}^{-1}(\sigma)$.

Another nice property of bisectors is that just as they decompose naturally into complex hyperplanes, as described by the Slice Decomposition Theorem above, they also decompose naturally into totally real geodesic subspaces and one has the corresponding Meridianal Decomposition Theorem (see [67, Theorem 5.1.10]).

2.3 The Siegel domain model

There is another classical model for complex hyperbolic geometry, the Siegel domain (or paraboloid) model. While the ball models describe complex projective space regarded from within, the Siegel domain model describes this space as regarded from a point at infinity. Thence it is in some sense analogous to the upper-half plane model for real hyperbolic geometry. The basic references for this section are [67], [71] and [169].

Consider $\mathbb{C}^n$ as $\mathbb{C}^{n-1} \times \mathbb{C}$ with coordinates (w', w_n), $w' \in \mathbb{C}^{n-1}$, and let $\langle\langle\,,\,\rangle\rangle$ be the usual Hermitian product in $\mathbb{C}^{n-1}$. So $\langle\langle w', w'\rangle\rangle = w'_1 \bar{w}'_1 + \cdots + w'_{n-1}\bar{w}'_{n-1}$. The *Siegel domain* $\mathfrak{S}^n$ consists of the points in $\mathbb{C}^n$ satisfying

$$2\mathrm{Re}(w_n) > \langle\langle w', w'\rangle\rangle \,.$$

Its boundary is a paraboloid in $\mathbb{C}^n$.

Now consider the embedding of $\mathfrak{S}^n$ in $\mathbb{P}^n_{\mathbb{C}}$ given by

$$(w', w_n) \overset{B}{\mapsto} [w', \frac{1}{2} - w_n, \frac{1}{2} + w_n]\,.$$

We claim that the image $B(\mathfrak{S}^n)$ is the ball $\mathbb{B}$ of negative points that serves as a model for complex hyperbolic $\mathbb{H}^n_{\mathbb{C}}$. To see this, notice that given points $w = (w', w_n)$, $z = (z', z_n) \in \mathbb{C}^n$, the Hermitian product of their image in $\mathbb{P}^n_{\mathbb{C}}$ (induced by the product $\langle\,,\,\rangle$ in $\mathbb{C}^{n,1}$) takes the form

$$\langle B(w), B(z)\rangle = \langle\langle w', z'\rangle\rangle - w_n - \bar{z}_n\,.$$

In particular $\langle B(w), B(w)\rangle = \langle\langle w', w'\rangle\rangle - \mathrm{Re}\,w_n$. Hence $B(w)$ is a negative point in $\mathbb{P}^n_{\mathbb{C}}$ if and only if w is in $\mathfrak{S}^n$.

Now consider the null vector $\tilde{p}_\infty := (0', -1, 1) \in \mathbb{C}^{n,1}$ and its image p_∞ in $\mathbb{P}^n_{\mathbb{C}}$. Let H_∞ be the unique complex projective hyperplane in $\mathbb{P}^n_{\mathbb{C}}$ which is tangent to $\mathbb{H}^n_{\mathbb{C}}$ at p_∞. This hyperplane consists of all points $[z] \in \mathbb{P}^n_{\mathbb{C}}$ whose homogeneous coordinates $[z_1 : \ldots : z_{n+1}]$ satisfy $z_n = z_{n+1}$. Therefore the image of B does not meet H_∞ and B provides an affine coordinate chart for $\mathbb{P}^n_{\mathbb{C}} \setminus H_\infty$, carrying the Siegel domain $\mathfrak{S}^n$ into the ball $\mathbb{H}^n_{\mathbb{C}}$.

Notice that the boundary of $\mathfrak{S}^n$ is the paraboloid $\{2\mathrm{Re}(w_n) = \langle\langle w', w'\rangle\rangle\}$, and its image under B is the sphere $\partial\mathbb{H}^n_{\mathbb{C}}$ minus the null point $p_\infty := [0', -1, 1] \in \mathbb{P}^n_{\mathbb{C}}$, where $0' := (0, \ldots, 0) \in \mathbb{C}^{n-1}$. That is:

$$\partial\mathfrak{S}^n \cong \partial\mathbb{H}^n_{\mathbb{C}} \setminus \{q_\infty\}\,.$$

One thus has an induced complex hyperbolic metric on $\mathfrak{S}^n$, induced by the Bergman metric on $\mathbb{H}^n_{\mathbb{C}}$ (see [71, p. 520] for the explicit formula).

Definition 2.3.1. For each positive real number u, the *horosphere* in $\mathfrak{S}^n$ (centred at q_∞) of level u is the set

$$\mathfrak{H}_u := \{w = (w', w_n) \in \mathfrak{S}^n \,\big|\, \mathrm{Re}\,w_n - \langle\langle w', w'\rangle\rangle = u\}\,.$$

We also set $\mathfrak{H}_0 := \partial\mathfrak{S}^n$.

Hence, horospheres in $\mathfrak{S}^n$ are paraboloids, the translates of the boundary. When regarded in $\mathbb{H}^n_{\mathbb{C}}$ they become spheres, tangent to $\partial\mathbb{H}^n_{\mathbb{C}}$ at $q_\infty = [0', -1, 1$. This definition can be easily adapted horospheres in $\mathbb{H}^n_{\mathbb{C}}$ centred at every point in $\partial\mathbb{H}^n_{\mathbb{C}}$.

2.3.1 Heisenberg geometry and horospherical coordinates

Recall that the classical Heisenberg group is the group of 3×3 triangular matrices of the form

$$H = \begin{pmatrix} 1 & a & t \\ 0 & 1 & b \\ 0 & 0 & 1 \end{pmatrix}$$

where the coefficients are real numbers. This is a 3-dimensional nilpotent Lie group diffeomorphic to $\mathbb{R}^3$. Its group structure, coming from the multiplication of matrices, is a semi-direct product $\mathbb{C} \ltimes \mathbb{R}$ where $\mathbb{C}$ and $\mathbb{R}$ are being regarded as additive groups:

$$\big((a,b),t\big) \cdot \big((a',b'),t'\big) \mapsto \big((a+a',b+b'),t+t'+a\cdot b'\big),$$
$$\big(v,t\big) \cdot \big(v',t'\big) \mapsto \big(v+v',t+t'+a\cdot b'\big),$$

where $v = (a,b)$ and $v' = (a',b')$.

More generally, let V be a finite-dimensional real vector space, equipped with a symplectic form ω. This means that ω is a nondegenerate skew symmetric bilinear form on V. One has a Heisenberg group $H = H(V,\omega)$ associated with the pair (V,ω). This group is a semi-direct product $V \ltimes \mathbb{R}$, where the group structure is given by the following law:

$$(v_1,t_1) \cdot (v_2,t_2) = (v_1 + v_2, t_1 + t_2 + 2\omega(v_1,v_2));$$

the factor 2 is included for conventional reasons. This group is a central extension of the additive group V and there is an exact sequence

$$0 \longrightarrow \mathbb{R} \longrightarrow H(V,\omega) \longrightarrow V \longrightarrow 0.$$

A *Heisenberg space* is a principal $H(V,\omega)$-homogeneous space N, say a smooth manifold, for some Heisenberg group as above. In other words, H acts (say by the left) transitively on N with trivial isotropy. So N is actually parametrised by H and it can be equipped with a Lie group structure coming from that in H.

Consider now the isotropy subgroup G_∞ of the null point $p_\infty := [0', -1, 1] \in \partial\mathbb{H}^n_{\mathbb{C}}$ under the action of $\mathrm{PU}(n,1)$. Let $\mathfrak{N}$ be the set of unipotent elements in G_∞. It is proved in [67, §4.2] (see also [71]) that $\mathfrak{N}$ is isomorphic to a Heisenberg group as above. This group is a semidirect product $\mathbb{C}^{n-1} \ltimes \mathbb{R}$ and consists of the so-called *Heisenberg translations* $\{T_{\zeta,t}\}$. These are more easily defined in $\mathfrak{S}^n$. For each $\zeta \in \mathbb{C}^{n-1}$, $t \in \mathbb{R}$ and $(w', w_n) \in \mathfrak{S}^n \subset \mathbb{C}^n$ one has:

$$(w', w_n) \xrightarrow{T_{\zeta,u}} \left(w' + \zeta,\ w_n + \langle\langle w',\zeta\rangle\rangle + \frac{1}{2}\langle\langle \zeta,\zeta\rangle\rangle - \frac{1}{2}it\right).$$

Notice that each orbit in $\mathfrak{S}^n$ is a horosphere $\mathfrak{H}_u$, and $\mathfrak{N}$ acts simply transitively on $\mathfrak{H}_0 := \partial\mathfrak{S}^n$.

One also has the one-parameter group $\mathfrak{D} = \{D_u\}$ of *real Heisenberg dilatations*: For each $u > 0$ define

$$(w', w) \xrightarrow{D_t} \left(\sqrt{u}\,w', u\,w_n\right).$$

Each dilatation carries horospheres into horospheres.

These two groups of transformations were used in [71] to equip the complex hyperbolic space with horospherical coordinates, obtained by identifying $\mathfrak{S}^n$ with the orbit of a "marked point" under the action of the group $\mathfrak{N} \cdot \mathfrak{D}$ generated by Heisenberg translations and Heisenberg dilatations. The choice of "marked point" is the obvious one: $(0', \frac{1}{2}) \in \mathfrak{S}^n$, the inverse image under the map B of the centre $\underline{0} = [0', 0, 1] \in \mathbb{H}^n_{\mathbb{C}}$.

We thus get an identification $(\mathbb{C}^{n-1} \times \mathbb{R} \times \mathbb{R}_+) \xrightarrow{\cong} \mathbb{H}^n_{\mathbb{C}}$ given by

$$(\zeta, t, u) \mapsto \left(\zeta, \frac{1}{2}\big(1 - \langle\langle\zeta, \zeta\rangle\rangle - u + it\big), \frac{1}{2}\big(1 - \langle\langle\zeta, \zeta\rangle\rangle - u + it\big)\right).$$

Following [71] we call $(\zeta, t, u) \in \mathbb{C}^{n-1} \times \mathbb{R} \times \mathbb{R}_+$ the *horospherical coordinates* of the corresponding point in $\mathbb{H}^n_{\mathbb{C}}$.

From the previous discussion we also get a specific identification of $\partial\mathbb{H}^n_{\mathbb{C}} \setminus q_\infty \cong \mathfrak{N}$. Thence the sphere at infinity can be thought of as being "the horosphere" of level 0, and if we remove from it the point q_∞, then it carries the structure of a Heisenberg space $\mathbb{C}^{n-1} \times \mathbb{R}$ where the group operation is

$$(\zeta, t) \cdot (\zeta', t') = \big(\zeta + \zeta', t + t' + 2\Im(\langle\langle\zeta, \zeta'\rangle\rangle)\big).$$

We now consider a point $q \in \partial\mathbb{H}^n_{\mathbb{C}}$, the family $\{\mathfrak{H}_u(q)\}$ of horospheres centred at q, and the pencil of all real geodesics in $\mathbb{H}^n_{\mathbb{C}}$ ending at q. Just as in real hyperbolic geometry, one has (see [67, §4.2] or [71, §1.3]) that every such geodesic is orthogonal to every horosphere $\mathfrak{H}_u(q)$. Thence, for every $u, u' \geq 0$, the geodesic from q to a point $x \in \mathfrak{H}_u(q)$ meets the horosphere $\mathfrak{H}_{u'}(q)$ at exactly one point. This gives a canonical identification $\Pi : \mathfrak{H}_u(q) \to \mathfrak{H}_{u'}(q)$ called *the geodesic perspective map*. In particular we get an identification between $\partial\mathbb{H}^n_{\mathbb{C}} = \mathfrak{H}_0(q)$ and every other horosphere.

2.3.2 The geometry at infinity

Just as there is a deep relation between real hyperbolic geometry and the conformal geometry on the sphere at infinity (as described in Chapter 1), so too there is a deep relation between the geometry of complex hyperbolic space and a geometry on its sphere at infinity. In this case the relevant geometry is the spherical CR (or Heisenberg) geometry. In both cases (real and complex hyperbolic geometry) this relation can be explained by means of the geodesic perspective introduced above.

In real hyperbolic geometry, geodesic perspective from a point q in the sphere at infinity identifies the various horospheres centred at q, and the corresponding maps between these spheres are conformal. This yields to the fact (explained in Chapter 1) that real hyperbolic geometry in the open ball degenerates into conformal geometry in the sphere at infinity. There is an analogous phenomenon in complex hyperbolic space, which we now briefly explain.

Recall that a CR-structure on a manifold means a codimension 1 sub-bundle of its tangent bundle which is complex and satisfies certain integrability conditions. In general, CR-manifolds arise as boundaries of complex manifolds, as for instance real hypersurfaces in complex manifolds, which carry a natural CR-structure.

In our setting, every horosphere, including $\partial \mathbb{H}_\mathbb{C}^n = \mathfrak{H}_0$, is a real hypersurface in $\mathbb{P}_\mathbb{C}^n$ and therefore carries a natural CR-structure. This is determined at each point by the unique complex $(n-1)$-dimensional subspace subspace of the bundle tangent to the corresponding horosphere. In fact, since real geodesics are orthogonal to all horospheres, the CR-structure can be regarded as corresponding to the Hermitian orthogonal complement of the line field tangent to the geodesics emanating from q, the centre of the horosphere in question. Therefore, it is clear that the geodesic perspective maps preserve the CR-structures on horospheres.

We are in fact interested in a refinement of this notion: spherical CR-structures:

Definition 2.3.2. A manifold M of dimension $2n - 1$ has a *spherical CR-structure* if it has an atlas which is locally modeled on the sphere $\mathbb{S}^{2n-1}$ with coordinate changes lying in the group $\mathrm{PU}(n, 1)$.

Amongst CR-manifolds, the spherical CR-manifolds are characterised as being those for which the Cartan connection on the CR-bundle has vanishing curvature.

Now consider the sphere $\mathbb{S}^{2n-1} \cong \partial \mathbb{H}_\mathbb{C}^n$. Then one has that $\mathrm{PU}(n, 1)$ acts on it by automorphisms that preserve the spherical CR-structure, and one actually has that every CR-automorphism of the sphere is an element in $\mathrm{PU}(n, 1)$. Thus one has that every CR-automorphism of the sphere at infinity corresponds to a holomorphic isometry of $\mathbb{H}_\mathbb{C}^n$.

In particular, the Heissenberg group $\mathfrak{N}$ acts by left multiplication on the sphere at infinity, and by holomorphic isometries on complex hyperbolic space.

2.4   Isometries of the complex hyperbolic space

This section is based on [67]. We consider the ball model for $\mathbb{H}_\mathbb{C}^n \subset \mathbb{P}_\mathbb{C}^n$, equipped with the Bergman metric, and we recall that $\mathrm{PU}(n, 1)$ is the group of holomorphic isometries of the complex hyperbolic n-space $\mathbb{H}_\mathbb{C}^n$. When $n = 1$ the space $\mathbb{H}_\mathbb{C}^1$ coincides with $\mathbb{H}_\mathbb{R}^2$, the group $\mathrm{PU}(n, 1)$ is $\mathrm{PSL}(2, \mathbb{R})$, and we know that its elements are classified into three types: elliptic, parabolic and hyperbolic. This classification is determined by their dynamics (the number and location of their fixed points), and also according to their trace. We will see that a similar classification holds for isometries of $\mathbb{H}_{\mathbb{C}}^2$, and to some extent also for those of $\mathbb{H}_{\mathbb{C}}^n$.

2.4.1 Complex reflections

Recall that in Euclidean geometry reflections play a fundamental role. Given a hyperplane $H \subset \mathbb{R}^n$, a way for defining the reflection on H is to look at its orthogonal complement $H^\perp$, and consider the orthogonal projections,

$$\pi_H : \mathbb{R}^n \to H \quad \text{and} \quad \pi_{H^\perp} : \mathbb{R}^n \to H^\perp \,.$$

Then the reflection on H is the map

$$x \mapsto \pi_H(x) - \pi_{H^\perp}(x) \,.$$

This notion extends naturally to complex geometry in the obvious way. Yet, in complex geometry this definition is too rigid and it is convenient to make it more flexible. For instance a hyperplane in $\mathbb{C}$ is just a point, say the origin 0, and the reflection above yields to the antipodal map, while we would like to get the full group $\mathrm{U}(1)$. Thus, more generally, given a complex hyperplane $H \subset \mathbb{C}^n$, to define a reflection on H we consider the orthogonal projections

$$\pi_H : \mathbb{R}^n \to H \quad \text{and} \quad \pi_{H^\perp} : \mathbb{R}^n \to H^\perp \,,$$

as before, but now with respect to the usual Hermitian product in $\mathbb{C}^n$. Then a reflection on H is any map of the form

$$x \mapsto \pi_H(x) + \zeta \cdot \pi_{H^\perp}(x) \,,$$

where ζ is a unit complex number. In this way we get, for instance, that the group of such reflections in $\mathbb{C}^2$ is $\mathrm{U}(2) \cong \mathrm{SU}(2) \times \mathrm{U}(1) \cong \mathbb{S}^3 \times \mathbb{S}^1$.

Sometimes the name "complex reflection" requires also that the complex number ζ be a root of unity, so that the corresponding automorphism has finite order. And it is also usual to extend this concept so that complex reflection means an automorphism of $\mathbb{C}^n$ that leaves a hyperplane fix-point invariant. This includes for instance, automorphisms constructed as above but considering different Hermitian products, and this brings us closer to the subject we envisage here.

Denote by $\langle\,,\,\rangle$ the Hermitian product on $\mathbb{C}^{n,1}$ corresponding to the quadratic form Q of signature $(n,1)$. Let F be a complex linear subspace of $\mathbb{C}^{n,1}$ such that the restriction of the product $\langle\,,\,\rangle$ to F is nondegenerate. Then there is an orthogonal direct-sum decomposition

$$\mathbb{C}^{n,1} = F \oplus F^\perp \,,$$

where $F^\perp$ is the Q-orthogonal complement of F:

$$F^\perp := \{ z \in \mathbb{C}^{n,1} \,|\, \langle z, f \rangle = 0 \quad \forall\, f \in F \} \,.$$

Let $\pi_F, \pi_{F^\perp}$ be the corresponding orthogonal projections of $\mathbb{C}^{n,1}$ into F and $F^\perp$ respectively, so one has $\pi_{F^\perp}(z) = z - \pi_F(z)$. Then following [67, p. 68] we have:

Definition 2.4.1. For each unit complex number ζ, define *the complex reflection in F with reflection factor ζ* to be the element in $\mathrm{U}(n,1)$ defined by

$$\varrho_F^\zeta(z) = \pi_F(x) + \zeta\,\pi_{F^\perp}(z)\,.$$

When $\zeta = -1$ the complex reflection is said to be an *inversion*.

For instance, if F is 1-dimensional and V is a nonzero vector in this linear space, then

$$\pi_F(z) = \frac{\langle z, V \rangle}{\langle V, V \rangle}\,V\,,$$

and therefore the complex reflection in F with reflection factor ζ is

$$\varrho_F^\zeta(z) = \zeta z + (1 - \zeta)\frac{\langle z, V \rangle}{\langle V, V \rangle}\,V\,.$$

In particular, every complex reflection is conjugate to an inversion and in this case the formula is

$$I_F(z) = -z + 2\,\frac{\langle z, V \rangle}{\langle V, V \rangle}\,V\,. \tag{2.4.1}$$

Notice that classically reflections are taken with respect to a hyperplane, which is not required here: We now have reflections with respect to points, lines, etc., which is in fact a concept coming from classical geometry, where one speaks of symmetries with respect to points, lines, planes, etc.

Of course our interest is in looking at the projectivisations of these maps, that we call complex reflections as well.

Example 2.4.2. Consider in $\mathbb{C}^{2,1}$ the vector $v = (-1, 1, 0)$, which is positive for the Hermitian product

$$\langle z, w \rangle = z_1\bar{w}_1 + z_2\bar{w}_2 - z_3\bar{w}_3\,.$$

The inversion on the line F spanned by v is the map

$$I_v(z_1, z_2, z_3) = (-z_2, -z_1, -z_3).$$

Notice that $\mathbb{H}_\mathbb{C}^2$ is contained in the coordinate patch of $\mathbb{P}_\mathbb{C}^2$ with homogeneous coordinates $[u_1 : u_2 : 1]$. In these coordinates the automorphism of $\mathbb{H}_\mathbb{C}^2$ determined by I_v is the map $[u_1 : u_2 : 1] \mapsto [u_2 : u_1 : 1]$. Notice also that in this example the points in $\mathbb{H}_\mathbb{C}^2$ with homogeneous coordinates $[u : u : 1]$ are obviously fixed points of the inversion I_v. Indeed these points form a complex geodesic in $\mathbb{H}_\mathbb{C}^2$, which is the projectivisation of the orthogonal complement of F:

$$F^\perp = \{(w_1, w_2, w_3) \in \mathbb{C}^{2,1} \,|\, w_1 = -w_2\}\,.$$

Now consider in $\mathbb{C}^2$ the inversion with respect to the negative vector $w = (0, 0, 1)$. The inversion is: $I_w(z) = (-1 + 2z_3)\,z\,.$ Its fixed point set in $\mathbb{P}_\mathbb{C}^2$ consists of the point $[0:0:1]$, which is the centre of the ball $\mathbb{H}^2_{\mathbb{C}}$, together with the complex line in $\mathbb{P}^2_{\mathbb{C}}$ consisting of points with homogeneous coordinates $[u_1:u_2:0]$. These points are the projectivisation of the orthogonal complement of the line spanned by w; they are all positive vectors (or the origin).

Thence in the first case, the complex reflection has a complex geodesic of fixed points in $\mathbb{H}^2_{\mathbb{C}}$ and the corresponding projective map has a circle of fixed points in $\mathbb{S}^3 = \partial \mathbb{H}^2_{\mathbb{C}}$, plus another fixed point far from $\mathbb{H}^2_{\mathbb{C}}$. In the second example the reflection has a single fixed point in $\mathbb{H}^2_{\mathbb{C}}$ and the remaining fixed points form a complex projective line in $\mathbb{P}^2_{\mathbb{C}} \setminus (\overline{\mathbb{H}}^2_{\mathbb{C}})$. This illustrates the two types of complex reflections one has in $\mathbb{H}^2_{\mathbb{C}}$. Of course similar considerations apply in higher dimensions, but in that case there is a larger set of possibilities.

2.4.2 Dynamical classification of the elements in $\mathrm{PU}(2,1)$

Every automorphism γ of $\mathbb{H}^n_{\mathbb{C}}$ lifts to a unitary transformation $\tilde{\gamma} \in \mathrm{SU}(n,1)$. Just as for classical Möbius transformations in $\mathrm{PSL}(2,\mathbb{C})$ their geometry and dynamics is studied by looking at their liftings to $\mathrm{SL}(2,\mathbb{C})$, here too we study the geometry and dynamics of γ by looking at their liftings $\tilde{\gamma} \in \mathrm{U}(n,1)$. The fixed points of γ correspond to eigenvectors of $\tilde{\gamma}$. By the Brouwer fixed point theorem, every automorphism of the compact ball $\overline{\mathbb{H}}^n_{\mathbb{C}} := \mathbb{H}^n_{\mathbb{C}} \cup \partial \mathbb{H}^n_{\mathbb{C}}$ has a fixed point.

The following definition generalises to complex hyperbolic spaces the corresponding notions from the classical theory of Möbius transformations.

Definition 2.4.3. An element $g \in \mathrm{PU}(n,1)$ is called *elliptic* if it has a fixed point in $\mathbb{H}^n_{\mathbb{C}}$; it is *parabolic* if it has a unique fixed point in $\partial \mathbb{H}^n_{\mathbb{C}}$, and *loxodromic* (or *hyperbolic*) if it fixes a unique pair of points in $\partial \mathbb{H}^n_{\mathbb{C}}$.

In fact this classification can be refined as follows. Recall that a square matrix is unipotent if all its eigenvalues are 1.

Definition 2.4.4. An elliptic transformation in $\mathrm{PU}(n,1)$ is: *regular* if it can be represented by an element in $\mathrm{SU}(n,1)$ whose eigenvalues are pairwise different, or a *complex reflection* otherwise (and this can be either with respect to a point or to a complex geodesic). There are two classes of parabolic transformations in $\mathrm{PU}(n,1)$: *unipotent* if it can be represented as a unipotent element of $\mathrm{PU}(n,1)$, and *ellipto-parabolic* otherwise. A loxodromic element is strictly hyperbolic if it has a lifting whose eigenvalues are all real.

One has (see [67, p. 201]) that if γ is ellipto-parabolic, then there exists a unique invariant complex geodesic in $\mathbb{H}^n_{\mathbb{C}}$ on which γ acts as a parabolic element of $\mathrm{PSL}(2,\mathbb{R}) \cong \mathrm{Iso}\,\mathbb{H}^1_{\mathbb{C}} \cong \mathrm{Iso}\,\mathbb{H}^2_{\mathbb{R}}$. Furthermore, around this geodesic, γ acts as a nontrivial unitary automorphism of its normal bundle.

From now on in this section, we restrict to the case $n = 2$ and we think of $\mathbb{H}^2_{\mathbb{C}}$ as being the ball in $\mathbb{P}^2_{\mathbb{C}}$ consisting of points whose homogeneous coordinates satisfy $|z_1|^2 + |z_2|^2 < |z_3|^2$.

Let $g \in \mathrm{PU}(2,1)$ be an elliptic element. Since $\mathrm{PU}(2,1)$ acts transitively on $\mathbb{H}_\mathbb{C}^2$, we can assume that $[0:0:1]$ is fixed by g. If $\tilde{g}$ denotes a lift to $\mathrm{SU}(2,1)$ of g then $(0,0,1)$ is an eigenvector of $\tilde{g}$, so it is of the form

$$\tilde{g} = \begin{pmatrix} A & 0 \\ 0 & \lambda \end{pmatrix},$$

where $A \in \mathrm{U}(2)$ and $\lambda \in \mathbb{S}^1$. Then every eigenvector of $\tilde{g}$ has module 1 and g generates a cyclic group with compact closure.

Conversely, if $\tilde{g}$ is as above (an element in $\mathrm{U}(2) \times \mathbb{S}^1$), or a conjugate of such an element, then the transformation g induced by $\tilde{g}$ is elliptic.

When g is regular elliptic, then it has precisely three fixed points in $\mathbb{P}_\mathbb{C}^2$, which correspond to $(0,0,1)$ and two other distinct eigenvectors, both being positive vectors with respect to the Hermitian product $\langle \cdot, \cdot \rangle$. If g is elliptic but not regular, then there exist two cases: if g is a reflection with respect to a point x in $\mathbb{H}_\mathbb{C}^2$, then the set of fixed points of g is the polar to x which does not meet $\mathbb{H}_\mathbb{C}^2 \cup \partial\mathbb{H}_\mathbb{C}^2$. If g is a reflection with respect to a complex geodesic, then g has a whole circle of fixed points contained in $\partial\mathbb{H}_\mathbb{C}^2$. Therefore the definitions of elliptic, loxodromic and parabolic elements are disjoint.

Now we assume $g \in \mathrm{PU}(2,1)$ is a loxodromic element. Let $\tilde{g} \in \mathrm{SU}(2,1)$ be a lift of g, we denote by x and y the fixed points of g and by $\tilde{x}$ and $\tilde{y}$ some respective lifts to $\mathbb{C}^{2,1}$. Let Σ be the complex geodesic determined by x and y, and let σ be the geodesic determined by x and y. We can assume that $\Sigma = \mathbb{H}_\mathbb{C}^1 \times 0$; in other words, we can assume $x = [-1:0:1]$, $y = [1:0:1]$. To see this, notice that if $z \in \sigma$, then there exists $h \in \mathrm{PU}(2,1)$ such that $h(z) = (0,0)$, so $h(L)$ contains the origin of $B^2 = \mathbb{H}_\mathbb{C}^2$. The stabiliser of the origin is $\mathrm{U}(2)$ and it acts transitively on the set of complex lines through the origin, so there exists $h_1 \in \mathrm{PU}(2,1)$ such that $h_1(\Sigma) = \mathbb{H}_\mathbb{C}^1 \times 0$ and $h_1(\sigma)$ contains the origin. Composing with a rotation in $\mathbb{H}_\mathbb{C}^1$, we prove the statement.

We see that any vector c polar to Σ is an eigenvector of $\tilde{g}$. In fact, if $v \in \mathbb{C}^{2,1}$ and $\mathbb{P}_\mathbb{C}(v) \in \Sigma$, then $\langle \tilde{g}(c), \tilde{g}(v) \rangle = \langle c, v \rangle = 0$, then $\tilde{g}(c)$ is polar to $g(\Sigma) = \Sigma$, but we know that the complex dimension of the orthogonal complement (respect to $\langle \cdot, \cdot \rangle$) of the vector subspace inducing Σ is 1.

Now, $c = (0,1,0)$ is a vector polar to Σ, then we can assume $[0:1:0]$ is a fixed point of g. Thus $\tilde{g}$ has the form

$$\begin{pmatrix} a & 0 & b \\ 0 & e^{-2i\theta} & 0 \\ b & 0 & a \end{pmatrix},$$

and the transformation of $\mathrm{PSL}(2,\mathbb{C})$ induced by the matrix

$$e^{-i\theta} \begin{pmatrix} a & b \\ b & a \end{pmatrix},$$

is a hyperbolic transformation preserving the unitary disc of $\mathbb{C}$ and with fixed points $1, -1$. With this information it is not hard to see that we can take $\tilde{g}$ of the form

$$\begin{pmatrix} e^{i\theta}\cosh u & 0 & e^{i\theta}\sinh u \\ 0 & e^{-2i\theta} & 0 \\ e^{i\theta}\sinh u & 0 & e^{i\theta}\cosh u \end{pmatrix}.$$

We notice that $\tilde{g}$ has an eigenvalue in the open unitary disc, one in $\mathbb{S}^1$ and another outside the closed unitary disc. Since they are parabolic, all of these have a unique fixed point in $\partial\mathbb{H}^2_\mathbb{C}$, but they have a different behaviour in $\mathbb{P}^2_\mathbb{C} \setminus \mathbb{H}^2_\mathbb{C}$.

Example 2.4.5. a) The transformation induced by the matrix

$$\tilde{g} = \begin{pmatrix} 1+i/2 & 0 & 1/2 \\ 0 & 1 & 0 \\ 1/2 & 0 & 1-i/2 \end{pmatrix}$$

is a unipotent transformation with a whole line of fixed points in $\mathbb{P}^2_\mathbb{C}$, tangent to $\partial\mathbb{H}^2_\mathbb{C}$.

b) The transformation in $SU(2,1)$ given by the matrix

$$\tilde{g} = \begin{pmatrix} \frac{1}{3} & \frac{2\sqrt{2}}{3} & 0 \\ -\frac{4\sqrt{2}}{3} & \frac{2}{3} & \sqrt{3} \\ -\frac{2\sqrt{6}}{3} & \frac{\sqrt{3}}{3} & 2 \end{pmatrix}$$

induces a unipotent automorphism of $\mathbb{H}^2_\mathbb{C}$ with one single fixed point in $\mathbb{P}^2_\mathbb{C}$.

c) The transformation in $SU(2,1)$ given by

$$\begin{pmatrix} \frac{2+\epsilon i}{2}e^{i\theta} & 0 & \frac{\epsilon i}{2}e^{i\theta} \\ 0 & e^{-2i\theta} & 0 \\ -\frac{\epsilon i}{2}e^{i\theta} & 0 & \frac{2-\epsilon i}{2}e^{i\theta} \end{pmatrix},$$

where $\epsilon \neq 0$, induces an ellipto-parabolic automorphism of $\mathbb{H}^2_\mathbb{C}$, with fixed points $[-1:0:1] \in \partial\mathbb{H}^2_\mathbb{C}$ and $[0:1:0]$.

Remark 2.4.6. We recall that complex hyperbolic space has (nonconstant) negative sectional curvature. As such, the classification given above of its isometries, into elliptic, parabolic and loxodromic (or hyperbolic), actually fits into a similar classification given in the general setting of the isometries of spaces of nonpositive curvature, and even more generally, for CAT(0)-spaces. We refer to [13], [32] for thorough accounts of this subject. The concept of CAT(0)-spaces captures the essence of nonpositive curvature and allows one to reflect many of the basic properties of such spaces, as for instance $\mathbb{H}^n_\mathbb{R}$, and $\mathbb{H}^n_\mathbb{C}$, in a much wider setting. The origins of CAT(0)-spaces, and more generally CAT(κ)-spaces, are in the work of A. D. Alexandrov, where he gives several equivalent definitions of what it means

for a metric space to have curvature bounded above by a real number κ. The
terminology "CAT(κ)" was coined by M. Gromov in 1987 and the initials are in
honour of E. Cartan, A. Alexandrov and V. Toponogov, each of whom considered
similar conditions.

Yet, we notice that our main focus in this work concerns automorphisms of
$\mathbb{P}^n_{\mathbb{C}}$, whose sectional curvature, when we equip it with the Fubini-Study metric,
is strictly positive, ranging from $1/4$ to 1. Moreover, the action of $\mathrm{PSL}(n+1, \mathbb{C})$
is not by isometries with respect to this metric. Thence, these transformations
do not fit in the general framework of isometries of CAT(0)-spaces. Even so, we
will see in the following chapter that the elements of $\mathrm{PSL}(3, \mathbb{C})$ are also naturally
classified into elliptic, parabolic and hyperbolic, both in terms of their geometry
and dynamics, and also algebraically.

2.4.3   Traces and conjugacy classes in $\mathrm{SU}(2,1)$

We now describe Goldman's classification of the elements in $\mathrm{PU}(2,1)$ by means of
the trace of their liftings to $\mathrm{SU}(2,1)$.

Let $\tau : \mathrm{SU}(2,1) \to \mathbb{C}$ be the function mapping an element in $\mathrm{SU}(2,1)$ to
its trace. Notice that a holomorphic automorphism of $\mathbb{H}^2_{\mathbb{C}}$ has a trace which is
well-defined up to multiplication by a cubic root of unity. The following result of
linear algebra is well known:

Lemma 2.4.7. *Let E be a Hermitian vector space, and let A be a unitary automorphism of E. The set of eigenvalues of A is invariant under the inversion $\imath$ in the
unitary circle of $\mathbb{C}$:*

$$\imath : \mathbb{C}^* \to \mathbb{C}^*$$

$$z \mapsto \frac{1}{\bar{z}}.$$

Proof. We assume the Hermitian form on E is given by a Hermitian matrix M.
The automorphism A is unitary if and only if

$$\bar{A}^t M A = M.$$

In other words,

$$A = M^{-1}(\bar{A}^t)^{-1} M.$$

Thus A has the same eigenvalues as $(\bar{A}^t)^{-1}$, which means that λ is an eigenvalue
of A if and only if $(\bar{\lambda})^{-1}$ is an eigenvalue. $\square$

Notice that when $\tilde{g} \in \mathrm{SU}(2,1)$, then $\tilde{g}$ has at least an eigenvalue of module
1. Moreover, the eigenvalues not lying in the unitary circle are given in an $\imath$-
invariant pair. Particularly, if $\tilde{g}$ has two eigenvalues of the same module, then
every eigenvalue has module 1.

Lemma 2.4.8. *The monic polynomial* $\chi(t) = t^3 - xt^2 + yt - 1$ *with complex coefficients has repeated roots if and only if*

$$\tilde{f}(x, y) = -x^2 y^2 + 4(x^3 + y^3) - 18xy + 27$$

$$= \begin{vmatrix} 1 & -x & y & -1 & 0 \\ 0 & 1 & -x & y & -1 \\ 3 & -2x & y & 0 & 0 \\ 0 & 3 & -2x & y & 0 \\ 0 & 0 & 3 & -2x & y \end{vmatrix}$$

$$= 0.$$

Proof. We assume χ has repeated roots, then χ and its derivative χ' have a common root. We suppose

$$\chi(t) = (t - a_1)(t - a_2)(t - a_3)$$

and

$$\chi'(t) = 3(t - a_1)(t - a_4).$$

Then

$$3(t - a_4)\chi(t) = (t - a_2)(t - a_3)\chi'(t),$$

or which is the same,

$$3t\chi(t) - 3a_4\chi(t) - t^2\chi'(t) + (a_2 + a_3)t\chi'(t) - a_2a_3\chi'(t) \equiv 0.$$

This means that the vectors obtained from the polynomials $t\chi(t)$, $\chi(t)$, $t^2\chi'(t)$, $t\chi'(t)$, $\chi'(t)$, taking the coefficients of the terms with degree ≤ 4, are linearly independent. But such vectors are precisely those row vectors of the determinant defining $\tilde{f}(x, y)$. Thence $\tilde{f}(x, y) = 0$.

Conversely, if we assume that $\tilde{f}(x, y) = 0$, then there exist complex numbers $c_1, \ldots, c_5$, not all zero, such that

$$c_1 t\chi(t) + c_2\chi(t) + c_3 t^2\chi'(t) + c_4 t\chi'(t) + c_5\chi'(t) \equiv 0.$$

Then $(c_1 t + c_2)\chi(t) = -(c_3 t^2 + c_4 t + c_5)\chi'(t)$, which implies that $\chi(t)$ and $\chi'(t)$ have a common root because $deg(\chi) = 3$. Therefore $\chi(t)$ has a repeated root. $\square$

We denote by $C_3 = \{\omega, \omega^2, 1\} \subset \mathbb{C}$ the set of cubic roots of unity, and $3C_3$ denotes the set $\{3\omega, 3\omega^2, 3\}$. We observe that there is a short exact sequence

$$1 \to C_3 \to \mathrm{SU}(2, 1) \to \mathrm{PU}(2, 1) \to 1.$$

Let $\tau : \mathrm{SU}(2, 1) \to \mathbb{C}$ be the function which assigns to an element of $\mathrm{SU}(2, 1)$ its trace. Goldman's classification theorem involves the real polynomial $f : \mathbb{C} \to \mathbb{R}$ defined by $f(z) = |z|^4 - 8\mathrm{Re}(z^3) + 18|z|^2 - 27$. In other words $f(z) = -\tilde{f}(z, \bar{z})$, where

$$\tilde{f}(x, y) = -x^2 y^2 + 4(x^3 + y^3) - 18xy + 27,$$

is the discriminant of the (characteristic) polynomial $\chi(t) = t^3 - xt^2 + yt - 1$.

Theorem 2.4.9. [67, Theorem 6.2.4] *The map* $\tau : \mathrm{SU}(2,1) \to \mathbb{C}$ *defined by the trace is surjective, and if* $A_1, A_2 \in \mathrm{SU}(2,1)$ *satisfy* $\tau(A_1) = \tau(A_2) \in \mathbb{C} - f^{-1}(0)$, *then they are conjugate. Furthermore, supposing* $A \in \mathrm{SU}(2,1)$ *one has:*

1) *A is regular elliptic if and only if* $f(\tau(A)) < 0$.

2) *A is loxodromic if and only if* $f(\tau(A)) > 0$.

3) *A is ellipto-parabolic if and only if A is not elliptic and* $\tau(A) \in f^{-1}(0) - 3C_3$.

4) *A is a complex reflection if and only if A is elliptic and* $\tau(A) \in f^{-1}(0) - 3C_3$.

5) $\tau(A) \in 3C_3$ *if and only if A represents a unipotent parabolic element.*

Proof. Let $\chi_A(t)$ be the characteristic polynomial of A:

$$\chi_A(t) = t^3 - xt^2 + yt - 1.$$

The eigenvalues $\lambda_1, \lambda_2, \lambda_3$ of A are the roots of $\chi_A(t)$. We have

$$x = \tau(A) = \lambda_1 + \lambda_2 + \lambda_3,$$

and

$$\lambda_1 \lambda_2 \lambda_3 = \det(A) = 1. \tag{2.4.10}$$

The coefficient y of χ_A is equal to

$$y = \lambda_1\lambda_2 + \lambda_2\lambda_3 + \lambda_3\lambda_1 = \overline{\lambda_1} + \overline{\lambda_2} + \overline{\lambda_3} = \overline{\tau(A)}.$$

Thus,

$$\chi_A(t) = t^3 - \tau(A)t^2 + \overline{\tau(A)}\,t - 1.$$

If $A \in \mathrm{SU}(2,1)$, then its eigenvalues satisfy equation 2.4.10 and the set

$$\tilde{\lambda} = \{\lambda_1, \lambda_2, \lambda_3\},$$

of these eigenvalues satisfies:

$$\lambda \in \tilde{\lambda} \;\Rightarrow\; \imath(\lambda) = \bar{\lambda}^{-1} \in \tilde{\lambda}. \tag{2.4.11}$$

Let $\tilde{\Lambda}$ (respectively Λ) be the set of unordered triples of complex numbers satisfying (2.4.10) (respectively (2.4.11)). Then $\imath$ induces an involution in $\tilde{\Lambda}$ (denoted by the same) whose set of fixed points is Λ. The function

$$\chi : \tilde{\Lambda} \to \mathbb{C}^2$$
$$\tilde{\lambda} \mapsto (\lambda_1 + \lambda_2 + \lambda_3, \lambda_1\lambda_2 + \lambda_2\lambda_3 + \lambda_3\lambda_1)$$

is bijective, with inverse function

$$\mathbb{C}^2 \to \tilde{\Lambda}$$
$$(x, y) \mapsto \{t \in \mathbb{C} : t^3 - xt + yt - 1 = 0\}.$$

The involution j on $\mathbb{C}^2$ defined as

$$j(x,y) = (\bar{y}, \bar{x})$$

satisfies

$$\chi \circ \imath = j \circ \chi$$

and χ restricted to the set of fixed points of $\imath$ is a bijection on the set of fixed points of j in $\mathbb{C}^2$, which is the image of

$$e : \mathbb{C} \to \mathbb{C}^2$$
$$z \mapsto (z, \bar{z}).$$

One has $e^{-1} \circ \chi|_\Lambda = \tau$. Let $\tilde{\Lambda}_{\mathrm{sing}} \subset \tilde{\Lambda}$ be the set of unordered triples $\{\lambda_1, \lambda_2, \lambda_3\}$, not all the λ_j being different. Lemma 2.4.8 implies that χ restricted to $\tilde{\Lambda}_{\mathrm{sing}}$ is a bijection on the set

$$\{(x,y) \in \mathbb{C}^2 : \tilde{f}(x,y) = 0\},$$

where

$$\tilde{f}(x,y) = \begin{vmatrix} 1 & -x & y & -1 & 0 \\ 0 & 1 & -x & y & -1 \\ 3 & -2x & y & 0 & 0 \\ 0 & 3 & -2x & y & 0 \\ 0 & 0 & 3 & -2x & y \end{vmatrix}$$
$$= -x^2 y^2 + 4(x^3 + y^3) - 18xy + 27.$$

We define $\Lambda_0 = \Lambda - \tilde{\Lambda}_{\mathrm{sing}}$ and notice that $\tau|_{\Lambda \cap \tilde{\Lambda}_{\mathrm{sing}}} : \Lambda \cap \tilde{\Lambda}_{\mathrm{sing}} \to f^{-1}(0)$ and $\tau|_{\Lambda_0} : \Lambda_0 \to \mathbb{C} - f^{-1}(0)$ are bijections. In fact, in order to prove that $\tau|_{\Lambda \cap \tilde{\Lambda}_{\mathrm{sing}}}$ is injective it is enough to see that $\tau = e^{-1} \circ \chi$ is the composition of injective functions. Now, if $z \in f^{-1}(0)$, then $0 = f(z) = -\tilde{f}(z, \bar{z}) = -\tilde{f}(e(z))$, but there exists $\lambda \in \Lambda$ such that $\chi(\lambda) = e(z)$, then $\tilde{f}(\chi(\lambda)) = 0$, which implies that $\lambda \in \Lambda \cap \tilde{\Lambda}_{\mathrm{sing}}$. Moreover $\tau(\lambda) = e^{-1} \circ \chi(\lambda) = z$. Therefore, $\tau|_{\Lambda \cap \tilde{\Lambda}_{\mathrm{sing}}}$ is onto. The proof of the fact that $\tau|_{\Lambda_0}$ is bijective is straightforward.

We have proved that $\tau : \mathrm{SU}(2,1) \to \mathbb{C}$ is onto. However we have implicitly assumed that Λ is the image of a correspondence defined in $\mathrm{SU}(2,1)$. Such correspondence is defined by

$$L : \mathrm{SU}(2,1) \to \Lambda$$
$$A \mapsto \{\lambda_1, \lambda_2, \lambda_3\},$$

where $\lambda_1, \lambda_2, \lambda_3$ are the eigenvalues of A. We must check that L is onto. We define $\lambda = \{\lambda_1, \lambda_2, \lambda_3\} \in \Lambda$. Notice there are only two possibilities:

(i) $|\lambda_i| = 1$ for $i = 1, 2, 3$; in this case

$$A = \begin{pmatrix} \lambda_1 & & \\ & \lambda_2 & \\ & & \lambda_3 \end{pmatrix} \in \text{SU}(2, 1),$$

and $L(A) = \lambda$.

(ii) $\lambda_3 = re^{i\theta}$, $0 < r < 1$ and we can take $\lambda_2 = e^{-2i\theta}$, $\lambda_1 = (1/r)e^{i\theta}$. We take $r = e^{-u}$ for some $u \in \mathbb{R}^+$, then

$$A = \begin{pmatrix} \cosh(u)e^{i\theta} & 0 & \sinh(u)e^{i\theta} \\ 0 & e^{-2i\theta} & 0 \\ \sinh(u)e^{i\theta} & 0 & \cosh(u)e^{i\theta} \end{pmatrix} \in \text{SU}(2, 1),$$

and $L(A) = \lambda$.

If $A_1, A_2 \in \text{SU}(2, 1)$ satisfy $\tau(A_1) = \tau(A_2) \in \mathbb{C} - f^{-1}(0)$, then the characteristic polynomials of A_1 and A_2 are equal, which implies they have the same eigenvalues (they are different amongst them). Then A_1 and A_2 are conjugate of the same diagonal matrix, so A_1 and A_2 are conjugate.

We define $\Lambda_0^l = \{\lambda \in \Lambda : \lambda \cap \mathbb{S}^1 \text{ has one single element }\}$. We can suppose that

$$|\lambda_1| > 1, \quad |\lambda_2| < 1, \quad |\lambda_3| = 1. \tag{2.4.12}$$

In particular, $\Lambda_0^l \subset \Lambda_0$, and given that λ_1 satisfies 2.4.12 , we obtain unique λ_2, λ_3 by means of the relations

$$\lambda_2 = (\overline{\lambda_1})^{-1}, \qquad \lambda_3 = \frac{\overline{\lambda_1}}{\lambda_1}$$

which proves that Λ_0^l is homeomorphic to the exterior of the unitary disc of $\mathbb{C}$ and therefore it is connected. [The topology on $\tilde{\Lambda}$ is the topology induced by the bijective function $\chi : \tilde{\Lambda} \to \mathbb{C}^2$ and the topology on Λ is the subspace topology.]

We denote by $\Lambda_0^e = \Lambda_0 - \Lambda_0^l$. If $\lambda \in \mathbb{R} \setminus \{0, 1, -1\}$, then

$$\tau(\lambda, \lambda^{-1}, 1) = \lambda + \lambda^{-1} + 1$$

which shows that $\tau|_{\Re(\Lambda_0^l)} : \Re(\Lambda_0^l) \to \mathbb{R} \setminus [-1, 3]$ is a bijection, where $\Re(\Lambda_0^l)$ denotes the set of triples in Λ_0^l: We note that $\tau|_{\Re(\Lambda_0^l)}$ is injective because it is the composition of injective functions and it is not hard to check that it is onto. Notice that f is positive in $\mathbb{R} \setminus [-1, 3]$. In fact, if $x \in \mathbb{R}$, then

$$f(x) = x^4 - 8x^3 + 18x^2 - 27 = (x + 1)(x - 3)^3.$$

We claim that $\tau|_{\Lambda_0^l} : \Lambda_0^l \to f^{-1}(\mathbb{R}^+)$ is bijective. In fact, first notice that $\tau(\Lambda_0^l) \subset f^{-1}(\mathbb{R}^+)$, for otherwise, using that Λ_0^l is connected, we can find a triplet

$\lambda \in \Lambda_0^l$ such that $f(\tau(\lambda)) = 0$, which means $\lambda \in \Lambda_0 \cap \tilde{\Lambda}_{\text{sing}} = \varnothing$, a contradiction. That τ is injective follows from the fact that it is a composition of the injective functions e^{-1} and χ.

In order to prove $\tau(\Lambda_0^l) = f^{-1}(\mathbb{R}^+)$, we take $z \in f^{-1}(\mathbb{R}^+)$, we know that $z = \tau(\lambda)$, for some $\lambda \in \Lambda$. We assume that $\lambda \notin \Lambda_0^l$, then every element in λ has module 1, which implies that

$$|z| = |\tau(\lambda)| \leq 3,$$

and then

$$|\Re(z^3)| \leq |z^3| \leq 27,$$

therefore,

$$f(z) \leq (3)^4 - 8(-27) + 18(3)^2 - 27 = 0,$$

which is a contradiction.

We now claim that the function $\tau|_{\Lambda_0^e}$ is a bijection on $f^{-1}(\mathbb{R}^-)$. We prove first that $\tau(\Lambda_0^e) \subset f^{-1}(\mathbb{R}^-)$. In fact, if $\lambda \in \Lambda_0^e$, then the elements in λ are different and they have module 1. We know that $\tau|_{\Lambda_0^e}$ is injective, because it is the restriction of the injective function $\tau|_{\Lambda_0}$. Finally, $\tau(\Lambda_0^e) = f^{-1}(\mathbb{R}^-)$, because $\tau(\Lambda_0^l) = f^{-1}(\mathbb{R}^+)$.

Hence $f(\tau(A)) < 0$ if and only if the eigenvalues of A are distinct unitary complex numbers, which happens if and only if A is regular elliptic. One has $f(\tau(A)) > 0$ if and only if A has exactly one eigenvalue in $\mathbb{S}^1$, if and only if A is loxodromic.

Now we consider the case when $f(\tau(A)) = 0$. Clearly $A \in \text{PU}(2,1)$ is unipotent if and only if it has a lift to $\text{SU}(2,1)$ having equal eigenvalues, and therefore $\frac{1}{3}\tau(A) \in C_3$. Conversely, if $\frac{1}{3}\tau(A) = \omega \in C_3$, then

$$\chi_A(t) = t^3 - 3\omega t^2 + 3\omega^2 t - 1 = (t - \omega)^3,$$

thus A has three repeated eigenvalues and it is projectively equivalent to a unipotent matrix.

Finally we consider the case when $\tau(A) \in f^{-1}(0) - 3C_3$. Then A has an eigenvalue $\zeta \in \mathbb{S}^1$ of multiplicity 2 and the other eigenvalue is equal to ζ^{-2}. Given that $\tau(A) \notin 3C_3$, we have $\zeta \neq \zeta^{-2}$. There are two cases depending on the Jordan canonical form of A: if A is diagonalizable, then A is elliptic (a complex reflection). This case splits in two cases, depending on whether the ζ-eigenspace V_ζ is positive or indefinite: if V_ζ is indefinite, then A is a complex reflection with respect to the complex geodesic corresponding to V_ζ; if V_ζ is positive, then A is a complex reflection with respect to the point corresponding to the ζ^{-2}-eigenspace. If A is not diagonalizable, then A has a repeated eigenvalue of module 1 and it has Jordan canonical form

$$\begin{pmatrix} \lambda & 1 & 0 \\ 0 & \lambda & 0 \\ 0 & 0 & \lambda^{-2} \end{pmatrix}.$$

In this case the eigenvector corresponding to λ is e_1 and A is ellipto-parabolic. $\square$

Corollary 2.4.13. *Let $\imath : \mathbb{C}^* \to \mathbb{C}^*$ be the inversion on the unitary circle; i.e., $\imath(z) = 1/\bar{z}$. If the set of eigenvalues of $A \in \mathrm{SL}(3, \mathbb{C})$ is invariant under $\imath$, then:*

(i) *The eigenvalues of A are unitary complex numbers, and pairwise different if and only if $f(\tau(A)) < 0$.*

(ii) *Precisely one eigenvalue is unitary if and only if $f(\tau(A)) > 0$.*

This corollary can be proved using the proof of Lemma 2.4.8. This is useful for proving Theorem 4.3.3, which is an extension to the elements in $\mathrm{PSL}(3, \mathbb{C})$ of the classification Theorem 2.4.9.

Remark 2.4.14. In [44] the authors look at the group of quaternionic Möbius transformations that preserve the unit ball in the quaternionic space $\mathcal{H}$, and they use this to classify the isometries of the hyperbolic space $\mathbb{H}^4_{\mathbb{R}}$ into six types, in terms of their fixed points and whether or not they are conjugate in $\mathrm{U}(1, 1; \mathcal{H})$ to an element of $\mathrm{U}(1, 1; \mathbb{C})$. In [72] the author uses also quaternionic transformations to refine the classification of the elements in $\mathrm{Conf}_+(\mathbb{S}^4)$ using algebraic invariants, and in [73] the authors characterize algebraically the isometries of quaternionic hyperbolic spaces (see also [43]). In Chapter 10 we follow [202] and use Ahlfor's characterization of the group $\mathrm{Conf}_+(\mathbb{S}^4) \cong \mathrm{Iso}_+(\mathbb{H}^5_{\mathbb{R}})$ as quaternionic Möbius transformations in order to describe the canonical embedding $\mathrm{Conf}_+(\mathbb{S}^4) \hookrightarrow \mathrm{PSL}(4, \mathbb{C})$ that appears in twistor theory, and use this to construct discrete subgroups of $\mathrm{PSL}(4, \mathbb{C})$.

2.5   Complex hyperbolic Kleinian groups

As before, let $\mathrm{U}(n, 1) \subset \mathrm{GL}(n + 1, \mathbb{C})$ be the group of linear transformations that preserve the quadratic form $|z_1|^2 + \cdots + |z_n|^2 - |z_{n+1}|^2$. We let V_- be the set of negative vectors in $\mathbb{C}^{n,1}$ for this quadratic form. We know already that the projectivization $[V_-]$ is a $2n$-ball that serves as a model for complex hyperbolic space $\mathbb{H}^n_{\mathbb{C}}$ and $\mathrm{PU}(n, 1)$ is its group of holomorphic isometries (see Section 2.2).

Definition 2.5.1. A subgroup $\Gamma \subset \mathrm{PSL}(n + 1, \mathbb{C})$ is a *complex hyperbolic Kleinian group* if it is conjugate to a discrete subgroup of $\mathrm{PU}(n, 1)$.

A fundamental problem in complex hyperbolic geometry is the construction and understanding of complex hyperbolic Kleinian groups.

This problem goes back to the work of Picard and Giraud (see the appendix in [67]), and it has been subsequently addressed by many authors, as for instance Mostow, Deligne, Hirzebruch and many others, as a means of generalising the classical theory of automorphic forms and functions, and also as a means of constructing complex manifolds with a rich geometry. Complex hyperbolic Kleinian groups can also be regarded as being discrete faithful representations of a group Γ into $\mathrm{PU}(n, 1)$.

Two basic questions are the construction of lattices in $\mathrm{PU}(n, 1)$, and the existence of lattices which are not commensurable with arithmetic lattices. Let us explain briefly what this means.

A discrete subgroup Γ of $\mathrm{PU}(n,1)$ (and more generally of a locally compact group G), equipped with a Haar measure, is said to be a *lattice* if the quotient $\mathrm{PU}(n,1)/\Gamma$ has finite volume. The lattice is said to be *uniform (or cocompact)* if the quotient $\mathrm{PU}(n,1)/\Gamma$ is actually compact.

A subgroup Γ of $\mathrm{U}(n,1)$ is *arithmetic* if there is an embedding $\mathrm{U}(n,1) \overset{\iota}{\hookrightarrow} \mathrm{GL}(N,\mathbb{C})$, for some N, such that the image of Γ is *commensurable* with the intersection of $\iota(\mathrm{U}(n,1))$ with $\mathrm{GL}(N,\mathbb{Z})$. That is, $\iota(\Gamma) \cap \mathrm{GL}(N,\mathbb{Z})$ has finite index in $\iota(\Gamma)$ and in $\mathrm{GL}(N,\mathbb{Z})$.

Complex hyperbolic space is, like real hyperbolic space, a noncompact symmetric space of rank 1, and an important problem in the study of noncompact symmetric spaces is the relationship between arithmetic groups and lattices: while all arithmetic groups are lattices (by [29]), the question of whether or not all lattices are arithmetic is rather subtle. We know from Margulis' work that for symmetric spaces of rank ≥ 2, all irreducible lattices are arithmetic. The rank 1 symmetric spaces of noncompact type come in three infinite families, real, complex and quaternionic hyperbolic spaces: $\mathbb{H}^n_{\mathbb{R}}, \mathbb{H}^n_{\mathbb{C}}, \mathbb{H}^n_{\mathcal{H}}$, and one has also the Cayley (or octonionic) hyperbolic plane $\mathbb{H}^2_{\mathbb{O}}$. We know by [81] (and work by K. Corlette) that all lattices on $\mathbb{H}^n_{\mathcal{H}}$ and $\mathbb{H}^2_{\mathbb{O}}$ are arithmetic, and we also know by [80] that there are nonarithmetic lattices acting on real hyperbolic spaces of all dimensions. In complex hyperbolic space $\mathbb{H}^n_{\mathbb{C}}$, we know that there are nonarithmetic lattices for $n = 2, 3$, by [50] respectively. The question of existence of nonarithmetic lattices on $\mathbb{H}^n_{\mathbb{C}}$ is open for $n \geq 4$ and this is one of the major open problems in complex hyperbolic geometry (see [50]).

There is a huge wealth of knowledge in literature about complex hyperbolic Kleinian groups published the last decades by various authors, as for instance P. Deligne, G. Mostow, W. Goldman, J. Parker, R. Schwartz, N. Gusevskii, E. Falbel, P. V. Koseleff, D. Toledo, M. Kapovich, E. Z. Xia, among others. Here we mention just a few words that we hope will give a taste of the richness of this branch of mathematics.

2.5.1 Constructions of complex hyperbolic lattices

We particularly encourage the reader to look at the beautiful articles [105] and [168] for wider and deeper surveys related to the topic of complex hyperbolic Kleinian groups. Kapovich's article is full of ideas and provides a very deep understanding of real and complex hyperbolic Kleinian groups. Parker's article explains the known ways and sources for producing complex hyperbolic lattices (see also [169]). Below we briefly mention some of Parker's explanations for lattices, and in the section below we give other interesting examples of constructions that produce complex hyperbolic groups which are not lattices.

Parker classifies the methods for constructing complex hyperbolic lattices in the following four main types, which of course overlap. All these are also present in the monograph [50], whose main goal is to investigate commensurability among lattices in $\mathrm{PU}(n, 1)$.

i) Arithmetic lattices: The natural inclusion of the integers in the real numbers is the prototype of an arithmetic group. This yields naturally to the celebrated modular group $\mathrm{PSL}(2, \mathbb{Z})$ in $\mathrm{PSL}(2, \mathbb{R})$. That construction was generalised to higher-dimensional complex hyperbolic lattices by Picard in 1882, and then studied by a number of authors. For instance, let d be a positive square-free integer, let $\mathbb{Q}(i\sqrt{d})$ be the corresponding quadratic number field and $\mathcal{O}_d$ its ring of integers, a discrete subgroup of $\mathbb{C}$. Let H be a Hermitian matrix with signature $(2, 1)$ and entries in $\mathcal{O}_d$. Let $\mathrm{SU}(H)$ be the group of unitary matrices that preserve H, and let $\mathrm{SU}(H; \mathcal{O}_d)$ be the subgroup of $\mathrm{SU}(H)$ consisting of matrices whose entries are in $\mathcal{O}_d$. Then $\mathrm{SU}(H; \mathcal{O}_d)$ is a lattice in $\mathrm{SU}(H)$. These type of arithmetic groups are known as Picard modular groups. We refer to Parker's article for a wide bibliography concerning these and other arithmetic constructions in the context of complex hyperbolic geometry.

ii) The second major technique mentioned by Parker for producing lattices in complex hyperbolic spaces is to consider objects that are parametrised by some $\mathbb{H}_{\mathbb{C}}^n$, with the property that the corresponding group of automorphisms is a complex hyperbolic lattice. For instance we know that the modular group $\mathrm{PSL}(2, \mathbb{Z})$ can be regarded as being the monodromy group of elliptic functions. Similar results were known to Poincaré, Schwartz and others for real hyperbolic lattices. In complex hyperbolic geometry, the first examples of this type of lattices were again given by Picard in 1885. He considered the moduli space of certain hypergeometric functions and showed that their monodromy groups were lattices in $\mathrm{PU}(2, 1)$, though his proof of the discreteness of the groups was not complete. This was settled and extended in [50], where the authors study the monodromy groups of a certain type of integrals, generalising the classical work of Schwarz and Picard. Under a certain integrality condition that they call (INT), they prove that the monodromy group Γ is a lattice in $\mathrm{PU}(n, 1)$; yet, for $d > 5$ this condition is never satisfied. They also give criteria for Γ to be arithmetic. Further research along similar lines was developed in [50], as well as by various other authors, as for instance Le Vavasseur, Terada, Thurston, Parker and others. Alternative approaches along this same general line of research have been followed by Allcock, Carlson, Toledo and others. Thurston's approach in [224] is particularly interesting and gives an alternative way of interpreting the (INT)-condition. We refer to Section 3 in [168] for an account on the construction of lattices arising as monodromy groups of hypergeometric functions.

iii) A third way for constructing discrete groups is by looking at lattices generated by complex reflections, or more generally by finding appropriate fundamental domains. This approach was introduced by Giraud (see Appendix A in [67]). Typically, a fundamental domain is a locally finite polyhedron P with some combinatorial structure that tells us how to identify its faces, called the sides, by maps in $\mathrm{PU}(n, 1)$. Given this information, Poincare's polyhedron theorem gives conditions under which the group generated by the sides pairing maps is discrete, and it gives a presentation of the group. One way for doing so is to construct the Dirichlet fundamental domain $D_\Gamma(z_0)$ as in Chapter 1: Assume $z_0 \in \mathbb{H}^n_\mathbb{C}$ is not fixed by any nontrivial element in a given group Γ, then $D_\Gamma(z_0)$ is the set of points in $\mathbb{H}^n_\mathbb{C}$ that are closer to z_0 than to any other point in its orbit. Its sides are contained in bisectors. Mostow used this approach in [155] to give the first examples of nonarithmetic lattices in $\mathrm{PU}(n,1)$ ($n \le 3$). Alternative methods for constructing lattices in this way have been given by Deraux, Falbel, Paupert, Parker, Goldman and others. Again, we refer to [168] for more on this topic.

iv) A fourth way for constructing complex hyperbolic lattices is using algebraic geometry. In fact, the Yau-Miyaoka uniformization theorem ([151]) states that if M is a compact complex 2-manifold whose Chern classes satisfy $c_1^2 = 3c_2$, then M is either $\mathbb{P}^2_\mathbb{C}$ or a complex hyperbolic manifold, i.e., the quotient of $\mathbb{H}^2_\mathbb{C}$ by some cocompact lattice. Thence the fundamental groups of such surfaces with $c_1^2 = 3c_2$ are uniform lattices in $\mathrm{PU}(2,1)$. Yet, techniques for having a direct geometric construction of such surfaces were not available until the Ph. D. Thesis of R. A. Livné [Harvard, Cambridge, Mass., 1981]. A variant on Livné's technique, using abelian branched covers of surfaces, was subsequently used in [93] to construct an infinite sequence of noncompact surfaces satisfying $\bar{c}_1^2 = 3\bar{c}_2$. See also [94] and [195].

2.5.2 Other constructions of complex hyperbolic Kleinian groups

In the previous subsection we briefly explained methods for constructing complex hyperbolic lattices. In the case of $\mathbb{H}^1_\mathbb{C} \cong \mathbb{H}^2_\mathbb{R}$ these are the so-called Fuchsian groups of the first kind, i.e., discrete subgroups of $\mathrm{PSL}(2,\mathbb{R}) \cong \mathrm{PU}(1,1)$ whose limit set is the whole sphere at infinity. And we know that Fuchsian groups of the second kind are indeed very interesting. It is thus natural to ask about discrete subgroups of $\mathrm{PU}(n,1)$ which are not lattices. This is in itself a whole area of research, that we will not discuss here, and we refer for this to the bibliography, particularly to [67], [105].

One of the classical ways of doing so is by taking discrete subgroups of $\mathrm{PSL}(2,\mathbb{R})$, considering representations of these in $\mathrm{PU}(2,1)$ and then looking at their deformations. More generally (see [67, Section 4.3.7]) one may consider $\Gamma \subset \mathrm{U}(n,1)$ a discrete subgroup and consider the natural inclusion $\mathrm{U}(n,1) \hookrightarrow \mathrm{U}(n+1,1)$. The composition

$$\Gamma \hookrightarrow \mathrm{U}(n,1) \hookrightarrow \mathrm{U}(n+1,1) \longrightarrow \mathrm{PU}(n+1,1)$$

defines a representation of Γ as a group of isometries of $\mathbb{H}^{n+1}_\mathbb{C}$. If Γ is a lattice, then there are strong local rigidity theorems, due to Goldman, Goldman-Millson, Toledo (for n=1) and Corlette (for n=2). If we now consider Γ to be discrete but not a lattice, then there is a rich deformation theory, and there are remarkable contributions by various authors.

We now give two interesting examples along this line of research, with references that can guide the interested reader into further reading.

Example 2.5.2 (Complex hyperbolic triangle groups). In Chapter 1 we discussed the classical hyperbolic triangle groups. These are discrete subgroups of $\mathrm{PSL}(2, \mathbb{R})$ generated by inversions on the sides (edges) of a triangle in the hyperbolic space $\mathbb{H}_{\mathbb{R}}^2$, bounded by geodesics, with inner angles $\pi/p, \pi/q$ and π/r for some integers $p, q, r > 1$; in fact some of these integers can be ∞, corresponding to triangles having one or more vertices on the visual sphere. When all vertices of the triangle are at infinity, this is called an ideal triangle.

In their article [70], W. Goldman and J. Parker give a method of constructing and studying complex hyperbolic ideal triangle groups. These are representations in $\mathrm{PU}(2, 1)$ of hyperbolic ideal triangle groups, such that each standard generator of the triangle group maps to a complex reflection, taking good care of the way in which products of pairs of generators are mapped. The fixed point set of a complex reflection is a complex slice (see Section 2.4.1).

Roughly speaking, the technique for constructing such groups begins with the embedding of a Fuchsian subgroup $\Gamma_0 \hookrightarrow \mathrm{PSL}(2, \mathbb{R})$, and deforms the representation inside $\mathrm{Hom}(\Gamma, \mathrm{PU}(2, 1))$. Thus Γ_0 preserves the real hyperbolic plane $\mathbb{H}^2$, but in general the deformed groups will not preserve any totally geodesic 2-plane (which are either complex lines or totally real 2-planes, intersected with the ball).

More precisely, Goldman and Parker look at the space of representations for a given triangle group, and for this they consider a triple of points (u_1, u_2, u_3) in ∂B^2, the boundary of the complex ball B^2. Let C_1, C_2, C_3 be the corresponding complex geodesics they span. Let Σ be the free product of three groups of order 2, and let $\phi : \Sigma \to \mathrm{PU}(2, 1)$ be the homomorphism taking the generators of Σ into the inversions (complex reflections) on C_1, C_2, C_3. Conjugacy classes of such homomorphisms correspond to $\mathrm{PU}(2, 1)$-equivalence classes of triples (u_1, u_2, u_3), and such objects are parametrised by their Cartan angular invariant $\phi(C_1, C_2, C_3)$, $-\frac{\pi}{2} \leq \phi \leq \frac{\pi}{2}$. The problem they address is: When is the subgroup $\Gamma \subset \mathrm{PU}(2, 1)$, obtained in this way, discrete? They prove that if the embedding of Γ is discrete, then $|\phi(x_0, x_1, x_2)| \leq \tan^{-1} \sqrt{125/3}$, and they conjectured that the condition $|\phi(x_0, x_1, x_2)| \leq \tan^{-1} \sqrt{125/3}$ was also sufficient to have a discrete embedding. This was referred to as the Goldman-Parker conjecture, and this was proved by R. Schwartz in [197] (see also [199]).

In this same vein one has R. Schwartz' article [198], where he discusses triangle subgroups of $\mathrm{PU}(2, 1)$ obtained also by complex reflections. Recall there is a simple formula (2.4.1) for the general complex reflection: Let V_+ and V_- be as above and choose a vector $c \in V_+$. For every nondegenerate vector $u \in \mathbb{C}^{2,1}$ the inversion on u is

$$I_c(u) = -u + \frac{2\langle u, c \rangle}{\langle c, c \rangle} \, c, \qquad (2.5.3)$$

and every complex reflection is conjugate to a map of this type. One may also consider the Hermitian cross product (see [67, p. 45])

$$u \boxtimes v = (u_3 v_2 - u_2 v_3 \, , \, u_1 v_3 - u_3 v_1 \, , \, u_1 v_2 - u_2 v_1) \qquad (2.5.4)$$

which satisfies: $\langle u, u \boxtimes v \rangle = \langle v, u \boxtimes v \rangle = 0$.

The two equations 2.5.3 and 2.5.4 enable us to generate discrete groups defined by complex reflections as follows: Take three vectors $V_1, V_2, V_3 \in V_-$ and set $c_j = V_{j-1} \boxtimes V_{j+1}$ (indices are taken modulo 3). For simplicity set $I_j = I_{c_j}$. Then the complex reflection I_j leaves invariant the complex geodesic determined by the points $[V_{j-1}]$ and $[V_{j+1}]$. The group $\Gamma := \langle I_1, I_2, I_3 \rangle$ is a complex-reflection triangle group determined by the triangle with vertices $[V_1], [V_2], [V_3]$, which is discrete if the vertices are chosen appropriately, as mentioned above. These groups furnish some of the simplest examples of complex hyperbolic Kleinian groups having a rich deformation theory (see for instance [198], [197], [56]).

This construction gives rise to manifolds with infinite volume obtained as quotient spaces $M = \mathbb{H}^2_{\mathbb{C}}/\Gamma$, which are the interior of a compact manifold-with-boundary, whose boundary ∂M is a real hyperbolic 3-manifold, quotient of a domain $\Omega \subset \partial \mathbb{H}^2_{\mathbb{C}} \cong \mathbb{S}^3$ by Γ (c.f. [198]).

Example 2.5.5 (Complex hyperbolic Kleinian groups with limit set a wild knot). As mentioned above, a way of producing interesting complex hyperbolic Kleinian groups is by taking a Fuchsian group $\Gamma \subset \mathrm{PSL}(2, \mathbb{R})$ and considering a representation of it in $\mathrm{PU}(2, 1)$ (or more generally in $\mathrm{PU}(n, 1)$). A specially interesting case is when Γ is the fundamental group of a hyperbolic surface, say of finite area. There has been a lot of progress in the study and classification of complex hyperbolic Kleinian groups which are isomorphic to such a surface group, but this is yet a mysterious subject which is being explored by several authors. The most natural way for this is by considering the *Teichmüller space* $T(\Gamma)$ of discrete, faithful, type-preserving representations of Γ in $\mathrm{PU}(2, 1)$; type-preserving means that every element in Γ that can be represented by a loop enclosing a single puncture is carried into a parabolic element in $\mathrm{PU}(2, 1)$, i.e., an element having exactly one fixed point on the boundary sphere of the ball in $\mathbb{P}^2_{\mathbb{C}}$ that serves as a model for $\mathbb{H}^2_{\mathbb{C}}$.

In [53] the authors prove that if Γ is the fundamental group of a noncompact surface of finite area, then the Teichmüller space $T(\Gamma)$ is not connected. For this they construct a geometrically finite quasi-Fuchsian group Γ acting on $H^2_{\mathbb{C}}$ whose limit set is a wild knot, and they show that this group can be also embedded in $\mathrm{PU}(2, 1)$ in such a way that the two embeddings are in different components of $T(\Gamma)$. They also prove that in both cases the two representations have the same Toledo invariant, thence this invariant does not distinguish different connected components of $T(\Gamma)$ when the surface has punctures, unlike the case where the surface is compact.

Let us sketch the construction of Dutenhefner-Gusevskii (see their article for more details). Consider the 3-dimensional Heisenberg group $\mathfrak{N}$, which is diffeomorphic to $\mathbb{C} \times \mathbb{R}$. Notice that this group carries naturally the *Heisenberg norm*

$$\|(\zeta, t)\| = \left| \|\zeta\|^2 + it \right|^{\frac{1}{2}}.$$

The corresponding metric on $\mathfrak{N}$ is *the Cygan metric*

$$\rho_0\big((\zeta, t), (\zeta', t')\big) = \left\| (\zeta - \zeta', t - t' + 2\Im\langle\langle \zeta - \zeta'\rangle\rangle) \right\|.$$

We use horospherical coordinates $\{(\zeta, t)\}$ for $\mathbb{H}^2_\mathbb{C}$, which allow us to identify $\mathfrak{N}$ with the horospheres in $\mathbb{H}^2_\mathbb{C}$ as explained in Subsection 2.3.1; these are the sets of points in $\mathbb{H}^2_\mathbb{C}$ of a constant "height".

Now consider a knot K and a finite collection $S = \{S_k, S'_k\}, k = 1\ldots n$, of Heisenberg spheres (in the Cygan metric) placed along K, satisfying the following condition: there is an enumeration $T_1; \ldots; T_{2n}$ of the spheres of this family such that each T_k lies outside all the others, except that T_k and T_{k+1} are tangent, for $k = 1, \ldots, 2n - 1$, and T_{2n} and T_1 are tangent. Such a collection S of Heisenberg spheres is called a *Heisenberg string of beads*, see Figure 2.1. Let g_k be elements from $\mathrm{PU}(2,1)$ such that:

(i) $g_k(S_k) = S'_k$,

(ii) $g_k(\mathrm{Ext}(S_k)) \subset \mathrm{Int}(S'_k)$,

(iii) g_k maps the points of tangency of S_k to the points of tangency of S'_k.

Let Γ be the group generated by g_k. Suppose now that Γ is Kleinian and the region D lying outside all the spheres of the family S is a fundamental domain for Γ. Then one can show that, under these conditions, the limit set of the group Γ is a wild knot.

The main difficulties in that construction are in finding a suitable knot, and appropriate spheres and pairing transformations g_k, so that the region D one gets is a *special Ford fundamental domain*, so that one can use Poincaré's polyhedron theorem (proved in [84]) for complex hyperbolic space, to ensure among other things that the group is Kleinian.

In order to construct a knot and a family of spheres as above, having the properties we need, they consider the granny knot K in $\mathcal{H}$; this is the connected sum of two right-handed trefoil knots (could also be left-handed). This has the property that it can be placed in $\mathcal{H}$ so as to be symmetric with respect to the reflection in the y-axis $\{v, (x + iy) \in \mathcal{H} \mid x = v = 0\}$, and also with respect to reflection in the vertical axis $\{x = y = 0\}$. These reflections are restrictions of elements in $\mathrm{PU}(2,1)$. Then they further choose K to be represented by a polygonal knot L, with the same symmetry properties and so that the edges of L are either segments of "horizontal" lines or segments of "vertical chains". Using this knot, they can show that there is a family of spheres and transformations with the required properties.

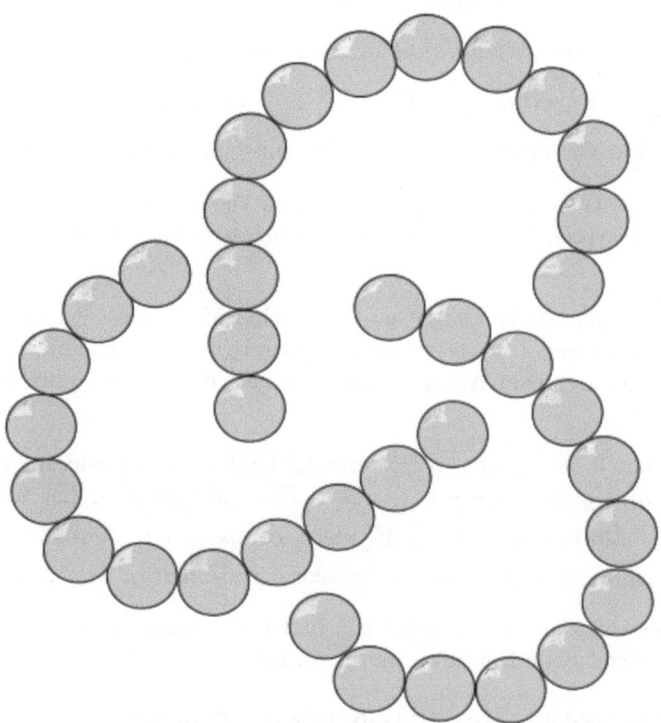

Figure 2.1: A Heisenberg string of beads

2.6   The Chen-Greenberg limit set

Consider now a discrete subgroup G of $\mathrm{PU}(n,1)$. As before, we take as a model for complex hyperbolic n-space $\mathbb{H}^n_{\mathbb{C}}$ the ball $\mathbb{B} \cong \mathbb{B}^{2n}$ in $\mathbb{P}^n_{\mathbb{C}}$ consisting of points with homogeneous coordinates satisfying

$$|z_1|^2 + \cdots + |z_n|^2 < |z_{n+1}|^2$$

whose boundary is a sphere $\mathbb{S} := \partial \mathbb{H}^n_{\mathbb{C}} \cong \mathbb{S}^{2n-1}$, and we equip $\mathbb{B}$ with the Bergman metric ρ to get $\mathbb{H}^n_{\mathbb{C}}$.

The following theorem can be found in [45]) and it is essentially a consequence of Arzelà-Ascoli's theorem, since G is acting on $\mathbb{H}^n_{\mathbb{C}}$ by isometries (see [185]).

Theorem 2.6.1. *Let G be a subgroup of $\mathrm{PU}(n,1)$. The following four conditions are equivalent:*

(i) *The subgroup $G \subset \mathrm{PU}(n,1)$ is discrete.*

(ii) G acts properly discontinuously on $\mathbb{H}_{\mathbb{C}}^n$.

(iii) The region of discontinuity of G in $\mathbb{H}_{\mathbb{C}}^n$ is all of $\mathbb{H}_{\mathbb{C}}^n$.

(iv) The region of discontinuity of G in $\mathbb{H}_{\mathbb{C}}^n$ is nonempty.

It follows that the orbit of every $x \in \mathbb{H}_{\mathbb{C}}^n$ must accumulate in $\partial \mathbb{H}_{\mathbb{C}}^n$.

Definition 2.6.2. If G is a subgroup of $\mathrm{PU}(n,1)$, the *Chen-Greenberg limit set* of G, denoted by $\Lambda_{CG}(G)$, is the set of accumulation points of the G-orbit of any point in $\mathbb{H}_{\mathbb{C}}^n$.

The following lemma is a slight generalisation of lemma 4.3.1 in [45]. This is essentially the *convergence property* for complex hyperbolic Kleinian groups (see page 15), and it implies that $\Lambda_{CG}(G)$ does not depend on the choice of the point in $\mathbb{H}_{\mathbb{C}}^n$.

Lemma 2.6.3. Let p be a point in $\mathbb{H}_{\mathbb{C}}^n$ and let (g_n) be a sequence of elements in $\mathrm{PU}(n,1)$ such that $g_n(p) \xrightarrow[m\to\infty]{} q \in \partial\mathbb{H}_{\mathbb{C}}^n$. Then for all $p' \in \mathbb{H}_{\mathbb{C}}^n$ we have that $g_n(p') \xrightarrow[m\to\infty]{} q$. Moreover, if $K \subset \mathbb{H}_{\mathbb{C}}^n$ is a compact set, then the sequence of functions $g_n|_K$ converges uniformly to the constant function with value q.

The expression $B_\rho(x, C)$ denotes the ball with centre at $x \in \mathbb{H}_{\mathbb{C}}^n$ and radius $C > 0$ with respect to the Bergman metric in $\mathbb{H}_{\mathbb{C}}^n$.

Proof. Assume the sequence $(g_n(p))$ converges to a point $q \in \partial\mathbb{H}_{\mathbb{C}}^n$, and we assume there exists $p' \in \mathbb{H}_{\mathbb{C}}^n$ such that the sequence $(g_n(p'))$ does not converge to q. Then there is a subsequence of $(g_n(p'))$ converging to a point $q' \in \overline{\mathbb{H}}_{\mathbb{C}}^n$, $q' \neq q$. So we may suppose that $g_n(p) \xrightarrow[m\to\infty]{} q$ and $g_n(p') \xrightarrow[m\to\infty]{} q'$. Let us denote by $[p, p']$ and $[q, q']$ the geodesic segments (with respect to the Bergman metric) joining p with p' and q with q', respectively. The distance from p to p' in the Bergman metric is equal to the length of $[p, p']$, and to $[g_n(p), g_n(p')]$, but this length goes to ∞ as $n \to \infty$, a contradiction. Therefore $g_n(p') \to q$ as $n \to \infty$.

Now we prove the convergence is uniform. If K is a compact subset of $\mathbb{H}_{\mathbb{C}}^n$, then there exists $C > 0$ such that $K \subset B_\rho(\mathbf{0}, C)$, where $\mathbf{0}$ denotes the origin in the ball model for $\mathbb{H}_{\mathbb{C}}^n$. The first part of this proof implies that $g_n(\mathbf{0}) \to q$. Given that g_n is an isometry of $\mathbb{H}_{\mathbb{C}}^n$, we have that $g_n(B_\rho(\mathbf{0}, C)) = B_\rho(g_n(\mathbf{0}), C)$ for each n, and the result follows from the lemma below. $\square$

Lemma 2.6.4. If (x_n) is a sequence of elements in $\mathbb{H}_{\mathbb{C}}^n$ such that $x_n \to q \in \partial\mathbb{H}_{\mathbb{C}}^n$ as $n \to \infty$, and $C > 0$ is a fixed positive number, then the Euclidean diameter of the ball $B_\rho(x_n, C)$ goes to zero as $n \to \infty$.

Proof. We use the model of the ball $B^n \subset \mathbb{C}^n$ for $\mathbb{H}_{\mathbb{C}}^n$. Let $x \in B^n$. Every complex geodesic through x has the form

$$\Sigma_y = \{x + \zeta y \mid \zeta \in \mathbb{C}, \ |x + \zeta y| < 1\}, \ y \in \mathbb{C}^n, \ |y| = 1.$$

The Euclidean distance of Σ_y to the origin of B^n is equal to

$$r(y) := |x - \langle\!\langle x, y\rangle\!\rangle y|,$$

where the symbol $\langle\!\langle x, y\rangle\!\rangle$ means the classical Hermitian product of the vectors $x, y \in \mathbb{C}^n$. Also, Σ_y is a Euclidean disc of Euclidean radius

$$(1 - r(y)^2)^{1/2} = (1 - |x|^2 + |\langle\!\langle x, y\rangle\!\rangle|^2)^{1/2} =: R(y).$$

Moreover, the Bergman metric in Σ_y has the form

$$\frac{4R(y)^2 dz d\bar{z}}{(R(y)^2 - |z|^2)^2}.$$

The intersection $B_\rho(x, C) \cap \Sigma_y$ is a hyperbolic disk of hyperbolic radius equal to C in Σ_y, and its hyperbolic centre $x \in \Sigma_y$ has Euclidean distance $|\langle\!\langle x, y\rangle\!\rangle|$ from the Euclidean centre of Σ_y, which is the point $x - \langle\!\langle x, y\rangle\!\rangle y$. Then $B_\rho(x, C) \cap \Sigma_y$ is a Euclidean disc of Euclidean radius

$$C_e(y) = R(y) \tanh(C/2) \frac{R(y)^2 - |\langle\!\langle x, y\rangle\!\rangle|^2}{R(y)^2 - \tanh^2(C/2)|\langle\!\langle x, y\rangle\!\rangle|^2}.$$

When x is fixed, the radius $C_e(y)$ is a continuous function of $y \in \mathbb{S}^{2n-1}$. Let y_M be such that $C_e(y_M) \geq C_e(y)$ for all $y \in \mathbb{S}^{2n-1}$. Let $X_1, X_2 \in B_\rho(x, C)$, then there exist $y_1, y_2 \in \mathbb{S}^{2n-1}$ such that $X_k \in \Sigma_{y_k} \cap B_\rho(x, C)$, $k = 1, 2$. If σ_k denotes the Euclidean centre of the disc $\Sigma_{y_k} \cap B_\rho(x, C)$, one has that

$$|X_1 - X_2| \leq |X_1 - \sigma_1| + |X_2 - \sigma_2| \leq 2C_e(y_1) + 2C_e(y_2) \leq 4C_e(y_M).$$

Finally we see that, for each $y \in \mathbb{S}^{2n-1}$, $C_e(y) \to 0$ as $x \to \partial\mathbb{H}^n_{\mathbb{C}}$. $\qquad\square$

It is clear from the definitions that the limit set $\Lambda_{CG}(G)$ is a closed, G-invariant set, and it is empty if and only if G is finite (since every sequence in a compact set contains convergent subsequences). Moreover the following proposition says that the action on $\Lambda_{CG}(G)$ is minimal (cf. [45]).

Proposition 2.6.5. *Let G be a nonelementary subgroup of* $\mathrm{PU}(n, 1)$. *If $X \subset \partial\mathbb{H}^n_{\mathbb{C}}$ is a G-invariant closed set containing more than one point, then $\Lambda_{CG}(G) \subset X$. Thence every orbit in $\Lambda_{CG}(G)$ is dense in $\Lambda_{CG}(G)$.*

Proof. Let q be a point in $\Lambda_{CG}(G)$. There exists a sequence (g_n) of elements of G such that $g_n(p) \to q$ for all $p \in \mathbb{H}^n_{\mathbb{C}}$. Let $x, y \in X$, $x \neq y$. We assume, taking subsequences if necessary, that $g_n(x) \to \hat{x}$ and $g_n(y) \to \hat{y}$, where $\hat{x} \neq q$, $\hat{y} \neq q$. Let $p' \in \mathbb{H}^n_{\mathbb{C}}$ be a point in the geodesic determined by x and y. Then $g_n(p')$ is in the geodesic determined by $g_n(x)$ and $g_n(y)$. Given that $q \neq \hat{x}$ and $q \neq \hat{y}$, we have that $q = \lim_{n\to\infty} g_n(p')$ belongs to the geodesic determined by $\lim_{n\to\infty} g_n(x) = \hat{x}$ and $\lim_{n\to\infty} g_n(y) = \hat{y}$, a contradiction. Thus, $g_n(x) \to q$ or $g_n(y) \to q$, so $q \in X$. Therefore $\Lambda_{CG}(G) \subset X$. $\qquad\square$

Theorem 2.6.6. *Let G be a discrete group such that $\Lambda_{CG}(G)$ has more than two points, then it has infinitely many points.*

Proof. Assume that $\Lambda_{CG}(G)$ is finite with at least three points. Then

$$\widetilde{G} = \bigcap_{x \in \Lambda_{CG}(G)} \mathrm{Isot}(x, G)$$

is a normal subgroup of G with finite index. Then for each $\gamma \in \widetilde{G}$, it follows that $\gamma(x) = x$, for each $x \in \Lambda_{CG}(\Gamma)$. Then, the classification of elements in $\mathrm{PU}(n,1)$ yields that each element in $\widetilde{G}$ is elliptic. Hence $\widetilde{G}$ is finite, which is a contradiction. $\square$

Definition 2.6.7. (cf. Definition 3.3.9) The group G is *elementary* if $\Lambda_{CG}(G)$ has at most two points.

Notice that Proposition 2.6.5 implies that if G is nonelementary and $\Lambda_{CG}(G)$ is not all of $\partial \mathbb{H}^n_{\mathbb{C}}$, then $\Lambda_{CG}(G)$ is a nowhere dense perfect set. In other words, $\Lambda_{CG}(G)$ has empty interior and every orbit in $\Lambda_{CG}(G)$ is dense in $\Lambda_{CG}(G)$.

The following corollary is an immediate consequence of Proposition 2.6.5:

Corollary 2.6.8. *If $G \subset \mathrm{PU}(1,n)$ is a nonelementary discrete group, then $\Lambda(G)$ is the unique closed minimal G-invariant set, for the action of G in $\overline{\mathbb{H}^n}$.*

Chapter 3

Complex Kleinian Groups

In this chapter we introduce some fundamental concepts in the theory of complex Kleinian groups that we study in the sequel. We begin with an example in $\mathbb{P}^2_{\mathbb{C}}$ that illustrates the diversity of possibilities one has when defining the notion of "limit set". In this example we see that there are several nonequivalent such notions, each having its own interest. We then define the Kulkarni limit set of discrete subgroups of $\mathrm{PSL}(n+1, \mathbb{C})$ and the regions of discontinuity and equicontinuity. Finally we outline the various constructions and families of complex Kleinian groups that we study in later chapters.

Notice that $\mathrm{PSL}(n+1, \mathbb{C})$, the group of automorphisms of $\mathbb{P}^n_{\mathbb{C}}$, contains as subgroups $\mathrm{PU}(n,1)$ and $\mathrm{Aff}\,(\mathbb{C}^n) \cong \mathrm{GL}\,(n, \mathbb{C}) \ltimes \mathbb{C}^n$, the latter being the group of affine transformations, the former being that of holomorphic isometries of complex hyperbolic space. Hence these are two natural sources of discrete subgroups of $\mathrm{PSL}(n+1, \mathbb{C})$ with a rich geometry. Yet this latter group is much larger than the others, so there is more in it, and that is one of the topics we explore in this monograph.

In this chapter, besides talking about the limit set, we also give an overview of the various means we know for constructing discrete subgroups of $\mathrm{PSL}(n+1, \mathbb{C})$. Of course this includes complex hyperbolic groups, discussed in the previous chapter, and complex affine groups, which we now discuss briefly.

3.1 The limit set: an example

In Chapter 1 we defined the limit set of a Kleinian group in the classical way, as the set of accumulation points of the orbits. This is indeed a good definition in that setting in all possible ways: its complement Ω is the maximal region of discontinuity for the action of the group on the sphere (see Definition 3.2.1 below), and Ω is also the region of equicontinuity, i.e., the set of points where the group forms a normal family, see Chapter 1.

It could be nice to have such a "universal" concept in the setting we envisage in this book, that of groups of automorphisms of $\mathbb{P}^n_\mathbb{C}$. Alas this is not possible in general and there is not a "correct" concept of limit set. This is illustrated by the example below.

Indeed the question of giving "the definition" of limit set can be rather subtle, as pointed out by R. Kulkarni in the general setting of discrete group actions [132], and in [203] for the particular setting we envisage here. As we will see, there can be several definitions of "limit set", each having its own interest, its own characteristics and leading to interesting results. Yet, one has that for complex dimension 2, "generically" all limit sets we envisage in this work coincide (See Chapter 6).

Let us start with an example from [160]. Let $\gamma \in \mathrm{PSL}(3,\mathbb{C})$ be the projectivisation of the linear map $\tilde{\gamma}$ given by

$$\tilde{\gamma} = \begin{pmatrix} \alpha_1 & 0 & 0 \\ 0 & \alpha_2 & 0 \\ 0 & 0 & \alpha_3 \end{pmatrix}$$

where $\alpha_1\alpha_2\alpha_3 = 1$ and $|\alpha_1| < |\alpha_2| < |\alpha_3|$. We denote by Γ the cyclic subgroup of $\mathrm{PSL}(3,\mathbb{C})$ generated by γ; we choose the α_i so that Γ is contained in $\mathrm{PU}(2,1)$. This is a loxodromic element in the notation of [67] (see Chapter 2 above). Each α_i corresponds to an eigenvector, hence to a fixed point of γ that we denote by e_i. The points $\{e_1, e_3\}$ are contained in the sphere at infinity $\mathbb{S}^3_\infty$, corresponding to the null vectors $[V_0]$; the first of these is a repulsor while the third is an attractor. The point e_2 is in $[V_+]$ and it is a saddle point. The projective lines $\overleftrightarrow{e_1, e_2}$ and $\overleftrightarrow{e_2, e_3}$ are both invariant lines, tangent to $\mathbb{S}^3_\infty$. The orbits of points in the line $\overleftrightarrow{e_1, e_2}$ accumulate in e_1 going backwards, and they accumulate in e_2 going forwards. Similar considerations apply to the line $\overleftrightarrow{e_2, e_3}$.

The orbit of each point in $\mathbb{P}^2_\mathbb{C} \setminus (\overleftrightarrow{e_1, e_2} \cup \overleftrightarrow{e_2, e_3})$ accumulates at the points $\{e_1, e_3\}$, and it is not hard to see that Γ forms a normal family at all points in $(\mathbb{H}^2_\mathbb{C} \cup \mathbb{S}^3_\infty) \setminus \{e_1, e_3\}$. In fact one has:

Proposition 3.1.1. *If $\Gamma = \langle \gamma \rangle$ is the cyclic group generated by γ, then:*

(i) *Γ acts discontinuously on $\Omega_0 = \mathbb{P}^2_\mathbb{C} - (\overleftrightarrow{e_1, e_2} \cup \overleftrightarrow{e_3, e_2})$, $\Omega_1 = \mathbb{P}^2_\mathbb{C} - (\overleftrightarrow{e_1, e_2} \cup \{e_3\})$ and $\Omega_3 = \mathbb{P}^2_\mathbb{C} - (\overleftrightarrow{e_3, e_2} \cup \{e_1\})$.*

(ii) *Ω_1 and Ω_2 are the maximal open sets where Γ acts (properly) discontinuously; and Ω_1/Γ and Ω_2/Γ are compact complex manifolds. (In fact they are Hopf manifolds.)*

(iii) *Ω_0 is the largest open set where Γ forms a normal family.*

Thence, unlike the conformal case, now different orbits accumulate at different points, while the set of accumulation points of all orbits consists of the points $\{e_1, e_2, e_3\}$. However, it follows from the proposition above that Γ is not acting discontinuously on the complement of this set.

The main problem resides in the fact that the invariant lines $\overleftrightarrow{e_1, e_2}$, $\overleftrightarrow{e_3, e_2}$ are attractive sets for the iterations of γ (in one case) or γ^{-1} (in the other case).

Notice that a similar example was studied in [223], showing that the quotient space $(\mathbb{P}_{\mathbb{C}}^2 - \{e_1, e_2, e_3\})/\Gamma$ is non-Hausdorff, hence the action cannot be discontinuous.

Proof. (i) To prove this assertion it is enough to observe that $\Omega_0 = \Omega_1 \cap \Omega_2$, and that the action of Γ restricted to Ω_1 or to Ω_2 corresponds to the action of the cyclic groups generated, respectively, by the contracting maps

$$(z, w) \mapsto (z\,\alpha_1\,\alpha_3^{-1}, \, w\,\alpha_3\,\alpha_3^{-1}),$$
$$(z, w) \mapsto (z\,\alpha_2^{-1}\,\alpha_1, \, w\,\alpha_3^{-1}\,\alpha_1).$$

(ii) Let Ω be an open set where Γ acts (properly) discontinuously, then we claim that either $\overleftrightarrow{e_1, e_2} \subset \mathbb{P}_{\mathbb{C}}^2 - \Omega$, or else $\overleftrightarrow{e_2, e_3} \subset \mathbb{P}_{\mathbb{C}}^2 - \Omega$. For this, suppose there is $z \in \mathbb{C}$ such that $[z; 1; 0] \in \Omega$, then

$$[z\,w\,\alpha_3^n : w\,\alpha_3^n : \alpha_2^n] \xrightarrow[n \to \infty]{} [z : 1 : 0],$$

for each $w \in \mathbb{C}^*$, and

$$\gamma^n([z\,w\,\alpha_3^n : w\,\alpha_3^n : \alpha_2^n]) = [z\,w\,\alpha_1^n\,\alpha_2^{-n} : w : 1] \xrightarrow[n \to \infty]{} [0 : w : 1],$$

for all $w \in \mathbb{C}^*$. Therefore $\overleftrightarrow{e_2, e_3} \subset \mathbb{P}_{\mathbb{C}}^2 - \Omega$. The analogous statement for the other line follows similarly.

(iii) Follows easily from (i) and (ii). $\qquad\square$

From the preceding example we see that even in simple cases, when we look at actions on higher-dimensional projective spaces, there is no definition of the limit set having all the properties one has in the conformal setting. Here one might take as "limit set":

- The points $\{e_1, e_2, e_3\}$ where all orbits accumulate. But the action is not properly discontinuous on all of its complement. Yet, this definition is good if we restrict the discussion to the "hyperbolic disc" $\mathbb{H}_{\mathbb{C}}^2$ contained in $\mathbb{P}_{\mathbb{C}}^2$. This corresponds to taking the Chen-Greenberg limit set of Γ, that we define below, which is contained in the "visual sphere" at infinity.

- The two lines $\overleftrightarrow{e_1, e_2}$, $\overleftrightarrow{e_3, e_2}$, which are attractive sets for the iterations of γ (in one case) or γ^{-1} (in the other case). This corresponds to Kulkarni's limit set of Γ, that we define below, and it has the nice property that the action on its complement is discontinuous and also, in this case, equicontinuous. And yet, the proposition above says that away from either one of these two lines (and a point) the action of Γ is properly discontinuous. So each of these is a "maximal" region where the action is properly discontinuous.

- We may be tempted to take as limit set the complement of the "largest region where the action is properly discontinuous", but there is no such region: there are two of them, the complements of each of the two invariant lines, so which one do we choose?

- Similarly we may want to define the limit set as the complement of "the equicontinuity region". In this particular example, that definition may seem appropriate. The problem is that, by [116], the Hopf manifolds one gets as quotients of Ω_1 and Ω_2 by the action of Γ cannot be obtained as the quotient of the region of equicontinuity of some subgroup of $\mathrm{PSL}(3, \mathbb{C})$. That is, these manifolds can not be written in the form U/G where G is a discrete subgroup of $\mathrm{PSL}(3, \mathbb{C})$ acting equicontinuously on an open set U of $\mathbb{P}^2_{\mathbb{C}}$. Moreover, there are examples where Γ is the fundamental group of a certain Inoue surface (those of type Sol^4_0 in Remark 8.7.2, see Chapter 8), and the action of Γ on $\mathbb{P}^2_{\mathbb{C}}$ has no points of equicontinuity.

Thus one has different definitions with nice properties in different settings. For example, we will see that for Schottky groups in higher odd-dimensional projective spaces, there is yet another definition of limit set which seems appropriate.

We will say more about limit sets in the rest of this chapter, and one of the main problems we study are the relations amongst these different concepts of "a limit set".

3.2   Complex Kleinian groups: definition and examples

Even if there is not a "well-defined" concept of limit set for discrete subgroups of $\mathrm{PSL}(n + 1, \mathbb{C})$, there are well-defined concepts of discontinuous and properly discontinuous actions, as explained already in Chapter 1. We recall these:

Definition 3.2.1. Let Γ be a subgroup of $\mathrm{PSL}(n + 1, \mathbb{C})$ and let Ω be a Γ-invariant open subset of $\mathbb{P}^n_{\mathbb{C}}$. The action of Γ is *discontinuous* on Ω if each point $x \in \Omega$ has a neighbourhood U_x which intersects at most finitely many copies of its Γ-orbit. The action is *properly discontinuous* if for every compact set $K \subset \Omega$ we have that K intersects at most a finite number of copies of its Γ-orbit.

Notice that if a subgroup $\Gamma \subset \mathrm{PSL}(n + 1, \mathbb{C})$ acts discontinuously on some nonempty open subset $\Omega \subset \mathbb{P}^n_{\mathbb{C}}$, then Γ is necessarily discrete. However the converse is not true in general, because the action of $\mathrm{PSL}(n + 1, \mathbb{C})$ on $\mathbb{P}^n_{\mathbb{C}}$ is not isometric. In fact there is an example in [207] of a discrete subgroup of $\mathrm{PSL}(3, \mathbb{C})$ acting on $\mathbb{P}^2_{\mathbb{C}}$ with dense orbits, and in [203] it is proved that the fundamental group of every closed, hyperbolic manifold of dimension 5 acts on $\mathbb{P}^3_{\mathbb{C}}$ so that every orbit is dense.

Definition 3.2.2. A discrete subgroup Γ of $\mathrm{PSL}(n + 1, \mathbb{C})$ is *complex Kleinian* if it acts properly discontinuously on some open subset of $\mathbb{P}^n_{\mathbb{C}}$. More generally, if $A \subset \mathbb{P}^n_{\mathbb{C}}$ is an invariant set for the action of a subgroup $\Gamma \subset \mathrm{PSL}(n + 1, \mathbb{C})$, we say that the action on A is *Kleinian* if A contains a nonempty invariant open (relative to A) subset on which the action is properly discontinuous.

Let us name some examples and sources of constructing complex Kleinian groups. These are studied in the sequel.

Example 3.2.3. (i) **Complex hyperbolic Kleinian groups.** Every discrete subgroup of $\mathrm{PU}(n,1)$ is a discrete subgroup of $\mathrm{PSL}(n+1,\mathbb{C})$ that acts properly discontinuously on the open ball that serves as a model for the hyperbolic n-space $\mathbb{H}^n_\mathbb{C}$. Hence it is a complex Kleinian group. The corresponding orbit spaces $\mathbb{H}^n_\mathbb{C}/\Gamma$ are by definition *complex hyperbolic orbifolds*. These are the groups we studied in Chapter 2, and these will appear also in Chapters 4, 7 and 8.

(ii) **Schottky groups.** Consider an arbitrary collection $\mathcal{L} := \{L_1, \ldots, L_r\}$ of projective n-spaces in $\mathbb{P}^{2n+1}_\mathbb{C}$, all of them pairwise disjoint. Given arbitrary neighbourhoods $U_1, \ldots, U_r$ of the L_i's, pairwise disjoint, one can easily show that there exist compact tubular neighbourhoods V_i of the L_i contained in U_i, and projective transformations T_i of $\mathbb{P}^{2n+1}_\mathbb{C}$, $i = 1, \ldots, r$, interchanging the interior with the exterior of each V_i, leaving invariant its boundary $E_i = \partial(V_i)$. The E_i's are *mirrors* that play the same role in $\mathbb{P}^{2n+1}_\mathbb{C}$ that $(n-1)$-spheres play in $\mathbb{S}^n$ to define the classical Schottky groups. The group of automorphisms of $\mathbb{P}^{2n+1}_\mathbb{C}$ generated by the T_i's is complex Kleinian.

These groups were introduced in [201], [203] and they are our focus of study in Chapter 9.

(iii) **Subgroups of the affine group.** As noted earlier, the affine group $\mathrm{Aff}(\mathbb{C}^n)$ is a subgroup of $\mathrm{PSL}(n+1,\mathbb{C})$, and every discrete subgroup of $\mathrm{Aff}(\mathbb{C}^n)$ with a nonempty region of discontinuity gives rise to a projective group with a nonempty region of discontinuity. Let us name some explicit examples that we discuss in more detail in Section 3.4 below.

(a) Fundamental groups of Hopf Manifolds.

(b) Fundamental groups of Complex Tori.

(c) The Suspension or Cone construction.

(d) Fundamental groups of Inoue surfaces.

(e) Complex reflections groups.

The suspension is a construction that allows us to construct a discrete subgroups of $\mathrm{PSL}(n+1,\mathbb{C})$ out from subgroups of $\mathrm{PSL}(n,\mathbb{C})$. The groups obtained in this way are the prototype of the *groups with a control group* that we study in Chapter 5. These are all affine groups.

(iv) **Groups constructed via twistor theory.** The so-called Calabi-Penrose fibration, which is the most basic twistor fibration, describes $\mathbb{P}^3_\mathbb{C}$ as a fibre bundle over the 4-sphere $\mathbb{S}^4$ with fibre $\mathbb{P}^1_{\mathbb{C}} \cong \mathbb{S}^2$. A basic fact of twistor theory is that the group $\mathrm{Conf}_+(\mathbb{S}^4)$, of orientation preserving conformal maps of $\mathbb{S}^4$, has a canonical embedding in $\mathrm{PSL}(4,\mathbb{C})$. Thus one has a canonical action of $\mathrm{Conf}_+(\mathbb{S}^4)$ on $\mathbb{P}^3_{\mathbb{C}}$, and this action carries "twistor lines" (the fibres of the fibration $\mathbb{P}^3_{\mathbb{C}} \to \mathbb{S}^4$) into twistor lines. In [202] it is proved that these maps on the fibres are isometries with respect to the usual round metric on $\mathbb{S}^2$ and therefore given $\Gamma \subset \mathrm{Conf}_+(\mathbb{S}^4)$ discrete, the maximal region of discontinuity for the induced action on $\mathbb{P}^3_{\mathbb{C}}$ is just the lifting of the corresponding region of discontinuity in $\mathbb{S}^4$.

In other words, every conformal Kleinian subgroup $\Gamma \subset \mathrm{Conf}_+(\mathbb{S}^4)$ is canonically a complex Kleinian subgroup of $\mathrm{PSL}(4,\mathbb{C})$, and the whole dynamics of Γ on the sphere can be read from its action on $\mathbb{P}^3_{\mathbb{C}}$, but in $\mathbb{P}^3_{\mathbb{C}}$ one can see facets of Γ that cannot be seen in $\mathbb{S}^4$. Analogous statements hold in higher dimensions.

These types of complex Kleinian groups were introduced in [201] and studied in [202]; they are our focus of study in Chapter 10.

Notice that if $\Gamma \subset \mathrm{Conf}_+(\mathbb{S}^4)$ is a lattice (i.e., the quotient $\mathrm{Conf}_+(\mathbb{S}^4)/\Gamma$ has finite volume), then Mostow's rigidity theorem says that Γ is rigid in $\mathrm{Conf}_+(\mathbb{S}^4)$, and therefore it is rigid in $\mathrm{PSL}(4,\mathbb{C})$ since the Lie algebra of this group is the complexification of the Lie algebra of $\mathrm{Conf}_+(\mathbb{S}^4)$. Yet, for discrete subgroups of $\mathrm{Conf}_+(\mathbb{S}^4)$ which are not lattices, there are deformations in $\mathrm{PSL}(4,\mathbb{C})$ that do not come from $\mathrm{Conf}_+(\mathbb{S}^4)$, and their study can be rather interesting.

(v) **The Join.** This is a construction introduced in [201] that allows us to construct a complex Kleinian group from two given complex Kleinian groups. Let $\Gamma_1 \subset \mathrm{GL}(n+1,\mathbb{C})$ be a discrete group acting on $\mathbb{C}^{n+1}$ which induces a Kleinian action on $\mathbb{P}^n_{\mathbb{C}}$; let $\Gamma_2 \subset \mathrm{GL}(m+1,\mathbb{C})$ be a discrete group acting on $\mathbb{C}^{m+1}$, which induces a Kleinian action on $\mathbb{P}^m_{\mathbb{C}}$. Then $\Gamma_1 \times \Gamma_2$ acts on $\mathbb{C}^{n+m+2}$ and induces a Kleinian action on $\mathbb{P}^{n+m+1}_{\mathbb{C}}$. The discontinuity set is the projective join of the corresponding discontinuity sets of the previous actions on $\mathbb{P}^n_{\mathbb{C}}$ and $\mathbb{P}^m_{\mathbb{C}}$. This construction can be iterated to any number of Kleinian actions.

It is clear that in this case $\mathbb{P}^n_{\mathbb{C}}$ and $\mathbb{P}^m_{\mathbb{C}}$ are both invariant subsets for the action on $\mathbb{P}^{n+m+1}_{\mathbb{C}}$. This motivates the following definitions from [201], [202]:

Definition 3.2.4. Let Γ be a discrete subgroup of $\mathrm{PSL}(n+1,\mathbb{C})$, $n > 1$.

(a) The action of Γ on $\mathbb{P}^n_{\mathbb{C}}$ is *reducible* if it is obtained either by suspending a complex Kleinian action on $\mathbb{P}^{n-1}_{\mathbb{C}}$, or as the join of two Kleinian actions on $\mathbb{P}^{n_1}_{\mathbb{C}}$ and $\mathbb{P}^{n_2}_{\mathbb{C}}$, $n = n_1 + n_2 + 1$. Otherwise we say that the action of Γ is *irreducible*.

(b) The action of Γ on $\mathbb{P}^n_{\mathbb{C}}$ is *(real or complex) algebraically-mixing* if there are no proper (real or complex, respectively) compact submanifolds of $\mathbb{P}^n_{\mathbb{C}}$ which are Γ-invariant.

It is clear that algebraically-mixing implies irreducible. In Chapter 10 we give examples from [202] of Kleinian actions which are algebraically mixing. As noticed in [202], the same arguments also prove that the fundamental group of every compact 5-dimensional hyperbolic manifold acts on $\mathbb{P}^3_\mathbb{C}$ with dense orbits; thence the action is algebraically-mixing.

In dimension 2, the example in [207], of a discrete subgroup of $\mathrm{PSL}(3, \mathbb{C})$ that acts on $\mathbb{P}^2_\mathbb{C}$ with dense orbits, has an empty region of discontinuity, so the action is algebraically mixing.

3.3 Limit sets: definitions and some basic properties

There are several types of possible "limit sets" relevant for this work, namely:

a) the Chen-Greenberg limit set of complex hyperbolic groups;

b) the limit set of Kulkarni;

c) the complement of the region of equicontinuity;

d) complements of maximal regions of discontinuity; and

e) the limit set of complex Schottky groups.

There are other possible definitions of a limit set. For instance, in the classical conformal case, the limit set can be regarded as the closure of the set of fixed points of loxodromic elements. This is analogous to what happens for rational maps on the Riemann sphere, where the Julia set is the closure of the repelling periodic points. One has a notion of "loxodromic elements" in $\mathrm{PSL}(3, \mathbb{C})$", which in some sense can be extended to higher dimensions. This may lead to another natural notion of limit set, as the closure of the fixed points of loxodromic elements.

A natural question is: what are the relations amongst all these types of limit sets? This is one of the questions we address in this work.

The Chen-Greenberg limit set of complex hyperbolic groups was already defined in the previous chapter. We define below the Kulkarni limit set. The case of Schottky groups is discussed in Chapter 9.

3.3.1 The Limit set of Kulkarni

Consider a locally compact Hausdorff space X with a countable base for its topology. Let G be a group acting on X and let $\Omega \subset X$ be a G-invariant subset. We recall from Chapter 1 that the action on Ω is *properly discontinuous* if for every pair of compact subsets C and D of Ω, the cardinality of the set $\{\gamma \in G \mid \gamma(C) \cap D \neq \emptyset\}$, is finite.

In [132] there is a concept of limit set, motivated through an example that inspired our example in Section 3.1. This definition of a limit set applies in a

very general setting of discrete group actions, and it has the important property
of assuring that its complement is an open invariant set where the group acts
properly discontinuously. For this, recall that given a family $\{A_\beta\}$ of subsets of
X, where β runs over some infinite indexing set B, a point $x \in X$ is a *cluster
(or accumulation) point* of $\{A_\beta\}$ if every neighbourhood of x intersects A_β for
infinitely many $\beta \in B$.

   Consider a space X and a group G as above. Then Kulkarni looks at the
following three sets.

   a) Let $L_0(G)$ be the closure of the set of points in X with infinite isotropy group.

   b) Let $L_1(G)$ be the closure of the set of cluster points of orbits of points in
      $X - L_0(G)$, i.e., the cluster points of the family $\{\gamma(x)\}_{\gamma \in G}$, where x runs
      over $X - L_0(G)$.

   c) Finally, let $L_2(G)$ be the closure of the set of cluster points of $\{\gamma(K)\}_{\gamma \in G}$,
      where K runs over all the compact subsets of $X - \{L_0(G) \cup L_1(G)\}$.

   We have ([132]):

Definition 3.3.1.   (i) Let X be as above and let G be a group of homeomorphisms
   of X. The *Kulkarni limit set* of G in X is the set

$$\Lambda_{\mathrm{Kul}}(G) := L_0(G) \cup L_1(G) \cup L_2(G).$$

(ii) The *Kulkarni region of discontinuity* of G is

$$\Omega_{\mathrm{Kul}}(G) \subset X := X - \Lambda_{\mathrm{Kul}}(G).$$

   Notice that the set $\Lambda_{\mathrm{Kul}}(G)$ is closed in X and it is G-invariant (it can be
empty). The set $\Omega_{\mathrm{Kul}}(G)$ (which also can be empty) is open, G-invariant, and G
acts properly discontinuously on it.

   It is not hard to show that when G is a Möbius (or conformal) group, the
Kulkarni limit set coincides with the classical limit set . In fact in that case there
are no points with infinite isotropy and the set $L_1(G)$ is the usual limit set $\Lambda(G)$.
The only thing to prove is that the set $L_2(G)$ does not add new points to the limit
set $\Lambda(G)$, and this is a consequence of the fact that in the conformal case, the
action on the complement of the limit set is properly discontinuous.

   In the example in Section 3.1 one has that the sets $L_0(G)$ and $L_1(G)$ are equal
and consist of the three points $\{e_1, e_2, e_3\}$, while $L_2(G)$ consists of the two lines
$\overleftrightarrow{e_1, e_2}$, $\overleftrightarrow{e_3, e_2}$. This example shows also that although Kulkarni's limit set has the
property of assuring that the action on its complement is properly discontinuous,
this region is not always the largest such region.

   In this work we are mostly interested in the case where the space X is a
complex projective space $\mathbb{P}^n_{\mathbb{C}}$, and this definition of a limit set will be essential in
several chapters. In Chapter 7 we will compare the limit sets of Chen-Greenberg
and that of Kulkarni in the case of discrete subgroups of $\mathrm{PU}(n, 1)$ acting on $\mathbb{P}^n_{\mathbb{C}}$.
The results in that chapter also show several other interesting facts, as for instance:

(i) There is no monotony regarding limit sets in the sense that if Γ_1, Γ_2 are Kleinian subgroups of some $\mathrm{PSL}(n, \mathbb{C})$, then $\Gamma_1 \subset \Gamma_2$ does not imply $\Lambda_{\mathrm{Kul}}(\Gamma_1) \subset \Lambda_{\mathrm{Kul}}(\Gamma_2)$ (see Remark 9.6.4), as it happens for conformal Kleinian groups and also for subgroups of $\mathrm{PU}(2, 1)$, see Corollary 7.3.11.

(ii) There is no relation amongst L_0, L_1 and L_2, in the sense that there are cases where two of them coincide and the other does not, or in some examples one has a strict contention $L_i \subset L_j$, and in other examples this is the other way round, $L_j \subset L_i$. We refer to Chapter 7 for details on these and other properties of the Kulkarni limit set for Kleinian actions of subgroups of $\mathrm{PU}(2n, 1)$.

The limit set also has the following properties that will be used in the sequel:

Proposition 3.3.2. *Let $G \subset \mathrm{PSL}(n, \mathbb{C})$ be a discrete group, then G is finite if and only if $\Lambda_{\mathrm{Kul}}(G)$ is empty.*

Proof. Obviously $\Lambda_{\mathrm{Kul}}(G)$ is empty when G is finite, so we assume that G is not finite. Then by Proposition 2.1.2 there is an element γ with infinite order. Let $\tilde{\gamma}$ be a lift of γ and x be an eigenvalue of $\tilde{g}$; such an eigenvector exists since $\mathbb{C}$ is complete as a field. Then $[x]$ is a fixed point of γ and therefore $[x] \in L_0(G)$. This completes the proof. □

Proposition 3.3.3. *Let $\Gamma \subset \mathrm{PSL}(n+1, \mathbb{C})$ be a group acting properly discontinuously on an open invariant set $\Omega \subset \mathbb{P}^n_{\mathbb{C}}$. Then $L_0(\Gamma) \cup L_1(\Gamma) \subset \mathbb{P}^n_{\mathbb{C}} - \Omega$, and for every compact set $K \subset \Omega$ one has that the set of cluster points of ΓK is contained in $\mathbb{P}^n_{\mathbb{C}} \setminus \Omega$.*

Proof. The proof follows easily from the fact that Γ acts discontinuously. □

Proposition 3.3.4. *Let $G \subset \mathrm{PSL}(n+1, \mathbb{C})$ be a discrete, infinite group, $n > 1$. Then the Kulkarni limit set of G always contains a projective subspace of dimension at least 1.*

Proof. For simplicity we assume $n = 2$; the general case is similar. By Proposition 2.1.2 there is an element $\gamma \in \Gamma$ with infinite order. By the Jordan Normal Form theorem, as in the example of Section 3.1, it is enough to consider the cases where $\gamma \in \mathrm{PSL}(3, \mathbb{C})$ has a lift $\tilde{\gamma}$ whose Jordan's normal form is either diagonal or one of the following. In the diagonal case the conclusion follows from the λ-lemma that we prove in Lemma 7.3.6.

Case 1. For all $m \in \mathbb{N}$ one has that:

$$\tilde{\gamma}^m = \begin{pmatrix} 1 & m & \frac{m(m-1)}{2} \\ 0 & 1 & m \\ 0 & 0 & 1 \end{pmatrix}. \tag{3.3.4}$$

We claim that in this case $\overleftrightarrow{e_1, e_2} \subset \mathbb{P}^2_{\mathbb{C}} - \Omega_{\mathrm{Kul}}(G)$. Otherwise there exists $z \in \mathbb{C}$ such that $[z; 1; 0] \in \Omega_{\mathrm{Kul}}(G)$. Let $\epsilon \in \mathbb{C}$, then $\left[z; 1; \frac{2(\epsilon - m)}{m(m-1)}\right] \xrightarrow{m \to \infty} [z; 1; 0]$. Thus for $m(\epsilon)$ large $(a_m(\epsilon) = \left[z; 1; \frac{2(\epsilon - m)}{m(m-1)}\right])_{m \geq m(\epsilon)} \subset \Omega_{\mathrm{Kul}}(G)$. By Proposition 3.3.4 we conclude that

$$\gamma^m(a_m) = \left[z + \epsilon; \frac{2\epsilon - (m+1)}{m-1}; \frac{2(\epsilon - m)}{m(m-1)}\right] \xrightarrow[m \to \infty]{} [-z - \epsilon; 1; 0] \text{ for each } \epsilon \in \mathbb{C}.$$

By Proposition 3.3.3 we conclude that $\overleftrightarrow{e_1, e_2} \subset \mathbb{P}^2_{\mathbb{C}} - \Omega_{\mathrm{Kul}}(G)$, which is a contradiction.

Case 2. There is a $\lambda \in \mathbb{C}$ such that for all $m \in \mathbb{N}$ one has that

$$\tilde{\gamma}^m = \begin{pmatrix} \lambda^m & m\lambda^{m-1} & 0 \\ 0 & \lambda^m & 0 \\ 0 & 0 & \lambda^{-2m} \end{pmatrix}.$$

If $|\lambda| = 1$, then the line $\overleftrightarrow{e_1, e_3}$ is contained in $L_0(G) \cup L_1(G)$ because the dynamics of the group restricted to this line is that of a rotation.

We assume now that $|\lambda| < 1$, the case $|\lambda| > 1$ being analogous. We claim that $\overleftrightarrow{e_1, e_2} \subset \mathbb{P}^2_{\mathbb{C}} - \Omega$ or $\overleftrightarrow{e_1, e_3} \subset \mathbb{P}^2_{\mathbb{C}} - \Omega$. If $\overleftrightarrow{e_1, e_2} \not\subset \mathbb{P}^2_{\mathbb{C}} - \Omega$, then there is $z \in \mathbb{C}$ such that $[z; 1; 0] \in \Omega$. Observe that for each $w \in \mathbb{C}^*$ we have $[wz; w; m\lambda^{3n-1}] \xrightarrow[m \to \infty]{} [z; 1; 0]$ and

$$\gamma^m \left(\left[z; 1; \frac{n\lambda^{3m-1}}{w}\right]\right) = \left[\frac{wz\lambda}{m} + w; \frac{w\lambda}{m}; 1\right] \xrightarrow[m \to \infty]{} [w; 0; 1],$$

for all $w \in \mathbb{C}^*$. As in the previous case we deduce that $\overleftrightarrow{e_1, e_3} \subset \mathbb{P}^2_{\mathbb{C}} - \Omega$. $\square$

From the proof of the previous proposition and the example in Section 3.1 we see that we have a matrix as in Case 2 above, with $|\lambda| \neq 1$, and we have two invariant lines which meet at a point. One of these lines is attractive while the other is repelling. Hence we conclude:

Corollary 3.3.5. (i) *Let $\gamma \in \mathrm{PSL}(3, \mathbb{C})$ be the transformation induced by the matrix:*

$$\tilde{\gamma}^m = \begin{pmatrix} \lambda^m & m\lambda^{m-1} & 0 \\ 0 & \lambda^m & 0 \\ 0 & 0 & \lambda^{-2m} \end{pmatrix},$$

with $|\lambda| \neq 1$. Then $\overleftrightarrow{e_1, e_2} \cup \overleftrightarrow{e_1, e_3} \subset \mathbb{P}^2_{\mathbb{C}} \setminus \mathrm{Eq}(\langle \gamma \rangle)$.

(ii) *Let $\gamma \in \mathrm{PSL}(3, \mathbb{C})$ be the transformation induced by the diagonal matrix $\tilde{\gamma} = (\gamma_{ij})_{i,j=1,3}$, where $|\gamma_{11}| < |\gamma_{22}| < |\gamma_{33}|$ and $\gamma_{11}\gamma_{22}\gamma_{33} = 1$. Then $\overleftrightarrow{e_1, e_2} \cup \overleftrightarrow{e_2, e_3} \subset \mathbb{P}^2_{\mathbb{C}} \setminus \mathrm{Eq}(\langle \gamma \rangle)$.*

Proposition 3.3.6. *Let $\mathcal{C}$ be a closed G invariant set which satisfies that for every compact set $K \subset \mathbb{P}^n_{\mathbb{C}} - \mathcal{C}$ the set of cluster points of GK lies in $\mathcal{C} \cap (L_0(G) \cup L_1(G))$, then $\Lambda_{\mathrm{Kul}}(G) \subset \mathcal{C}$.*

Proof. If this is not the case, we deduce that there is a point in $x \in \mathbb{P}_{\mathbb{C}}^n - \mathcal{C}$, a sequence $(k_m \subset \mathbb{P}_{\mathbb{C}}^n - (L_0(G) \cup L_1(G))$ and a sequence (g_m) of different elements in G, such that

$$k_m \xrightarrow[m \to \infty]{} k \in \mathbb{P}_{\mathbb{C}}^n - (L_0(G) \cup L_1(G)),$$

$$g_m(k_m) \xrightarrow[m \to \infty]{} x \in \mathbb{P}_{\mathbb{C}}^n - \mathcal{C}.$$

Thus, we can assume that $(g_m(k_m)) \subset \mathbb{P}_{\mathbb{C}}^n - \mathcal{C}$. Applying the hypothesis to the compact set $K = \{g_m(k_m) : m \in \mathbb{N}\} \cup \{x\}$ we get $k \in L_0(G) \cup L_1(G)$, which is a contradiction. Therefore $\Lambda(G) \subset \mathcal{C}$. $\qquad\square$

Remark 3.3.7. [**Complements of maximal regions**] We can see from the previous examples (in Section 3.1 and in Subsection 3.2.3) that there are cases for which neither of the two concepts of limit set defined above (Chen-Greenberg's and Kulkarni's) is "appropriate". For instance when considering Schottky groups, the limit set $\Lambda_S(\Gamma)$ is defined to be the set of accumulation points of the projective subspaces $\{L_i\}$ used to define the Schottky group. The limit set for other complex Schottky groups is defined similarly (see Chapter 9). We have not yet been able to determine whether or not for Schottky groups this limit set coincides with Kulkarni's limit set, nor if its complement coincides with the region of equicontinuity.

In the case of the fundamental group of a Hopf surface, it is easy to determine its Kulkarni limit set, but it has the disadvantage that its complement (the Kulkarni region of discontinuity) is not the largest region where the action is discontinuous. In fact, there is not such a "largest" region, but there are two "maximal" regions.

3.3.2 Elementary groups

We now introduce the notion of elementary complex Kleinian groups. Recall that for a conformal Kleinian group $\Gamma \subset \mathrm{Iso}(\mathbb{H}_{\mathbb{R}}^n)$, its limit set either has infinite cardinality or else its cardinality is at most 2, and in that case we say that the group is elementary.

In the case of complex Kleinian groups which are infinite, Proposition 3.3.4 says that the Kulkarni limit set always contains projective subspaces of positive dimension, so it has infinite cardinality. So it makes no sense to count the number of points in the limit set.

In fact, when $n = 2$ one can show that if the Kulkarni limit set has more than three projective lines, then it contains infinitely many lines. In Chapter 4 there are examples of cyclic groups having one, two and three lines in its Kulkarni limit set. Moreover, if $n = 2$ we have the following result from [14]:

Proposition 3.3.8. *Let $\Gamma \subset \mathrm{PSL}(3, \mathbb{C})$ be discrete. Assume it acts properly discontinuously on an open invariant set $\Omega \subset \mathbb{P}_{\mathbb{C}}^n$ whose complement $\mathbb{P}_{\mathbb{C}}^n \setminus \Omega$ consists of a finite union of complex projective subspaces. Then one of the following statements is verified:*

(i) $\mathbb{P}^2_{\mathbb{C}} \setminus \Omega$ is a complex line;

(ii) $\mathbb{P}^2_{\mathbb{C}} \setminus \Omega$ is the union of two complex lines;

(iii) $\mathbb{P}^2_{\mathbb{C}} \setminus \Omega$ is the union of 3 nonconcurrent lines;

(iv) $\mathbb{P}^2_{\mathbb{C}} \setminus \Omega$ is the union of a point and a complex line not containing it.

The previous discussion motivates the following definition:

Definition 3.3.9. Let $\Gamma \subset \mathrm{PSL}(n+1, \mathbb{C})$ be a discrete group. Then Γ is *elementary* if its Kulkarni limit set is a finite union of complex projective subspaces.

3.4 On the subgroups of the affine group

As mentioned before, every discrete subgroup of the affine group $\mathrm{Aff}\,(\mathbb{C}^n)$ is a subgroup of $\mathrm{PSL}(n+1, \mathbb{C})$. Let us look at some concrete examples:

3.4.1 Fundamental groups of Hopf manifolds

Let $\Gamma_g = \langle g \rangle$ be the cyclic group generated by an element $g \in \mathrm{GL}\,(n, \mathbb{C})$ which is a contraction. Then Γ_g acts properly discontinuously and freely on $\mathbb{C}^n - \{0\}$. The quotient $H = \mathbb{C}^n - \{0\}/\Gamma_g$ is compact and it is a Hopf manifold.

3.4.2 Fundamental groups of complex tori

Let $v_1, \ldots, v_{2n} \in \mathbb{C}^{2n} - \{0\}$ be points which are $\mathbb{R}$-linearly independent, and let g_i be the translation induced by v_i. Then $\Gamma = \langle g_1, \ldots, g_{2n} \rangle$ is a group isomorphic to $\mathbb{Z}^{2n}$ which acts properly discontinuously and freely on $\mathbb{C}^n$. The quotient $\mathbb{C}^n/\Gamma$ is homeomorphic to a torus $\mathbb{T}^{2n} \cong \mathbb{T}^2 \times \cdots \times \mathbb{T}^2$ with $\mathbb{T}^2 := \mathbb{S}^1 \times \mathbb{S}^1$. Thinking of $\mathbb{P}^n_{\mathbb{C}}$ as being $\mathbb{C}^n$ union a $\mathbb{P}^{n-1}_{\mathbb{C}}$ at ∞, one can easily show that $\mathbb{C}^n = \Omega_{\mathrm{Kul}}(\Gamma) = \mathrm{Eq}\,(\Gamma)$ is the largest open set where Γ acts properly discontinuously, and it coincides with the equicontinuity region.

3.4.3 The suspension or cone construction

This construction is studied in detail in Chapter 5. This allows us to construct complex Kleinian subgroups of $\mathrm{PSL}(n+1, \mathbb{C})$ from a complex Kleinian subgroup of $\mathrm{PSL}(n, \mathbb{C})$, and these are all affine. As noted before, $\mathbb{P}^{n-1}_{\mathbb{C}}$ is the space of lines in $\mathbb{C}^n$ and $\mathbb{P}^n_{\mathbb{C}}$ can be thought of as being $\mathbb{C}^n$ union the hyperplane at infinity.

Let Γ be a subgroup of $\mathrm{PSL}(n, \mathbb{C})$ and consider the covering map $[[\ ]]_n :$ $\mathrm{SL}(n, \mathbb{C}) \to \mathrm{PSL}(n, \mathbb{C})$. Let $\widehat{\Gamma} \subset \mathrm{SL}(n, \mathbb{C})$ be $[[\ ]]_n^{-1}(\Gamma)$, i.e., the inverse image of Γ in $\mathrm{SL}(n, \mathbb{C})$. Think of $\mathbb{P}^n_{\mathbb{C}}$ as being $\mathbb{C}^n \cup \mathbb{P}^{n-1}_{\mathbb{C}}$ and let $\widetilde{\Gamma}$ be the subgroup of $\mathrm{PSL}(n+1, \mathbb{C})$ that acts as $\widehat{\Gamma}$ on $\mathbb{C}^n$, so its action on $\mathbb{P}^{n-1}_{\mathbb{C}}$, the hyperplane at infinity, coincides with that of Γ. We call $\widetilde{\Gamma}$ the *full suspension group* of Γ, following [201].

If we can actually lift $\Gamma \subset \mathrm{PSL}(n, \mathbb{C})$ to a subgroup $\widehat{\Gamma} \subset \mathrm{SL}(n, \mathbb{C})$ that intersects the kernel of $[[\ ]]_n$ only at the identity, so that $\widehat{\Gamma}$ is isomorphic to Γ, then we can define $\widetilde{\Gamma}$ to be the subgroup of $\mathrm{PSL}(n + 1, \mathbb{C})$ that acts as $\widehat{\Gamma}$ on $\mathbb{C}^n$ and as Γ on $\mathbb{P}_{\mathbb{C}}^{n-1}$, the hyperplane at infinity. We now call $\widetilde{\Gamma}$ a *(simple) suspension group* of Γ.

As noted in [201], the obstruction to lifting a Kleinian subgroup Γ of $\mathrm{PSL}(n, \mathbb{C})$ to an isomorphic group in $\mathrm{SL}(n, \mathbb{C})$ is an element in $H^2(\Gamma, \mathbb{Z}_n)$. If this obstruction vanishes, then Γ can be regarded as a Kleinian group on $\mathbb{P}_{\mathbb{C}}^n$ via a suspension as above. This happens, for instance, if $H^2(\Gamma, \mathbb{Z}_n) \cong 0$.

For example, if Γ is the fundamental group of a complete (nonnecessarily compact), hyperbolic 3-manifold, so that $\Gamma \subset \mathrm{PSL}(2, \mathbb{C})$, then the obstruction in question can be identified with the second Stiefel-Whitney class ω_2 of the 3-manifold, as pointed out by Thurston, see [127]. This class is always 0, because every oriented 3-manifold is parallelisable. Hence Γ can always be lifted isomorphically to $\mathrm{SL}(2, \mathbb{C})$. Thus the fundamental group Γ of a hyperbolic 3-manifold acting on $\mathbb{H}^3$, whose action on the sphere at infinity is Kleinian, can be considered as a complex Kleinian group $\widetilde{\Gamma}$ acting on $\mathbb{P}_{\mathbb{C}}^2$ and leaving the line at infinity invariant. In this case the Kulkarni limit set of $\widetilde{\Gamma}$ is the cone, with vertex at 0, over the limit set of Γ on $\mathbb{P}_{\mathbb{C}}^1$; the equicontinuity region of $\widetilde{\Gamma}$ agrees with its discontinuity region and it is the largest open set where Γ acts properly discontinuously. By Ahlfors' Finiteness Theorem [3], the quotient $\Omega(\Gamma)/\Gamma$ is a Riemann surface of finite type. Hence the quotient $\Omega(\Gamma)/\widehat{\Gamma}$ is a complex line bundle over a Riemann surface of finite type; $\Omega(\Gamma)/\widehat{\Gamma}$ is homotopically equivalent to $\mathbb{H}^3/\Gamma$.

Remark 3.4.1. Recall that $\mathrm{PSL}(n, \mathbb{C}) \cong \mathrm{SL}(n, \mathbb{C})/\mathbb{Z}_n$. Then it can happen that a subgroup $\Gamma \subset \mathrm{PSL}(n, \mathbb{C})$ has a lifting to $\mathrm{SL}(n, \mathbb{C})$ that intersects the kernel of $[[\ ]]_n$ in a nontrivial subgroup, other than $\mathbb{Z}_n$. In this case we can consider suspensions similarly as before.

It is worth noting that one has a canonical embedding $\mathrm{SL}(n, \mathbb{C}) \to \mathrm{SL}(n + k, \mathbb{C})$, for all $n > 1, k > 0$, given by $A \mapsto \begin{pmatrix} A & 0 \\ 0 & I_{k \times k} \end{pmatrix}$. Therefore, if $\Gamma \subset \mathrm{PSL}(n, \mathbb{C})$ is a group with a region of discontinuity in $\mathbb{P}_{\mathbb{C}}^{n-1}$ that can be suspended to a linear group in $\mathrm{SL}(n, \mathbb{C})$, then Γ can be automatically suspended to a linear group in $\mathrm{SL}(n + k, \mathbb{C})$.

Thus, via this construction, Γ has a region of discontinuity in $\mathbb{P}_{\mathbb{C}}^{n+k}$. If Γ acts properly discontinuously on $\Omega \subset \mathbb{P}_{\mathbb{C}}^{n-1}$, then the corresponding region of discontinuity Ω^k of Γ in $\mathbb{P}_{\mathbb{C}}^{n+k}$ is a $(k + 1)$-dimensional complex bundle over $\mathbb{P}_{\mathbb{C}}^{n-1}$. Hence, in particular, *the fundamental group of every complete, connected, open, 3-dimensional hyperbolic manifold can be regarded in this way as a Kleinian group on $\mathbb{P}_{\mathbb{C}}^n$, for all $n > 0$.*

This construction of a "suspension" has been generalised in [41], leading to the concept of a *controllable Kleinian group*, that we study in Chapter 5 below. This means a Kleinian subgroup $\Gamma \subset \mathrm{PSL}(n + 1, \mathbb{C})$ that leaves invariant a hyperplane $\mathbb{P}_{\mathbb{C}}^{n-1} \subset \mathbb{P}_{\mathbb{C}}^{n}$ and a point p away from $\mathbb{P}_{\mathbb{C}}^{n-1}$. One thus has a natural holomorphic projection map $\mathbb{P}_{\mathbb{C}}^{n} \setminus \{p\} \to \mathbb{P}_{\mathbb{C}}^{n-1}$ and a group homomorphism $\Gamma \to \mathrm{PSL}(n, \mathbb{C})$ that allows us to study the geometry and dynamics of Γ from its image in $\mathrm{PSL}(n, \mathbb{C})$.

3.4.4   Example of elliptic affine surfaces

Let us describe an example which belongs to the important class of compact complex surfaces known as *elliptic affine surfaces*; for a precise definition see [18]. This shows how this kind of surfaces appear naturally in our topic.

Let $\Sigma \subset \mathrm{PSL}(2, \mathbb{C})$ be a Kleinian group and U a connected component of the discontinuity region $\Omega(\Sigma)$ such that $U/\mathrm{Isot}(U, \Sigma)$ is a compact manifold, where $\mathrm{Isot}(U, \Sigma)$ is the isotropy of U, i.e., the subgroup of Σ that leaves U invariant. Now let $G \subset \mathbb{C}^*$ be an infinite discrete group, and consider the suspension of $\mathrm{Isot}(U, \Sigma)$ with respect to G. Let Γ be a torsion free subgroup of $\mathrm{Susp}(\mathrm{Isot}(U, \Sigma), G)$ with finite index. Then the quotient $(U \times C^*)/\Gamma$ is a compact manifold, called an *elliptic affine manifolds*.

3.4.5   Fundamental groups of Inoue surfaces

The formal definition of Inoue surfaces can be seen in [18]. By a theorem of [231] all Inoue surfaces can be described as follows. There are three types (or families) of such surfaces, all of them being affine manifolds. Let us describe briefly each of these.

i) The S_M family.

Let $M \in \mathrm{SL}(3, \mathbb{Z})$ have eigenvalues $\alpha, \beta, \overline{\beta}$ with α a real number > 1 and $\beta \neq \overline{\beta}$. Choose a real eigenvector (a_1, a_2, a_3) belonging to α and an eigenvector (b_1, b_2, b_3) belonging to β. Now let G_M be the group of automorphisms of $\mathbb{H}^+ \times \mathbb{C}$ generated by

$$\gamma_0(w, z) = (\alpha w, \beta z),$$
$$\gamma_i(w, z) = (w + a_i, z + b_i) \, i = 1, 2, 3,$$

where $\mathbb{H}^+ \cong \mathbb{H}_{\mathbb{R}}^2$ denotes the upper half plane in $\mathbb{C}$ equipped with the hyperbolic metric. Then G_M is a subgroup of a certain solvable group Sol_0^4 (that we define in page 175), and G_M acts properly discontinuously and freely on $\mathbb{H}^+ \times \mathbb{C}$. The quotient $(\mathbb{H}^+ \times \mathbb{C})/G_M$ is a compact surface.

ii) The $\mathbb{S}_N^+$ family.

Let $N \in \mathrm{SL}(2, \mathbb{Z})$ have real eigenvalues α, α^{-1} with corresponding real eigenvectors (a_1, a_2), (b_1, b_2). Choose a nonzero integer r, a complex number t and complex numbers c_1, c_2 satisfying a certain integrability condition (that can be made precise). Consider the automorphisms of $\mathbb{H}^+ \times \mathbb{C}$ generated by

$$
\begin{aligned}
\gamma_0(w, z) &= (\alpha w, \alpha^{-1} z + t), \\
\gamma_i(w, z) &= (w + a_i, z + b_i w + c_i) \, i = 1, 2, \\
\gamma_3(w, z) &= (w, z + r^{-1}(b_1 a_2 - b_2 a_1)).
\end{aligned}
$$

These generate a group $G_M \subset \mathrm{Sol}_1^4$ and G_M acts properly discontinuously, freely and with compact quotient on $\mathbb{H}^+ \times \mathbb{C}$.

iii) The $\mathbb{S}_N^-$ family.

This family is defined by modifying the above construction as follows. Consider $N \in \mathrm{GL}(2, \mathbb{Z})$ with real eigenvalues $\alpha, -\alpha^{-1}$, and do as above but now setting $\gamma_0(w, z) = (\alpha w, -z)$.

In each of these cases the group G_M can be regarded as acting on $\mathbb{P}_{\mathbb{C}}^2$, and it can be proved (see Chapter 8) that its Kulkarni region of discontinuity is given by $(\mathbb{C} - \mathbb{R}) \times \mathbb{C}$. Also this set is the largest one where G_M acts properly discontinuously. However in the case $G_M \subset \mathrm{Sol}_0^4$ one has that $\mathrm{Eq}\,(G_M) = \emptyset$.

3.4.6 A group induced by a hyperbolic toral automorphism

Let M be the matrix $M = \begin{pmatrix} 3 & 5 \\ -5 & 8 \end{pmatrix}$. This matrix has eigenvalues $\alpha_\pm = \frac{-5 \pm \sqrt{21}}{2}$, and one can see that $v_+ = (1, \frac{-11 + \sqrt{21}}{10})$, $v_- = (\frac{-11 + \sqrt{21}}{10}, 1)$ are eigenvectors corresponding to α_+ and α_- respectively. Now consider the open set

$$
W = \bigcup_{i,j=0,1} (\mathbb{H}^{(-1)^i} \times H^{(-1)^j}),
$$

where $H^{\pm 1}$ are the upper and the lower half planes in $\mathbb{C}$. Let $\Gamma_M^\times$ be the group of automorphisms of: generated by:

$$
\begin{aligned}
\gamma_0(w, z) &= (\alpha_+ w, \alpha_- z); \\
\gamma_1(w, z) &= (w + 1, z + \frac{-11 + \sqrt{21}}{10}); \\
\gamma_2(w, z) &= (w + \frac{-11 + \sqrt{21}}{10}, z + 1); \\
\gamma_3(w, z) &= (z, w).
\end{aligned}
$$

Then it is easily seen that $\Gamma_M^\times$ acts properly discontinuously on $W = \Omega_{\mathrm{Kul}}(\Gamma_A^\times) = \mathrm{Eq}\,(\Gamma_A^\times)$. Moreover, this set is the largest open set where Γ acts properly discontinuously.

3.4.7 Crystallographic and complex affine reflection groups

A complex crystallographic group Γ is a discrete subgroup of $\text{Aff}\,(\mathbb{C}^n)$ with compact quotient. An element $g \neq 1$ of $\text{Aff}\,(\mathbb{C}^n)$ is called a reflection if it has finite order and it leaves point-wise fixed a hyperplane H. A crystallographic group Γ is called a reflection group if it is generated by finitely many reflections in $\text{Aff}\,(\mathbb{C}^n)$.

Crystallographic and complex affine reflection groups have been studied by various authors. See for instance [225] where the authors give a complete classification of 2-dimensional crystallographic reflection groups, or [178] where the author gives a classification of the discrete groups generated by affine complex reflections. (This classification is meant to be complete, though there is in [75] a reflection group which is not in Popov's list. In this and other articles, V. Goryunov gives interesting relations of complex affine reflection groups and singularity theory.)

Chapter 4

Geometry and Dynamics of Automorphisms of $\mathbb{P}^2_{\mathbb{C}}$

In this chapter we study and describe the geometry, dynamics and algebraic classification of the elements in $\mathrm{PSL}(3, \mathbb{C})$, extending Goldman's classification for the elements in $\mathrm{PU}(2, 1) \subset \mathrm{PSL}(3, \mathbb{C})$. Just as in that case, and more generally for the isometries of manifolds of negative curvature, the automorphisms of $\mathbb{P}^2_{\mathbb{C}}$ can also be classified into the three types of elliptic, parabolic and loxodromic (or hyperbolic) elements, according to their geometry and dynamics. This classification can be also done algebraically, in terms of their trace.

It turns out that elliptic and parabolic elements in $\mathrm{PSL}(3, \mathbb{C})$ are all conjugate to elliptic and parabolic elements in $\mathrm{PU}(2, 1)$. Also, the loxodromic elements in $\mathrm{PU}(2, 1)$ are loxodromic as elements in $\mathrm{PSL}(3, \mathbb{C})$, but in this latter group we get new types of loxodromic elements that cannot exist in $\mathrm{PU}(2, 1)$. In this chapter we study and describe in detail each of these types of automorphims of $\mathbb{P}^2_{\mathbb{C}}$.

Notice that when we look at subgroups of $\mathrm{PU}(2, 1)$ we have the stringent condition of preserving the corresponding quadratic form, and therefore one has an invariant ball. Then the elliptic elements are those having a fixed point in the interior of the ball, parabolics have a fixed point in the boundary of the ball and loxodromic elements have two fixed points in the boundary. Yet, when we think of automorphisms of $\mathbb{P}^2_{\mathbb{C}}$, this type of classification makes no sense, since in general there is not an invariant ball or sphere.

In $\mathbb{P}^2_{\mathbb{C}}$ we must think globally. Each such an automorphism γ has a lifting to $\mathrm{SL}(3, \mathbb{C})$ with three eigenvalues (possibly not all distinct). Each eigenvector gives rise to a fixed point of γ. All these fixed points, and their local properties, must be taken into account for a classification of the elements in $\mathrm{PSL}(3, \mathbb{C})$.

The material in this chapter is essentially contained in [160]. The first section gives a qualitative overview of the classification problem we address in this chapter. This somehow serves as an introduction to the topic. In the following sections we make precise the notions of elliptic, parabolic and loxodromic elements in PSL(3, $\mathbb{C}$), as well as the various subclasses of maps one has in each type.

We carefully describe the geometry and dynamics in each case. We determine in each case the corresponding Kulkarni limit set, the equicontinuity region and the maximal region of discontinuity. We also give an algebraic characterisation of the various types of transformations in terms of the trace of their liftings to SL(3, $\mathbb{C}$).

4.1   A qualitative view of the classification problem

We consider an element $g \in \mathrm{PSL}(3, \mathbb{C})$ and all its iterates $g^n := g \circ g^{n-1}$, for all $n \in \mathbb{Z}$ (with $g_1 := g$, $g_0 := \mathrm{Id}$ and $g^{-n} := (g^{-1})^n$). In other words, we are considering the cyclic group generated by g. The element g is represented by a matrix $\tilde{g}$ in GL $(3, \mathbb{C})$, unique up to multiplication by nonzero complex numbers.

Such a matrix $\tilde{g}$ has three eigenvalues, say $\lambda_1, \lambda_2, \lambda_3$, which may or may not be equal, and if they are distinct, they may or may not have equal norms: These facts make big differences in their geometry and dynamics, as we will see in the sequel. These, together with the corresponding Jordan canonical form of $\tilde{g}$, yield to the geometric and dynamical characterisations of the elements in PSL(3, $\mathbb{C}$) that we give in this chapter. Yet, in the case of PSL(3, $\mathbb{C}$) we give also an algebraic classification in terms of the trace, and this looks hard to do in higher dimensions.

Notice also that what really matters are the ratios amongst the λ_i, since multiplication of a matrix by a scalar multiplies all its eigenvalues by that same scalar. Recall also that each eigenvalue determines a one-dimensional space of eigenvectors in $\mathbb{C}^3$, so its projectivisation fixes the corresponding point in $\mathbb{P}_{\mathbb{C}}^2$. Distinct eigenvalues give rise to distinct fixed points in $\mathbb{P}_{\mathbb{C}}^2$. Also, every two points in $\mathbb{P}_{\mathbb{C}}^2$ determine a unique projective line; if the two points are fixed by g, then the corresponding line is g-invariant.

Let us use this information to have a closer look at the dynamics of g by considering a lifting $\tilde{g} \in \mathrm{SL}(3, \mathbb{C})$ and looking at its Jordan canonical form. One can check that this must be of one of the following three types:

$$\begin{pmatrix} \lambda_1 & 0 & 0 \\ 0 & \lambda_2 & 0 \\ 0 & 0 & \lambda_3 \end{pmatrix}, \qquad \text{where } \lambda_3 = (\lambda_1 \lambda_2)^{-1},$$

$$\begin{pmatrix} \lambda_1 & 1 & 0 \\ 0 & \lambda_1 & 0 \\ 0 & 0 & \lambda_3 \end{pmatrix}, \qquad \text{where } \lambda_3 = (\lambda_1)^{-2},$$

$$\begin{pmatrix} 1 & 1 & 0 \\ 0 & 1 & 1 \\ 0 & 0 & 1 \end{pmatrix}.$$

Let us see what happens in each case. In the first case, the images of e_1, e_2, e_3 are fixed by the corresponding map in $\mathbb{P}_{\mathbb{C}}^2$; for simplicity we denote the corresponding images by the same letters (to avoid having too many brackets $[e_i]$). One also has at least three invariant projective lines in $\mathbb{P}^2_{\mathbb{C}}$: $\mathcal{L}_1 := \overleftrightarrow{e_1, e_2}$, $\mathcal{L}_2 : \overleftrightarrow{e_2, e_3}$ and $\mathcal{L}_3 := \overleftrightarrow{e_1, e_3}$.

Up to re-numbering the eigenvalues there are three essentially different possibilities (though a closer look at them shows that there are actually certain subcases):

(i) $|\lambda_1| < |\lambda_2| < |\lambda_3|$.

(ii) $|\lambda_1| = |\lambda_2| < |\lambda_3|$ (could be $|\lambda_1| < |\lambda_2| = |\lambda_3|$, but this is similar).

(iii) $|\lambda_1| = |\lambda_2| = |\lambda_3| = 1$

Case (i) was essentially discussed in Chapter 3. The point e_1 is repelling, e_2 is a saddle and e_3 is an attractor. Notice that the restriction of g to each of the three lines $\mathcal{L}_i$ is a loxodromic transformation in the group of automorphisms of this line, that we can identify with $\mathrm{PSL}(2, \mathbb{C})$: it has two fixed points in the line, one is repelling and the other is attracting.

Each point in $\mathcal{L}_1$ determines a unique projective line passing through that point and e_3, and the union of all these lines fills up the whole space $\mathbb{P}^2_{\mathbb{C}}$. In other words, the points in $\mathcal{L}_1$ parametrise the pencil $\{\mathcal{L}_y\}_{e_3}$ of projective lines in $\mathbb{P}^2_{\mathbb{C}}$ passing through e_3. Since the line $\mathcal{L}_1$ is g-invariant, and e_3 is a fixed point of g, it follows that each element in this pencil is carried by g into another element of the pencil. Furthermore, we can say that this is happening in a "loxodromic" way in the following sense: if we start with a point x in one of these lines, then the g-orbit of x will travel from line to line, converging towards e_3 under the iterates of g, and getting closer and closer to the line $\mathcal{L}_1$ under the iterates of g^{-1}, thus converging to a fixed point in this line. This kind of transformations will correspond to the so-called subclass of **strongly-loxodromic** elements.

Notice that in this case, since e_3 is an attracting fix point, we can choose a small enough "round ball" U containing e_3 such that $g(\overline{U}) \subset U$. This is relevant because it is this property which characterises loxodromic elements (see Definition 4.2.14).

Now consider the case

$$|\lambda_1| = |\lambda_2| < |\lambda_3| \,.$$

We can assume $|\lambda_1| = 1$. As before, the three points e_i are fixed points, the three lines are g-invariant and g carries elements in the pencil $\{\mathcal{L}_y\}_{e_3}$ into elements of this same pencil in a "loxodromic" way, as in the previous case, since the eigenvalue λ_3 has larger norm. The difference with the previous case is that the restriction of g to the invariant line $\mathcal{L}_1$ is now elliptic, not loxodromic. Hence the orbits of points in $\mathcal{L}_1$, other than the two fixed points e_1, e_2, move rotating along circles. All other points approach e_3 when travelling forwards, doing "spirals", and they approach the line $\mathcal{L}_1$ when moving backwards. These transformations are therefore called **screws**, and we will see that they are also loxodromic as elements in $\mathrm{PSL}(3, \mathbb{C})$.

When $|\lambda_1| = |\lambda_2| = |\lambda_3| = 1$ the situation is quite different. Now the restriction of g to each of the three lines $\mathcal{L}_1$, $\mathcal{L}_2$ and $\mathcal{L}_3$ is an elliptic transformation, and g carries the elements of the pencil into elements of the pencil in an "elliptic way", that we will make precise. Notice one has in this case that

$$T^{(1)}(r) = \{[z_1 : z_2 : z_3] \in P_{\mathbb{C}}^2 : |z_2|^2 + |z_3|^2 = r|z_1|^2\}, \quad r > 0,$$

is a family of 3-spheres, each of these being invariant under the action of g.

Let us envisage now the second case considered above, that is matrices of the form

$$\begin{pmatrix} \lambda_1 & 1 & 0 \\ 0 & \lambda_1 & 0 \\ 0 & 0 & \lambda_3 \end{pmatrix},$$

with $\lambda_3 = (\lambda_1)^{-2}$. Notice that the top Jordan block determines a projective line $\mathcal{L}_1$ on which the transformation is parabolic. As a Möbius transformation in $\mathcal{L}_1$ this map is

$$z \mapsto z + \frac{1}{\lambda_1}. \tag{4.1.0}$$

So the map in $\mathcal{L}_1$ is parabolic. Now observe that the points e_1 and e_3 are the only fixed points of g. As before, we have the invariant pencil $\{\mathcal{L}_y\}_{e_1}$. Notice there are two cases: $|\lambda_1| = 1$ or $|\lambda_1| \neq 1$. In the first case, g carries each element in the pencil into another element in the pencil in an "elliptic way". One can show too that in this case there is a family of 3-spheres in $\mathbb{P}_{\mathbb{C}}^2$ which are invariant by g and they all meet at the point e_3. These type of maps belong to the class of **parabolic elements** in PSL$(3, \mathbb{C})$, and they belong to the sub-class of **ellipto-parabolic** transformations.

If we now take $|\lambda_1| \neq 1$, then the dynamics in $\mathcal{L}_1$ is as before, but away from this invariant line the dynamics is dominated by the eigenvalue λ_3. If we assume $|\lambda_1| > 1$, then all points in $\mathbb{P}_{\mathbb{C}}^2 \setminus \mathcal{L}_1$ escape towards e_3 when moving forwards, and they accumulate in the line $\mathcal{L}_1$ when moving backwards. If $|\lambda_1| < 1$ the dynamics just reverses and the backwards orbits accumulate at e_3. These maps are **loxodromic elements**, of the type called **loxo-parabolic**.

Finally consider the case

$$\begin{pmatrix} 1 & 1 & 0 \\ 0 & 1 & 1 \\ 0 & 0 & 1 \end{pmatrix}.$$

Now one has that all eigenvalues are equal to 1. There is only one fixed point, e_1, and an invariant line, $\mathcal{L}_1 := \overleftrightarrow{e_1, e_2}$, in which the transformation is parabolic. Moreover, one has in this case the following family of g-invariant 3-spheres, which are all tangent to the line $\mathcal{L}_1$ at the point e_1,

$$T_r = \{[z_1 : z_2 : z_3] \mid |z_2|^2 + r|z_3|^2 - (\overline{z_1}z_3 + z_1\overline{z_3}) - \frac{1}{2}(\overline{z_2}z_3 + z_2\overline{z_3}) = 0\}, \ r \in \mathbb{R}.$$

These maps are all **parabolic**, of the type called **unipotent**.

4.2   Classification of the elements in $\mathrm{PSL}(3, \mathbb{C})$

In this section we study and classify the elements in $\mathrm{PSL}(3, \mathbb{C})$ by means of their geometry and dynamics, following the ideas sketched in the previous section. As indicated above, these will be of three types: elliptic, parabolic and loxodromic, and this classification can be naturally refined into several subclasses. Elliptic elements can be, up to conjugation, of two types: regular elliptic or complex reflections. There are also two types of parabolic elements: unipotent and ellipto-parabolic. The loxodromic elements are of four types: loxo-parabolic, homotheties, screws and strongly-loxodromic. The two types of elliptic elements, as well as the two types of parabolic elements, appear in Goldman's classification of elements in $\mathrm{PU}(2, 1)$.

Regarding loxodromic we find that those coming from $\mathrm{PU}(2, 1)$ are all strong-ly loxodromic, and in fact only some types of the strongly loxodromic elements are conjugate to elements in $\mathrm{PU}(2, 1)$. So we have that there are several new types of transformations in $\mathrm{PSL}(3, \mathbb{C})$ that do not exist in $\mathrm{PU}(2, 1)$. These are the loxo-parabolics, screws, homotheties and a "large" set of strongly-loxodromic transformations (see Theorem 4.3.3). Notice that $\mathrm{PU}(2, 1)$ has dimension 8 while $\mathrm{PSL}(3, \mathbb{C})$ has dimension 15, so it is natural to expect having new types of trans-formations in the latter group that do not appear in the former.

We first define and study the elliptic elements, then the parabolics and finally the loxodromic elements.

4.2.1   Elliptic Transformations in $\mathrm{PSL}(3, \mathbb{C})$

Recall that up to conjugation, an elliptic element $g \in \mathrm{PU}(2, 1)$ can be represented by a matrix of the form

$$\tilde{g} = \begin{pmatrix} A & 0 \\ 0 & \lambda \end{pmatrix},$$

where $A \in \mathrm{U}(2)$ and $\lambda \in \mathbb{S}^1$. Actually $\mathrm{U}(2) \cong \mathrm{SU}(2) \times \mathbb{S}^1$ acts on $\mathbb{C}^2$ preserving the usual Hermitian product. So its action on $\mathbb{C}^2$ has a fixed point at 0 and preserves the foliation given by all 3-spheres centred at 0. The action of $\tilde{g}$ on $\mathbb{P}^2_{\mathbb{C}}$ essentially corresponds to extending the action of A to $\mathbb{P}^2_{\mathbb{C}} \cong \mathbb{C}^2 \cup \mathbb{P}^1_{\mathbb{C}}$, where $\mathbb{P}^1_{\mathbb{C}}$ is regarded as the line at infinity; this is an invariant line for the action of $\mathrm{U}(2)$ on $\mathbb{P}^2_{\mathbb{C}}$. Notice that a neighbourhood of $\mathbb{P}^1_{\mathbb{C}}$ in $\mathbb{P}^2_{\mathbb{C}}$ looks like the total space of the tautological (or universal) bundle over $\mathbb{P}^1_{\mathbb{C}}$; the above 3-spheres can each be regarded as the boundary of a tubular neighbourhood of $\mathbb{P}^1_{\mathbb{C}}$, and each of these corresponds to the usual Hopf bundle $\mathbb{S}^3 \to \mathbb{S}^2 \cong \mathbb{P}^1_{\mathbb{C}}$.

More generally:

Definition 4.2.1. The 3-*spheres in* $\mathbb{P}^2_{\mathbb{C}}$ are defined as the images of the set

$$T = \{[z_1 : z_2 : z_3] \in \mathbb{P}^2_{\mathbb{C}} : |z_1|^2 + |z_2|^2 - |z_3|^2 = 0\}$$

under the action of $\mathrm{PSL}(3, \mathbb{C})$.

Notice that if in the above discussion we take the origin of $\mathbb{C}^2$ as being the point e_3 and the line at infinity $\mathbb{P}^1_{\mathbb{C}}$ as being the line $\overleftrightarrow{e_1, e_2}$, then the above family of spheres actually is a foliation of $\mathbb{P}^2_{\mathbb{C}} \setminus (\overleftrightarrow{e_1, e_2} \cup \{e_3\})$, given by:

$$T(r) = \{[z_1 : z_2 : z_3] \in P^2 : |z_1|^2 + |z_2|^2 = r|z_3|^2\}, \quad r > 0,$$

where $\overleftrightarrow{e_1, e_2}$ denotes the complex line $\{[z_1 : z_2 : z_3] \in \mathbb{P}^2_{\mathbb{C}} \mid z_3 = 0\}$ and $e_3 = [0 : 0 : 1]$.

Each automorphism $h \in \mathrm{PSL}(3, \mathbb{C})$ carries the above foliation into another family of 3-spheres given by $h(T(r))$, $r > 0$. These are the leaves of a foliation of $\mathbb{P}^2_{\mathbb{C}} \setminus (h(\overleftrightarrow{e_1, e_2}) \cup \{h(e_3)\})$.

Definition 4.2.2. A transformation $\hat{g} \in \mathrm{PSL}(3, \mathbb{C})$ is called *elliptic* if it preserves each one of the leaves of a foliation as above. In other words, $\hat{g} \in \mathrm{PSL}(3, \mathbb{C})$ is elliptic if and only if there exists $\hat{h} \in \mathrm{PSL}(3, \mathbb{C})$ such that $\hat{h}^{-1}\hat{g}\hat{h}(T(r)) = T(r)$ for every $r > 0$.

Proposition 4.2.3. *The element* $\hat{g} \in \mathrm{PSL}(3, \mathbb{C})$ *is elliptic if and only if it is conjugate to an elliptic element of* $\mathrm{PU}(2, 1)$.

Proof. Assume $\hat{g} \in \mathrm{PSL}(3, \mathbb{C})$ is elliptic, then there is $\hat{h} \in \mathrm{PSL}(3, \mathbb{C})$ such that $\hat{h}^{-1}\hat{g}\hat{h}$ preserves every 3-sphere $T(r)$, $r > 0$. It follows that $\hat{f} := \hat{h}^{-1}\hat{g}\hat{h} \in \mathrm{PU}(2, 1)$ and $[0 : 0 : 1]$ is a fixed point of $\hat{f}$. Therefore $\hat{f}$ is elliptic in $\mathrm{PU}(2, 1)$. The converse follows from the fact that every elliptic element in $\mathrm{PU}(2, 1)$ has a conjugate in $\mathrm{PU}(2, 1)$ which is induced by a matrix of the form

$$\begin{pmatrix} A & 0 \\ 0 & \lambda \end{pmatrix},$$

where $A \in \mathrm{U}(2)$ and $\lambda \in \mathrm{U}(1)$. $\square$

The next corollary follows easily from the proposition above and the fact that every element in $\mathrm{U}(2)$ is diagonalizable and its eigenvalues are unitary complex numbers.

Corollary 4.2.4. *An element* $\hat{g} \in \mathrm{PSL}(3, \mathbb{C})$ *is elliptic if and only if* $\hat{g}$ *has a lift* $g \in \mathrm{SL}(3, \mathbb{C})$ *such that* g *is diagonalizable and every eigenvalue is a unitary complex number.*

Hence every elliptic element has a canonical Jordan form of the type

$$\begin{pmatrix} \lambda_1 & 0 & 0 \\ 0 & \lambda_2 & 0 \\ 0 & 0 & \lambda_3 \end{pmatrix}$$

with the λ_i satisfying $|\lambda_1| = |\lambda_2| = |\lambda_3| = |\lambda_1\lambda_2\lambda_3| = 1$.

By definition an elliptic element in $\mathrm{PSL}(3, \mathbb{C})$ preserves a foliation by "concentric" spheres. The proposition below says that such a transformation actually preserves three foliations by "concentric" 3-spheres, and each such sphere is itself foliated by invariant tori (with two circles as singular set of the foliation).

Proposition 4.2.5. *If $\hat{g} \in \mathrm{PSL}(3, \mathbb{C})$ is an elliptic transformation, then there are three families of invariant 3-spheres. Furthermore, each one of these 3-spheres has a $\hat{g}$-invariant foliation whose nonsingular leaves are torus, and it has two singular leaves, each one being a circle.*

In fact the foliation by tori one has on each sphere is the usual one: we decompose the sphere as the union of two (unknotted) solid tori $\mathbb{S}^1 \times \mathbb{D}^2$, glued along their boundary, and each torus is foliated by concentric tori. The singular fibres are the cores of the two tori.

Proof. Let $\hat{f}$ be an elliptic transformation, then there exists $\hat{h} \in \mathrm{PSL}(3, \mathbb{C})$ such that $\hat{g} = \hat{h}\tilde{g}\hat{h}^{-1}$ has a diagonal matrix with unitary eigenvalues as a lift to $SL(3, \mathbb{C})$. Then the three families of 3-spheres are given by:

$$T^{(1)}(r) = \{[z_1 : z_2 : z_3] \in P_\mathbb{C}^2 : |z_2|^2 + |z_3|^2 = r|z_1|^2\}, \quad r > 0;$$

$$T^{(2)}(r) = \{[z_1 : z_2 : z_3] \in P_\mathbb{C}^2 : |z_1|^2 + |z_3|^2 = r|z_2|^2\}, \quad r > 0;$$

$$T^{(3)}(r) = \{[z_1 : z_2 : z_3] \in P_\mathbb{C}^2 : |z_1|^2 + |z_2|^2 = r|z_3|^2\}, \quad r > 0.$$

Notice that each of these 3-spheres is $\hat{g}$-invariant. Therefore the intersection of members of these three families is also $\hat{g}$-invariant. So, we will describe the possible intersections amongst these 3-spheres. By symmetry, it is enough to consider the intersections of $T^{(3)}(1)$ with spheres of each family $T^{(1)}(r_1)$, $r_1 > 0$; $T^{(2)}(r_2)$, $r_2 > 0$.

We identify the sphere $T^{(3)}(1)$ with $\mathbb{S}^3 = \{(z_1, z_2) \in \mathbb{C}^2 : |z_1|^2 + |z_2|^2 = 1\}$. It is not difficult to see that the intersection with a 3-sphere of the family $T^{(1)}(r_1)$ is empty whenever $0 < r_1 < 1$; the intersection is the circle $z_2 = 0$ when $r_1 = 1$, and it is the torus $|z_1| = \sqrt{\frac{2}{1+r_1}}$ when $r_1 > 1$.

The intersection of $T^{(3)}(1)$ with a sphere of the family $T^{(2)}(r_2)$ is: empty if $0 < r_2 < 1$; it is the circle $z_1 = 0$ if $r_2 = 1$; and it is the torus $|z_2| = \sqrt{\frac{2}{1+r_2}}$ if $r_2 > 1$.

Every torus of the family $|z_1| = \sqrt{\frac{2}{1+r_1}}$, $r_1 > 1$, can be seen as one of the family $|z_2| = \sqrt{\frac{2}{1+r_2}}$, $r_2 > 1$, and vice versa. Therefore they define the same family. Moreover, this family of tori together with the two circles $z_1 = 0$, $z_2 = 0$ are the leaves of a foliation of $\mathbb{S}^3$, where the two circles are singular leaves.

Finally, the possible intersections for three 3-spheres, one of each family, are either the empty set or a torus as above. □

We now look at the Kulkarni limit set $\Lambda_{\mathrm{Kul}} = L_0 \cup L_1 \cup L_2$ of the cyclic group generated by an elliptic transformation (we refer to Chapter 3 for the definition of this limit set).

Lemma 4.2.6. *Let $\beta \in \mathbb{R} - \mathbb{Q}$, and consider an action on the 3-sphere*

$$\mathbb{S}^3(r) = \{(z_1, z_2) \in \mathbb{C}^2 : |z_1|^2 + |z_2|^2 = r\}, \quad r > 0,$$

given by $\psi : (z_1, z_2) \mapsto (e^{2\pi\alpha i}z_1, e^{2\pi\beta i}z_2)$, *where* α *is a real number. Then every point in* $\mathbb{S}^3(r)$ *is an accumulation point of some orbit. In other words, the Kulkarni set* $L_1(\psi)$ *is all of* $\mathbb{S}^3(r)$.

Proof. We prove the case when $r = 1$; the general case is analogous. We have a foliation of $\mathbb{S}^3$ whose leaves are the tori $\{(z_1, z_2) : |z_1| = c\}$, $0 < c < 1$, the circles $z_1 = 0$ and $z_2 = 0$ are singular leaves; and each leaf is invariant under the action. Now we restrict our attention to one single torus.

  Case 1. $\frac{\alpha}{\beta} = \gamma \in \mathbb{Q}$. The curves $t \mapsto (e^{2\pi t\alpha i}z_1, e^{2\pi t\beta i}z_2)$, $t \in [0, 1]$, are simple and closed for every $(z_1, z_2) \in T$. Moreover, the orbit of any point in this curve is dense in the curve, because $\beta \notin \mathbb{Q}$. Therefore $L_1(\psi) = \mathbb{S}^3$.

  Case 2. $\frac{\alpha}{\beta} = \gamma \notin \mathbb{Q}$. The curves $t \mapsto (e^{2\pi t\alpha i}z_1, e^{2\pi t\beta i}z_2)$, $t \in [0, 1]$, are simple, but not closed and the image of each such curve is dense in the corresponding torus T. As in Case 1, the orbit of any point in this curve is dense in it, therefore $L_1(\psi) = \mathbb{S}^3$. $\square$

Proposition 4.2.7. *Let* $\hat{g} \in \mathrm{PSL}(3, \mathbb{C})$ *be an elliptic transformation, and let* $\mathrm{Eq}\,(\langle\hat{g}\rangle)$ *be its equicontinuity region. Then the Kulkarni sets* L_0, L_1, L_2 *are as follows:*

(i) *If* $\hat{g}$ *has finite order, then*

$$\mathbb{P}_{\mathbb{C}}^2 \setminus \mathrm{Eq}\,(\langle\hat{g}\rangle) = L_0(\hat{g}) = L_1(\hat{g}) = L_2(\hat{g}) = \Lambda(\hat{g}) = \varnothing\,.$$

(ii) *If* $\hat{g}$ *has infinite order, then* $L_0(\hat{g}) = \{x : x \text{ is fixed point of } \hat{g}\}$ *and we have*

$$\mathrm{Eq}\,(\langle\hat{g}\rangle) = L_1(\hat{g}) = \mathbb{P}_{\mathbb{C}}^2, \qquad L_2(\hat{g}) = \varnothing.$$

Proof. Statement (i) is clear, let us prove (ii). It is easy to check that $L_0(\hat{g}) = \{x : x$ is fixed point of $\hat{g}\}$, since this set consists of the points with infinite isotropy. We can assume that $\hat{g}$ has a lift $\tilde{g}$ to $\mathrm{GL}\,(3, \mathbb{C})$ of the form

$$\begin{pmatrix} e^{2\pi i\alpha} & 0 & 0 \\ 0 & e^{2\pi i\beta} & 0 \\ 0 & 0 & 1 \end{pmatrix},$$

where $\beta \in \mathbb{R} - \mathbb{Q}$. Then $\hat{g}$ acts on the family of 3-spheres,

$$T(r) = \{[z_1 : z_2 : z_3] \in \mathbb{P}_{\mathbb{C}}^2 : |z_1|^2 + |z_2|^2 = r|z_3|^2\},$$

leaving invariant each member. The family $T(r)$, $r > 0$, can be considered as the family of 3-spheres in $\mathbb{C}^2$ with centre at the origin and positive radius. Then Lemma 4.2.6 implies that $L_1(\hat{g}) = \mathbb{P}_{\mathbb{C}}^2$. By definition we have that $L_2(\hat{g}) = \varnothing$. To conclude the proof, let $(\tilde{g}^{l_m}) \subset \langle\tilde{g}\rangle$ be a sequence of distinct elements. Since the set of unitary complex numbers is compact, we can assume that the sequences $(e^{2\pi i\alpha l_m})$, $(e^{2\pi i\beta l_m})$ are convergent, with limit points $\tilde{\alpha}$ and $\tilde{\beta}$ respectively. Thus

$$\tilde{g}^{l_m} = \begin{pmatrix} e^{2\pi i\alpha l_m} & 0 & 0 \\ 0 & e^{2\pi i\beta l_m} & 0 \\ 0 & 0 & 1 \end{pmatrix} \xrightarrow[m\to\infty]{} \left(\tilde{\gamma} = \begin{pmatrix} \tilde{\alpha} & 0 & 0 \\ 0 & \tilde{\beta} & 0 \\ 0 & 0 & 1 \end{pmatrix} \right).$$

That is $g^{lm} \xrightarrow[m \to \infty]{} [\gamma] \in \mathrm{PSL}(3, \mathbb{C})$ uniformly on $\mathbb{P}^2_{\mathbb{C}}$, which shows that $\mathrm{Eq}(\langle \hat{g} \rangle) = \mathbb{P}^2_{\mathbb{C}}$. $\square$

4.2.2 Parabolic Transformations in PSL(3, $\mathbb{C}$)

Recall that by definition a parabolic element $\hat{g} \in \mathrm{PU}(2, 1)$ has a fixed point p on $\mathbb{S}^3 = \partial \mathbb{H}^2_{\mathbb{C}} \subset \mathbb{P}^2_{\mathbb{C}}$. It is well-known (and it follows also from our proof below of Proposition 4.2.12) that the parabolic transformation Γ leaves invariant each horosphere at p in $\mathbb{H}^2_{\mathbb{C}} \subset \mathbb{P}^2_{\mathbb{C}}$. In real hyperbolic geometry, taking p to be infinity in the upper-half space model, we get that a parabolic element is essentially a translation. In the complex hyperbolic case this is not that simple, and yet, the transformation still preserves the horospheres.

Definition 4.2.8. The transformation $\hat{g} \in \mathrm{PSL}(3, \mathbb{C})$ is called *parabolic* if there exists a family $\mathcal{F}$ of $\hat{g}$-invariant 3-spheres and $Z_f \in \mathbb{P}^2_{\mathbb{C}}$ such that:

(i) For every pair of different elements $T_1, T_2 \in \mathcal{F}$ it follows that $T_1 \cap T_2 = \{Z_f\}$.

(ii) The set $\bigcup \mathcal{F}$ is a closed round ball.

Here by a round ball we mean the image by an element in $\mathrm{PSL}(3, \mathbb{C})$ of the ball consisting of points whose homogeneous coordinates satisfy $|z_1|^2 + |z_2|^2 < |z_3|^2$.

Proposition 4.2.9. *If the element $\hat{g} \in \mathrm{PSL}(3, \mathbb{C})$ is parabolic, then $\hat{g}$ is conjugate to a parabolic element of* $\mathrm{PU}(2, 1)$.

Proof. After conjugating with a projective transformation if necessary, we have that $\hat{g} \in \mathrm{PU}(2, 1)$. Since $T_1 \cap T_2 = \{Z_f\}$, for every pair of distinct elements $T_1, T_2 \in \mathcal{F}$, we conclude that Z_f is a fixed point of $\hat{g}$ and $Z_f \in \partial \mathbb{H}^n_{\mathbb{C}}$. Thus $\hat{g}$ is either loxodromic or parabolic. Let us assume that $\hat{g}$ is loxodromic, thus there is a pont $q \neq Z_f$ such that q is fixed by $\hat{g}$. Then $\ell = \overleftrightarrow{Z_f, q}$ is invariant under $\hat{g}$ and the restriction of $\hat{g}$ to ℓ is a loxodromic transformation. On the other hand, trivially we get that $\tilde{\mathcal{F}} = \{T \cap \ell : T \in \mathcal{F}\}$ is a family of $\hat{g}$-invariant circles tangent at $Z_f \in \mathbb{P}^2_{\mathbb{C}}$. Therefore $q \in \tilde{\mathcal{F}}$, which is a contradiction. $\square$

We now proceed to developing some tools we need for proving Proposition 4.2.12, which is the converse of Proposition 4.2.9.

As before, given $X, Y \in \mathbb{P}^2_{\mathbb{C}}$, we denote by $\overleftrightarrow{X, Y}$ the projective line in $\mathbb{P}^2_{\mathbb{C}}$ determined by X and Y.

Recall (see p. 16, equation 1.13, in [67]) that given $X, Y \in \mathbb{P}^2_{\mathbb{C}}$, the Fubini-Study metric $d(X, Y)$ between X and Y satisfies the equation

$$\cos^2(d(X, Y)) = \frac{\langle\!\langle x, y \rangle\!\rangle \langle\!\langle y, x \rangle\!\rangle}{\langle\!\langle x, x \rangle\!\rangle \langle\!\langle y, y \rangle\!\rangle},$$

where $x, y \in \mathbb{C}^3 - \{(0, 0, 0)\}$ are lifts of X and Y, respectively, and $\langle\!\langle \cdot, \cdot \rangle\!\rangle$ is the usual Hermitian product in $\mathbb{C}^3$.

Proposition 4.2.10. *Let $\hat{g} \in \mathrm{PSL}(3, \mathbb{C})$.*

(i) *Assume $\hat{g}$ has a lift*

$$\begin{pmatrix} 1 & 1 & 0 \\ 0 & 1 & 0 \\ 0 & 0 & 1 \end{pmatrix},$$

then $L_0(\hat{g}) = \overleftrightarrow{e_1, e_3}$, $L_1(\hat{g}) = L_2(\hat{g}) = \{e_1\}$ and $\mathrm{Eq}\,(\langle \hat{g} \rangle) = \Omega_{\mathrm{Kul}}(\langle \hat{g} \rangle)$.

(ii) *Assume $\hat{g}$ has a lift*

$$\begin{pmatrix} 1 & 1 & 0 \\ 0 & 1 & 1 \\ 0 & 0 & 1 \end{pmatrix},$$

then $\hat{g}$ has one single fixed point, which is e_1; $L_0(\hat{g}) = L_1(\hat{g}) = \{e_1\}$, and $L_2(\hat{g}) = \overleftrightarrow{e_1, e_2}$, with $\hat{g}$ acting as a classical parabolic transformation on this line. Moreover $\mathrm{Eq}\,(\langle \hat{g} \rangle) = \Omega_{\mathrm{Kul}}(\langle \hat{g} \rangle)$.

(iii) *If $\hat{g}$ has a lift*

$$\begin{pmatrix} 1 & 1 & 0 \\ 0 & 1 & 0 \\ 0 & 0 & \lambda \end{pmatrix},$$

where $\lambda = e^{2\pi i x} \neq 1$, then $\hat{g}$ has two fixed points e_1, e_3, and

$$L_0(\hat{g}) = \begin{cases} \overleftrightarrow{e_1, e_3} & \text{if x is rational,} \\ \{e_1, e_3\} & \text{if x is irrational,} \end{cases}$$

$$L_1(\hat{g}) = \begin{cases} \{e_1\} & \text{if x is rational,} \\ \overleftrightarrow{e_1, e_3} & \text{if x is irrational,} \end{cases}$$

$$L_2(\hat{g}) \hspace{5cm} = \{e_1\}.$$

Moreover $\mathrm{Eq}\,(\langle \hat{g} >) = \Omega_{\mathrm{Kul}}(\langle \hat{g} \rangle)$.

Proof. For (i), assume $\hat{g}$ has a lift to $\mathrm{SL}(3, \mathbb{C})$ of the form

$$\begin{pmatrix} 1 & 1 & 0 \\ 0 & 1 & 0 \\ 0 & 0 & 1 \end{pmatrix}.$$

Then it is clear one has $L_0(\hat{g}) = L_f = \{[z_1 : z_2 : z_3] \in P^2 \mid z_2 = 0\}$.
Let $Z = [z_1 : z_2 : z_3]$ be a point in $\mathbb{P}^2_{\mathbb{C}} - L_0(\hat{g})$. The equations

$$\cos^2(d(\hat{g}^n(Z), e_1)) = \frac{|z_1 + n z_2|^2}{|z_1 + n z_2|^2 + |z_2|^2 + |z_3|^2},$$

$$\cos^2(d(\hat{g}^{-n}(Z), e_1)) = \frac{|z_1 - n z_2|^2}{|z_1 - n z_2|^2 + |z_2|^2 + |z_3|^2},$$

imply that $L_1(\hat{g}) = \{e_1\}$.

For every $\epsilon > 0$, we define $K_\epsilon^{(i)}$ as the subset of $\mathbb{P}_\mathbb{C}^2$ given by $\pi(\tilde{K}_\epsilon^i)$, where $\pi : \mathbb{S}^5 \to \mathbb{P}_\mathbb{C}^2$ is the canonical projection and $\tilde{K}_\epsilon^{(i)} = \{(z_1, z_2, z_3) \in \mathbb{S}^5 : |z_i| \geq \epsilon\}$. We will prove that the sequence of functions $\hat{h}_n = \hat{g}^n|_{K_\epsilon^{(2)}}$ converges uniformly to the constant function with value e_1. We take a point $k = [z_1 : z_2 : z_3] \in K_\epsilon^{(2)}$, then

$$\cos^2(d(\hat{g}^n(k), e_1)) = \frac{|z_1 + nz_2|^2}{|z_1 + nz_2|^2 + |z_2|^2 + |z_3|^2} \leq 1.$$

Moreover, there exists $N \in \mathbb{N}$ such that $n\epsilon - 1 > 0$ for all $n \geq N$. It follows that

$$\frac{(n\epsilon - 1)^2}{(n\epsilon - 1)^2 + 1 + 1} \leq \frac{|z_1 + nz_2|^2}{|z_1 + nz_2|^2 + |z_2|^2 + |z_3|^2},$$

for all $n \geq N$. This proves our claim about uniform convergence.

Let K be a compact subset of $\mathbb{P}_\mathbb{C}^2 - \mathcal{L}_f$, then there exists $\epsilon > 0$ such that $K \subset K_\epsilon^{(2)}$. So, for every neighbourhood U of e_1, there is a number $N \in \mathbb{N}$ such that $\hat{g}^n(K) \subset U$ for all $n \geq N$, therefore e_1 is the unique cluster point of the family of compact sets $\{\hat{g}^n(K)\}_{n \in \mathbb{N}}$. An analogous proof shows that e_1 is the unique cluster point of the family $\{\hat{g}^{-n}(K)\}_{n \in \mathbb{N}}$. Therefore $L_2(\hat{g}) = \{e_1\}$. To conclude the proof observe that every sequence of distinct elements of $\langle \hat{g} \rangle$, has a subsequence which converges uniformly over compact sets of $\mathbb{P}_\mathbb{C}^2 - \overleftrightarrow{e_1, e_3}$ to the constant e_1, and every point in $\overleftrightarrow{e_1, e_3}$ is fixed by $\hat{g}$.

For (ii), we assume $\hat{g}$ has a lift to SL(3, ℂ) of the form

$$\begin{pmatrix} 1 & 1 & 0 \\ 0 & 1 & 1 \\ 0 & 0 & 1 \end{pmatrix}.$$

In this case we take $e_1 = [1 : 0 : 0]$, so it is clear that we have $L_0 = \{e_1\}$. We take $Z = [z_1 : z_2 : z_3] \in \mathbb{P}_\mathbb{C}^2$, then

$$\cos^2(d(\hat{g}^n(Z), e_1)) = \frac{|z_1 + nz_2 + \frac{n(n-1)}{2}z_3|^2}{|z_1 + nz_2 + \frac{n(n-1)}{2}z_3|^2 + |z_2 + nz_3|^2 + |z_3|^2},$$

$$\cos^2(d(\hat{g}^{-n}(Z), e_1)) = \frac{|z_1 - nz_2 + \frac{n(n+1)}{2}z_3|^2}{|z_1 - nz_2 + \frac{n(n+1)}{2}z_3|^2 + |z_2 - nz_3|^2 + |z_3|^2},$$

so $\hat{g}^n(Z)$ and $\hat{g}^{-n}(Z)$ converge to e_1 as $n \to \infty$. Therefore $L_1(\hat{g}) = \{e_1\}$.

For (iii) we first need:

Lemma 4.2.11. *One has $\hat{g}^{-n} \xrightarrow[n \to \infty]{} e_1$ uniformly on compact subsets of $\mathbb{P}_\mathbb{C}^2 \setminus \overleftrightarrow{e_1, e_2}$.*

Proof. Let K be a subset of $\mathbb{P}_\mathbb{C}^2 \setminus \overleftrightarrow{e_1, e_2}$, then there exists $\epsilon > 0$ such that $K \subset K_\epsilon^{(3)}$, where $K_\epsilon^{(3)}$ is the image, under the canonical projection $\pi : \mathbb{S}^5 \to \mathbb{P}_\mathbb{C}^2$, of the set

$\{(z_1, z_2, z_3) \in \mathbb{S}^5 : |z_3| \geq \epsilon\}$. There exists $N \in \mathbb{N}$ such that $\frac{n(n-1)}{2}\epsilon - (n+1) > 0$, whenever $n \geq N$. It follows that

$$\frac{(\frac{n(n-1)}{2}\epsilon - (n+1))^2}{(\frac{n(n-1)}{2}\epsilon - (n+1))^2 + (1+n)^2 + 1} \leq \cos^2(d(\hat{g}^n(Z), e_1)) \leq 1, \quad \text{whenever } n \geq N.$$

Therefore $\hat{g}^n(\cdot) \to e_1$ uniformly on K. An analogous argument is used when the exponents are negative. $\qquad\square$

Proposition 3.3.6 and Lemma 4.2.11 imply that $L_2(\hat{g}) \subset \overleftrightarrow{e_1, e_2}$. Conversely, the sequence $\{[\frac{z_1}{2} : -\frac{n-1}{2n} : \frac{1}{n}] : n \in \mathbb{N}\}$ converges to the point $[z_1 : -1 : 0] \neq e_1$, and the sequence $g^n([\frac{z_1}{2} : -\frac{n-1}{2n} : \frac{1}{n}]) = [\frac{z_1}{2} : \frac{1}{2} + \frac{1}{2n} : \frac{1}{n}]$, converges to the point $[z_1 : 1 : 0]$. Then $\overleftrightarrow{e_1, e_2} \subset L_2(\hat{g})$. To get $\Omega_{\text{Kul}}(\langle \hat{g} \rangle) = \text{Eq}(\langle \hat{g} \rangle)$, it is enough to apply Lemma 4.2.11 and the fact that $\hat{g}$ leaves invariant the line $\overleftrightarrow{e_1, e_2}$ where it acts as a parabolic transformation.

Now we prove (iii). We can assume that $\hat{g}$ has a lift to $\text{GL}(3, \mathbb{C})$ of the form

$$\begin{pmatrix} 1 & 1 & 0 \\ 0 & 1 & 0 \\ 0 & 0 & \lambda \end{pmatrix},$$

where $1 \neq \lambda = e^{2\pi i x}$. In this case we take, $e_1 = [1 : 0 : 0]$ and $e_3 = [0 : 0 : 1]$. Clearly

$$L_0(\hat{g}) = \begin{cases} \overleftrightarrow{e_1, e_3} & \text{if } x \text{ is rational,} \\ \{e_1, e_3\} & \text{if } x \text{ is not rational.} \end{cases}$$

If $Z = [z_1 : z_2 : z_3]$, then $\hat{g}^n(Z) = [z_1 + nz_2 : z_2 : \lambda^n z_3]$ and $\hat{g}^{-n}(Z) = [z_1 - nz_2 : z_2 : \lambda^{-n} z_3]$. It follows that $\hat{g}^n(Z) \to e_1$ and $\hat{g}^{-n}(Z) \to e_1$ whenever $Z \in \mathbb{P}_{\mathbb{C}}^2 - \overleftrightarrow{e_1, e_3}$. If x is rational, then $L_1(\hat{g}) = \{e_1\}$. If x is not rational, then $\hat{g}$ acts as an elliptic transformation on $\overleftrightarrow{e_1, e_3}$ which has infinite order, so $L_1(\hat{g}) = \overleftrightarrow{e_1, e_3}$.

A similar argument to the one used in i) shows that $\hat{g}^{\pm} \xrightarrow[n \to \infty]{} e_1$ uniformly on compact subsets of $\mathbb{P}_{\mathbb{C}}^2 \setminus \overleftrightarrow{e_1, e_2}$. Therefore one has $L_2(\hat{g}) = \{e_1\}$ and $\Omega_{\text{Kul}}(\langle \hat{g} \rangle) \subset \text{Eq}(\langle \hat{g} \rangle)$. Hence the claim $\Omega_{\text{Kul}}(\langle \hat{g} \rangle) = \text{Eq}(\langle \hat{g} \rangle)$ follows from Corollary 3.3.5. $\qquad\square$

Proposition 4.2.12. *If $\hat{g} \in \text{PSL}(3, \mathbb{C})$ is a transformation conjugate to a parabolic element of $\text{PU}(2, 1)$, then $\hat{g}$ is parabolic according to Definition 4.2.8.*

Proof. It suffices to specify the families T_r, $r \in \mathbb{R}$, when $\hat{g}$ is induced by one of the following matrices:

$$\begin{pmatrix} 1 & 1 & 0 \\ 0 & 1 & 0 \\ 0 & 0 & 1 \end{pmatrix}, \quad \begin{pmatrix} 1 & 1 & 0 \\ 0 & 1 & 1 \\ 0 & 0 & 1 \end{pmatrix}, \quad \begin{pmatrix} \zeta & 1 & 0 \\ 0 & \zeta & 0 \\ 0 & 0 & \zeta^{-2} \end{pmatrix}, \quad |\zeta| = 1.$$

In the first case,

$$T_r = \{[z_1 : z_2 : z_3] \mid r|z_2|^2 + |z_3|^2 + i(\overline{z_1}z_2 - z_1\overline{z_2}) = 0\}, \quad r \in \mathbb{R};$$

in the second case,

$$T_r = \{[z_1 : z_2 : z_3] \mid |z_2|^2 + r|z_3|^2 - (\overline{z_1}z_3 + z_1\overline{z_3}) - \frac{1}{2}(\overline{z_2}z_3 + z_2\overline{z_3}) = 0\}, \quad r \in \mathbb{R};$$

finally, in the third case,

$$T_r = \{[z_1 : z_2 : z_3] \mid r|z_2|^2 + |z_3|^2 + i(\overline{\zeta z_1}z_2 - \zeta z_1\overline{z_2}) = 0\}, \quad r \in \mathbb{R}.$$

Proposition 4.2.10 and an easy computation prove that each family T_r satisfies the conditions in Definition 4.2.8. $\qquad\square$

Definition 4.2.13. The parabolic transformation $\hat{g} \in \mathrm{PSL}(3,\mathbb{C})$ is called *rational* (respectively *irrational*) if it has a lift such that the quotient of two different eigenvalues is equal to $e^{2\pi i x}$ with x rational (respectively irrational). In either case, the transformation is

(i) *unipotent*, if it has a lift to $\mathrm{SL}(3,\mathbb{C})$ such that every eigenvalue is equal to 1;

(ii) *ellipto-parabolic* if it is not unipotent.

We remark that the element $\hat{g} \in \mathrm{PSL}(3,\mathbb{C})$ is parabolic if and only if $\hat{g}$ has a lift $g \in \mathrm{SL}(3,\mathbb{C})$, nondiagonalizable, such that every eigenvalue has module 1. It follows that a unipotent parabolic element has a Jordan normal form of the type

$$\begin{pmatrix} 1 & 1 & 0 \\ 0 & 1 & 0 \\ 0 & 0 & 1 \end{pmatrix}, \quad \begin{pmatrix} 1 & 1 & 0 \\ 0 & 1 & 1 \\ 0 & 0 & 1 \end{pmatrix}.$$

In the case of an ellipto-parabolic element, the normal form is:

$$\begin{pmatrix} \zeta & 1 & 0 \\ 0 & \zeta & 0 \\ 0 & 0 & \zeta^{-2} \end{pmatrix}, \quad |\zeta| = 1, \ \zeta \neq 1.$$

4.2.3 Loxodromic Transformations in $\mathrm{PSL}(3,\mathbb{C})$

Recall that a loxodromic element $\hat{g}$ in $\mathrm{PU}(2,1)$ by definition has two fixed points in $p, q \in \partial\mathbb{H}^2_\mathbb{C}$. One of these points is repelling, the other attracting. These two points determine a complex geodesic in $\mathbb{H}^2_\mathbb{C}$, which is the intersection of $\mathbb{H}^2_\mathbb{C}$ with the projective line in $\mathbb{P}^2_\mathbb{C}$ passing trough p, q. The restriction of $\hat{g}$ to this line is loxodromic as an element in $\mathrm{PSL}(2,\mathbb{C})$. Notice that the 3-sphere $\partial\mathbb{H}^2_\mathbb{C}$ splits $\mathbb{P}^2_\mathbb{C}$ in two connected components: one is the ball $\mathbb{H}^2_\mathbb{C}$, the other is $\mathbb{P}^2_\mathbb{C} \setminus \mathbb{H}^2_\mathbb{C}$, diffeomorphic to the total space of the Hopf bundle over $\mathbb{P}^1_\mathbb{C}$. Given a loxodromic element $\hat{g}$ in $\mathrm{PU}(2,1)$ one can always choose a small enough 3-sphere with centre at the attracting point such that $\hat{g}(\hat{W} \cup X) \subset X$, where X is the connected components of $\mathbb{P}^2_\mathbb{C} - \hat{W}$ containing the attracting point. We will see below that this property characterises the loxodromic elements.

Definition 4.2.14. The element $\hat{g} \in \mathrm{PSL}(3, \mathbb{C})$ is called *loxodromic* if there is a 3-sphere $\hat{W}$ in $\mathbb{P}^2_{\mathbb{C}}$, such that $\hat{g}(\hat{W} \cup \mathbb{X}_i) \subset \mathbb{X}_i$, for some $i = 1, 2$, where $\mathbb{X}_1$ and $\mathbb{X}_2$ are the connected components of $\mathbb{P}^2_{\mathbb{C}} - \hat{W}$.

We will show that, unlike the parabolic and the elliptic transformations which are all conjugate to elements in $\mathrm{PU}(2, 1)$, there are loxodromic elements in $\mathrm{PSL}(3, C)$ which are not conjugate to elements in $\mathrm{PU}(2, 1)$ (see Proposition 4.2.23 below). In fact we give necesary and sufficient conditions to decide whether or not a loxodromic element is conjugate to an element in $\mathrm{PU}(2, 1)$, see Theorem 4.3.3.

Proposition 4.2.15. *If $\hat{g} \in \mathrm{PSL}(3, \mathbb{C})$ is loxodromic, then $\hat{g}$ has infinite order and the set $L_0(\hat{g}) \cup L_1(\hat{g})$ is not connected.*

Proof. That $\hat{g}$ has infinite order is obvious since there is an invariant ball in which it is a contraction. In fact, let $\hat{W}$ be a 3-sphere as in Definition 4.2.14 and $\mathbb{X}_i$, $i = 1, 2$, the corresponding connected component of $\mathbb{P}^2_{\mathbb{C}} - \hat{W}$. Assume that $\hat{g}(\hat{W} \cup \mathbb{X}_1) \subset \mathbb{X}_1$, then $\hat{g}^n(\hat{W}) \subset \mathbb{X}_1$ for every $n \in \mathbb{N}$. This implies that $\hat{g}^n(w) \neq w$ for every $w \in \hat{W}$ and every $n \in \mathbb{N}$. It follows that $\hat{g}$ has infinite order.

We see that $\hat{W} \cap (L_0(\hat{g}) \cup L_1(\hat{g})) = \varnothing$, because there exists a neighbourhood of $\hat{W}$ on which $\langle \hat{g} \rangle$ acts discontinuously. The hypothesis $\hat{g}(\hat{W} \cup \mathbb{X}_1) \subset \mathbb{X}_1$, the equality $\hat{W} \cap (L_0(\hat{g}) \cup L_1(\hat{g})) = \varnothing$ and the fact that $\hat{W} \cup \mathbb{X}_1$ is compact imply that $\mathbb{X}_1 \cap (L_0(\hat{g}) \cup L_1(\hat{g})) \neq \varnothing$. An analogous reasoning, using the inclusion $\hat{g}^{-1}(\hat{W} \cup \mathbb{X}_2) \subset \mathbb{X}_2$, proves that $\mathbb{X}_2 \cap (L_0(\hat{g}) \cup L_1(\hat{g})) \neq \varnothing$. Hence $L_0(\hat{g}) \cup L_1(\hat{g})$ is not connected. $\qquad \square$

Example 4.2.16. The transformation $\hat{g} \in \mathrm{PSL}(3, \mathbb{C})$ induced by a diagonal matrix with entries $\lambda_1, \lambda_2, \lambda_3$, where $|\lambda_i| < |\lambda_3|$ for $i = 1, 2$, is a loxodromic transformation. In particular, the loxodromic transformations in $\mathrm{PU}(2, 1)$ are loxodromic transformations when considered as elements of $\mathrm{PSL}(3, \mathbb{C})$.

Proof. We take $\hat{W} = \{[z_1 : z_2 : z_3] \in \mathbb{P}^2_{\mathbb{C}} : |z_1|^2 + |z_2|^2 - |z_3|^2 = 0\} = \partial \mathbb{H}^2_{\mathbb{C}}$, then the ball $\mathbb{X}_1 = \{[v] \in P^2_{\mathbb{C}} : |z_1|^2 + |z_2|^2 - |z_3|^2 < 0\}$ is one of the components of $\mathbb{P}^2_{\mathbb{C}} \setminus \hat{W}$. If $Z = [z_1 : z_2 : z_3] \in \hat{W} \cap \mathbb{X}_1$, and $|\lambda| = \max\{|\lambda_1|, |\lambda_2|\}$, then:

$$|\lambda_1 z_1|^2 + |\lambda_2 z_2|^2 \leq |\lambda|^2(|z_1|^2 + |z_2|^2) \leq |\lambda|^2 |z_3|^2 < |\lambda_3 z_3|^2.$$

Therefore, $\hat{g}(Z) = [\lambda_1 z_1 : \lambda_2 z_2 : \lambda_3 z_3] \in \mathbb{X}_1$. $\qquad \square$

Example 4.2.17. The transformation $\hat{g} \in \mathrm{PSL}(3, \mathbb{C})$, induced by a matrix of the form

$$g = \begin{pmatrix} \lambda_1 & 1 & 0 \\ 0 & \lambda_1 & 0 \\ 0 & 0 & \lambda_2 \end{pmatrix},$$

where $|\lambda_1| \neq |\lambda_2|$, is a loxodromic transformation.

Proof. We can assume, conjugating and inverting if needed, that $\hat{g}$ is induced by
a matrix of the form

$$\begin{pmatrix} 1 & c & 0 \\ 0 & 1 & 0 \\ 0 & 0 & \lambda \end{pmatrix}, \quad |\lambda| > 1 \text{ and } 0 < c < 1 - |\lambda|.$$

We work with homogeneous coordinates. If $Z = [z_1 : z_2 : 1] \in \mathbb{H}_{\mathbb{C}}^2 \cup \partial\mathbb{H}_{\mathbb{C}}^2$, then
$|z_1|^2 + |z_2|^2 \le 1$, and $\hat{g}(Z) = [z_1 + cz_2 : z_2 : \lambda]$. Thus, $\hat{g}(Z) \in \mathbb{H}_{\mathbb{C}}^2$ if and only if
$|z_1 + cz_2|^2 + |z_2|^2 - |\lambda|^2 < 0$. Finally,

$$\begin{aligned}
|z_1 + cz_2|^2 + |z_2|^2 - |\lambda|^2 &= c^2|z_2|^2 + 2c\Re(z_1\overline{z_2}) + |z_1|^2 + |z_2|^2 - |\lambda|^2 \\
&\le c^2 + 2c + 1 - |\lambda|^2 \\
&< (|\lambda| - 1)^2 + 2(|\lambda| - 1) + 1 - |\lambda|^2 \\
&= 0.
\end{aligned}$$

Therefore g is loxodromic.   □

Definition 4.2.18. Every transformation in PSL(3, $\mathbb{C}$) conjugate to a transformation as in Example 4.2.17 is called a *loxo-parabolic* transformation.

Proposition 4.2.19. *If $\hat{g}$ is a loxo-parabolic transformation, with fixed points e_1, e_3,
then $L_0(\hat{g}) = \{e_1, e_3\} = L_1(\hat{g})$ and $L_2(\hat{g}) = \overleftrightarrow{e_1, e_2} \cup \overleftrightarrow{e_1, e_3}$, where $\overleftrightarrow{e_1, e_2}$ is a $\hat{g}$-
invariant line, and $\hat{g}$ acts as a classical parabolic transformation on $\overleftrightarrow{e_1, e_2}$. More-
over, one has* Eq$((\langle \hat{g} \rangle)) = \Omega_{\mathrm{Kul}}(\langle \hat{g} \rangle)$

Proof. We can assume that $\hat{g}$ has a lift to GL(3, $\mathbb{C}$) of the form

$$\begin{pmatrix} 1 & 1 & 0 \\ 0 & 1 & 0 \\ 0 & 0 & \lambda \end{pmatrix},$$

where $|\lambda| > 1$. In this case, $e_1 = [1 : 0 : 0]$, $e_2 = [0 : 1 : 0]$, $e_3 = [0 : 0 : 1]$; and
clearly $L_0(\hat{g}) = \{e_1, e_3\}$. We see that $\hat{g}^n(Z) \to e_3$ as $n \to \infty$, whenever $Z \notin \overleftrightarrow{e_1, e_2}$;
and $\hat{g}^n(Z) \to e_1$ as $n \to -\infty$, whenever $Z \in \mathbb{P}_{\mathbb{C}}^2 \setminus \{e_3\}$. When we restrict the action
to $\overleftrightarrow{e_1, e_2}$, the orbit of every point accumulates at e_1. Therefore $L_1(\hat{g}) = \{e_1, e_3\}$.

   One has that $\hat{g}^n(\cdot) \to e_3$ and $\hat{g}^{-n}(\cdot) \to e_1$ as $n \to \infty$, uniformly on compact
subsets of $\mathbb{P}_{\mathbb{C}}^2 \setminus (\overleftrightarrow{e_1, e_2} \cup \overleftrightarrow{e_1, e_3})$. Hence $\mathbb{P}_{\mathbb{C}}^2 \setminus (\overleftrightarrow{e_1, e_2} \cup \overleftrightarrow{e_1, e_3}) \subset$ Eq$((\langle \hat{g} \rangle))$.
   We now have to show that $\mathbb{P}_{\mathbb{C}}^2 \setminus (\overleftrightarrow{e_1, e_2} \cup \overleftrightarrow{e_1, e_3}) \supset$ Eq$((\langle \hat{g} \rangle))$, but this is an
immediate consequence of Corollary 3.3.5.
   Now observe that Proposition 3.3.6 implies that $L_2(\hat{g}) \subset \overleftrightarrow{e_1, e_2} \cup \overleftrightarrow{e_1, e_3}$, as
stated.
   Next we see that the sequence $\{[n : \lambda^n : nz_3]\}_{n \in \mathbb{N}}$ goes to $e_2 \notin L_0(\hat{g}) \cup L_1(\hat{g})$
as $n \to \infty$, and $\hat{g}^n([n : \lambda^n : nz_3]) = [1 + \frac{1}{\lambda^n} : \frac{1}{n} : z_3]$ goes to $[1 : 0 : z_3]$ as $n \to \infty$.
Therefore, $\overleftrightarrow{e_1, e_3} \subset L_2(\hat{g})$. Finally, $[1 + \frac{1}{\lambda^n} + \frac{1}{n} : \frac{z_2}{n} : 1] \to [1 : 0 : 1] \notin L_0(\hat{g}) \cup L_1(\hat{g})$
as $n \to \infty$ and $\hat{g}^-([1 + \frac{1}{\lambda^n} + \frac{1}{n} : \frac{z_2}{n} : 1]) = [\lambda^n + n : \lambda^n z_2 : n] \to [1 : z_2 : 0]$ as
$n \to \infty$. Therefore, $\overleftrightarrow{e_1, e_2} \subset L_2(\hat{g})$.   □

Proposition 4.2.20. *The element $\hat{g} \in \mathrm{PSL}(3, \mathbb{C})$ is loxodromic if and only if it has a lift $g \in \mathrm{SL}(3, \mathbb{C})$ with at least two eigenvalues of different module.*

Proof. Assume $\hat{g}$ has a lift $g \in \mathrm{SL}(3, \mathbb{C})$ with at least two eigenvalues of different module, then there are two cases according to whether g is diagonalizable or not.

If g is diagonalizable, then Example 4.2.16 shows $\hat{g}$ is loxodromic. If g is not diagonalizable, then Example 4.2.17 shows $\hat{g}$ is loxodromic.

Conversely, we assume $\hat{g}$ has a lift $g \in \mathrm{SL}(3, \mathbb{C})$ whose eigenvalues have the same module. We have two possibilities: If g is diagonalizable then $\hat{g}$ is elliptic and Proposition 4.2.7 implies $L_0(\hat{g}) \cup L_1(\hat{g}) = \varnothing$ or $\mathbb{P}^2_{\mathbb{C}}$. Then Proposition 4.2.15 implies $\hat{g}$ is not loxodromic. If g is not diagonalizable, then $\hat{g}$ is parabolic and Propositions 4.2.10 and 4.2.15 imply $\hat{g}$ is not loxodromic. $\square$

Definition 4.2.21. The element $\hat{g} \in \mathrm{PSL}(3, \mathbb{C})$ is called a *complex homothety* if it is conjugate to an element induced by a diagonal matrix

$$g = \begin{pmatrix} \lambda_1 & & \\ & \lambda_1 & \\ & & \lambda_2 \end{pmatrix},$$

where $|\lambda_1| \neq |\lambda_2|$.

Definition 4.2.22. The element $\hat{g} \in \mathrm{PSL}(3, \mathbb{C})$ is called a *rational screw* (respectively *irrational screw*) if it is not a complex homothety and it is conjugate to an element induced by a diagonal matrix

$$g = \begin{pmatrix} \lambda_1 & & \\ & \lambda_2 & \\ & & \lambda_3 \end{pmatrix},$$

where $|\lambda_1| = |\lambda_2| \neq |\lambda_3|$, $\lambda_1/\lambda_2 = e^{2\pi i x}$ with $x \in \mathbb{Q}$ (respectively $x \in \mathbb{R} - \mathbb{Q}$).

We remark that the complex homotheties and the screws are loxodromic transformations.

Proposition 4.2.23. (i) *If $\hat{g}$ is a complex homothety whose fixed point set consists of a line R and a point $Z_f \notin R$, then $L_0(\hat{g}) = L_1(\hat{g}) = L_2(\hat{g}) = R \cup \{Z_f\}$.*

(ii) *If $\hat{g}$ is a screw with fixed points e_1, e_2, e_3, then*

$$L_0(\hat{g}) = \begin{cases} \overleftrightarrow{e_1, e_2} \cup \{e_3\} & \text{if } \hat{g} \text{ is rational,} \\ \{e_1, e_2, e_3\} & \text{if } \hat{g} \text{ is irrational.} \end{cases}$$

In both cases we have $L_1(\hat{g}) = \overleftrightarrow{e_1, e_2} \cup \{e_3\} = L_2(\hat{g})$ and $\hat{g}$ acts as a classical elliptic transformation on the invariant line $\overleftrightarrow{e_1, e_2}$.

Proof. (i) We can assume $\hat{g}$ has a lift $g \in \mathrm{SL}(3, \mathbb{C})$ of the form

$$\begin{pmatrix} \lambda_1 & 0 & 0 \\ 0 & \lambda_1 & 0 \\ 0 & 0 & \lambda_2 \end{pmatrix},$$

where $|\lambda_1| < |\lambda_2|$. Then $R_f = \{[z_1 : z_2 : z_3] \in \mathbb{P}^2_{\mathbb{C}} \,|\, z_3 = 0\}$, $Z_f = [0 : 0 : 1]$. It follows that $L_0(\hat{g}) = R_f \cup \{Z_f\}$. If $Z = [z_1 : z_2 : z_3] \notin L_0(\hat{g})$, then $\hat{g}^n(Z) \to Z_f$ as $n \to \infty$, and $\hat{g}^{-n}(Z) \to [z_1 : z_2 : 0] \in R_f$ as $n \to \infty$. Therefore, $L_1(\hat{g}) = L_0(\hat{g}) = R_f \cup \{Z_f\}$.

We now prove that $L_2(\hat{g}) = R_f \cup \{Z_f\}$. Let K be a compact subset of $\mathbb{P}^2_{\mathbb{C}} \setminus (R_f \cup \{Z_f\})$. There exists $\epsilon > 0$ such that $K \subset K^{(3)}_\epsilon$, where $K^{(3)}_\epsilon$ is the image, under the canonical projection $\pi : \mathbb{S}^5 \to \mathbb{P}^2_{\mathbb{C}}$, of the set $\{(z_1, z_2, z_3) \in \mathbb{S}^5 \,|\, |z_3| \geq \epsilon\}$.

The inequality $\frac{|\lambda_2|^{2n}\epsilon^2}{|\lambda_1|^2 + |\lambda_1|^2 + |\lambda_2|^{2n}\epsilon^2} \leq \cos^2(d(\hat{g}^n(k), Z_f)) \leq 1$, for all $k \in K^{(3)}_\epsilon$, implies that $\hat{g}^n(\cdot) \to Z_f$ as $n \to \infty$ uniformly on $K^{(3)}_\epsilon$. Thus Z_f is the only cluster point of the family of compact sets $\{\hat{g}^n(K)\}_{n \in \mathbb{N}}$.

Let $p : \mathbb{P}^2_{\mathbb{C}} - \{Z_f\} \to R_f$, be the function defined by $p([z_1 : z_2 : z_3]) = [z_1 : z_2 : 0]$. Given that $K \subset \mathbb{P}^2_{\mathbb{C}} - (R_f \cup \{Z_f\})$, there exists $\delta > 0$ such that $K \subset C_\delta$, where C_δ is the compact subset $\pi(\{(z_1, z_2, z_3) \in \mathbb{S}^5 \,|\, |z_1|^2 + |z_2|^2 \geq \delta\})$.

The inequality $\frac{|\lambda_1|^{-2n}\delta}{|\lambda_1|^{-2n}\delta + |\lambda_2|^{-2n}} \leq \cos^2(d(g^{-n}(Z), p(Z))) \leq 1$, for all $Z \in C_\delta$, implies that $\hat{g}^{-n}|_{C_\delta} \to p|_{C_\delta}$ uniformly as $n \to \infty$. Hence $\mathbb{P}^2_{\mathbb{C}} \setminus (R_f \cup \{Z_f\}) = \text{Eq}(\langle \hat{g} \rangle)$. On the other hand the cluster points of the family of compact sets $\{\hat{g}^{-n}(K)\}_{n \in \mathbb{N}}$ are contained in R_f. Therefore $L_2(\hat{g}) \subset R_f \cup \{Z_f\}$ and it is easy to check that $R_f \cup \{Z_f\} \subset L_2(\hat{g})$. We conclude that $R_f \cup \{Z_f\} = L_2(\hat{g})$. This completes the proof of (i).

Let us now prove (ii). We can assume that $\hat{g}$ has a lift to SL(3, ℂ) of the form,

$$\begin{pmatrix} \lambda_1 & 0 & 0 \\ 0 & \lambda_2 & 0 \\ 0 & 0 & \lambda_3 \end{pmatrix},$$

where $|\lambda_1| = |\lambda_2| > |\lambda_3|$. In this case, $e_1 = [1 : 0 : 0]$, $e_2 = [0 : 1 : 0]$, $e_3 = [0 : 0 : 1]$. Clearly

$$L_0(\hat{g}) = \begin{cases} \overleftrightarrow{e_1, e_2} \cup \{e_3\} & \text{if } \hat{g} \text{ is rational,} \\ \{e_1, e_2, e_3\} & \text{if } \hat{g} \text{ is irrational.} \end{cases}$$

To complete the proof of statement (ii) we need the following lemma.

Lemma 4.2.24. *Let $\hat{g}$ be a screw with fixed points e_1, e_2, e_3. If $Z = [z_1 : z_2 : z_3] \in \mathbb{P}^2_{\mathbb{C}} \setminus \{e_1, e_2, e_3\}$, then:*

a) *Let $d(\,,\,)$ denote the distance with respect to the Fubini-Study metric. Then the sequence $d(\hat{g}^n(\cdot), \overleftrightarrow{e_1, e_2})$ converges to 0 uniformly on compact subsets of $\mathbb{P}^2_{\mathbb{C}} \setminus \{e_3\}$ as $n \to \infty$.*

b) *The sequence $\hat{g}^{-n}(\cdot)$ converges to e_3 uniformly on compact subsets of $\mathbb{P}^2_{\mathbb{C}} - \overleftrightarrow{e_1, e_2}$ as $n \to \infty$.*

Proof. Let $K \subset \mathbb{P}^2_{\mathbb{C}} \setminus \{e_3\}$ be a compact set and $Z = [z_1 : z_2 : z_3] \in K$. There exists $\delta > 0$ such that $K \subset C_\delta$, where

$$C_\delta = \pi(\{(z_1, z_2, z_3) \in \mathbb{S}^5 : |z_1|^2 + |z_2|^2 \geq \delta\}),$$

and $\pi : \mathbb{S}^5 \to \mathbb{P}^2_\mathbb{C}$ is the canonical projection. We have that

$$d(\hat{g}^n(Z), \overleftrightarrow{e_1, e_2}) \le d(\hat{g}^n(Z), [\lambda_1^n z_1 : \lambda_2^n z_2 : 0]).$$

Thus

$$\frac{|z_1|^2 + |z_2|^2}{|z_1|^2 + |z_2|^2 + |(\frac{\lambda_3}{\lambda_1})^n z_3|^2} \le \cos^2(d(\hat{g}^n(Z), \overleftrightarrow{e_1, e_2})) \le 1.$$

Hence

$$\frac{\delta}{\delta + |(\frac{\lambda_3}{\lambda_1})^n|^2} \le \cos^2(d(\hat{g}^n(Z), \overleftrightarrow{e_1, e_2})) \le 1,$$

which proves a).

In order to prove b), let $K \subset \mathbb{P}^2_\mathbb{C} \backslash \overleftrightarrow{e_1, e_2}$ be a compact subset and $Z \in K$. There exists $\epsilon > 0$ such that $K \subset K_\epsilon^{(3)}$, where $K_\epsilon^{(3)} = \pi(\{(z_1, z_2, z_3) \in \mathbb{S}^5 : |z_3| \ge \epsilon\})$. We have that

$$\cos^2(d(\hat{g}^n(Z), e_3)) = \frac{|\lambda_3^{-n} z_3|^2}{|\lambda_1^{-n} z_1|^2 + |\lambda_2^{-n} z_2|^2 + |\lambda_3^{-n} z_3|^2},$$

so

$$\frac{|\lambda_3|^{-2n} \epsilon^2}{|\lambda_1|^{-2n} + |\lambda_2|^{-2n} + |\lambda_3|^{-2n} \epsilon^2} \le \cos^2(d(\hat{g}^n(Z), e_3)) \le 1,$$

which proves b). $\square$

We now observe that Lemma 4.2.24 implies that $L_1(\hat{g}) \subset \overleftrightarrow{e_1, e_2} \cup \{e_3\}$ and $e_3 \in L_1(\hat{g})$. When $\hat{g}$ is a rational screw there exists $k \in \mathbb{N}$ such that $\lambda_1^k = \lambda_2^k$, then $\hat{g}^{nk}(Z) \to [z_1 : z_2 : 0]$ as $n \to \infty$, hence $\overleftrightarrow{e_1, e_2} \cup \{e_3\} = L_1(\hat{g})$. If $\hat{g}$ is an irational screw, then $\hat{g}\big|_{\overleftrightarrow{e_1, e_2}}$ is elliptic of infinite order. Hence in both cases we have $L_1(\hat{g}) = \overleftrightarrow{e_1, e_2} \cup \{e_3\}$.

Notice that Lemma 4.2.24 implies that $L_2(\hat{g}) \subset \overleftrightarrow{e_1, e_2} \cup \{e_3\}$, and it is easy to prove that $\overleftrightarrow{e_1, e_2} \cup \{e_3\} \subset L_2(\hat{g})$. $\square$

Corollary 4.2.25. *Let $g \in \mathrm{PSL}(3, \mathbb{C})$ be a screw or a homothety. Then $\mathrm{Eq}(\langle \hat{g} \rangle) = \overleftrightarrow{e_1, e_2} \cup \{e_3\}$.*

Definition 4.2.26. The transformation $\hat{g}$ is called *strongly loxodromic* if there exist two 3-spheres in $\mathbb{P}^2_\mathbb{C}$, say $T^{(1)}$ and $\mathbf{T}^{(2)}$, such that $\hat{g}(T^{(1)} \cup B_1) \subset B_1$, $\hat{g}^{-1}(T^{(2)} \cup B_2) \subset B_2$, and $B_1 \cap B_2 = \varnothing$, where B_i, $i = 1, 2$, is the connected component of $\mathbb{P}^2_\mathbb{C} \setminus T^{(i)}$ which is diffeomorphic to an open ball in $\mathbb{C}^2$.

Proposition 4.2.27. *The transformation $\hat{g} \in \mathrm{PSL}(3, \mathbb{C})$ is strongly loxodromic if and only if there exists a lift $g \in \mathrm{SL}(3, \mathbb{C})$ whose eigenvalues have pairwise different modules.*

Proof. Assume that $g \in \mathrm{SL}(3, \mathbb{C})$ is a lift of $\hat{g}$ whose eigenvalues have pairwise different modules. We can assume further that g has the form

$$
g = \begin{pmatrix} \lambda_1 & & \\ & \lambda_2 & \\ & & \lambda_3 \end{pmatrix},
$$

with $|\lambda_1| < |\lambda_2| < |\lambda_3|$.

We take the 3-spheres $T^{(1)} = \{[z_1 : z_2 : z_3] \in \mathbb{P}^2_{\mathbb{C}} : |z_1|^2 + |z_2|^2 = |z_3|^2\}$, and $T^{(2)} = \{[z_1 : z_2 : z_3] \in \mathbb{P}^2_{\mathbb{C}} : |z_2|^2 + |z_3|^2 = |z_1|^2\}$, then $\hat{g}(T^{(1)} \cup B_1) \subset B_1$ and $\hat{g}^{-1}(T^{(2)} \cup B_2) \subset B_2$, where B_i is the connected component of $\mathbb{P}^2_{\mathbb{C}} \setminus T^{(i)}$ which is diffeomorphic to an open ball.

Conversely, we assume $\hat{g}$ is strongly loxodromic. Since $\hat{g}$ is loxodromic, there exists a lift $g \in \mathrm{SL}(3, \mathbb{C})$ with at least two eigenvalues of different module.

We claim g must be diagonalizable. Suppose this does not happen, i.e., g is not diagonalizable, then $\hat{g}$ is loxo-parabolic. Thus $\hat{g}$ has two fixed points, two invariant complex lines, and $\hat{g}$ acts as a parabolic transformation on one of these lines. We take $T^{(1)}$ y $T^{(2)}$ such that

$$
\hat{g}(T^{(1)} \cup B_1) \subset B_1 , \tag{1}
$$

$$
\hat{g}^{-1}(T^{(2)} \cup B_2) \subset B_2 , \tag{2}
$$

where B_i, $i = 1, 2$, is the connected component of $\mathbb{P}^2_{\mathbb{C}} - T^{(i)}$ which is diffeomorphic to an open ball. Brouwer's fixed point theorem together with equations (1) and (2), imply that each set B_i, $i = 1, 2$, contains precisely one fixed point of $\hat{g}$. The $\hat{g}$-invariant line on which $\hat{g}$ acts as a parabolic transformation intersects one of the sets B_i, $i = 1, 2$, in a disc. This disc is mapped by $\hat{g}$ or $\hat{g}^{-1}$ to its interior, but this can only happen when $\hat{g}$ is loxodromic on that line, which is a contradiction. Therefore $\hat{g}$ is diagonalizable.

If g is diagonalizable with two eigenvalues having the same module, then $\hat{g}$ has an invariant complex line on which $\hat{g}$ acts as an elliptic transformation. A reasoning similar to that above proves that $\hat{g}$ is loxodromic and elliptic on the same line, a contradiction. Therefore g cannot have two eigenvalues with the same module. $\qquad \square$

Proposition 4.2.28. *Let $\hat{g} \in \mathrm{PSL}(3, \mathbb{C})$ be a strongly loxodromic transformation with fixed points e_1, e_2, e_3. Then $L_0(\hat{g}) = \{e_1, e_2, e_3\} = L_1(\hat{g})$, where e_3 is an attracting point in $\mathbb{P}^2_{\mathbb{C}} \setminus \overleftrightarrow{e_1, e_2}$ and e_1 is a repelling point in $\mathbb{P}^2_{\mathbb{C}} \setminus \overleftrightarrow{e_2, e_3}$. Furthermore $L_2(\hat{g}) = \overleftrightarrow{e_1, e_2} \cup \overleftrightarrow{e_2, e_3} = \mathbb{P}^2_{\mathbb{C}} \setminus \mathrm{Eq}(\langle \hat{g} \rangle)$.*

Proof. We can assume that $\hat{g}$ has a lift $h \in \mathrm{SL}(3, \mathbb{C})$ of the form

$$
\begin{pmatrix} \lambda_1 & 0 & 0 \\ 0 & \lambda_2 & 0 \\ 0 & 0 & \lambda_3 \end{pmatrix},
$$

where $0 < |\lambda_1| < |\lambda_2| < |\lambda_3|$. In this case $e_1 = [1 : 0 : 0]$, $e_2 = [0 : 1 : 0]$, $e_3 = [0 : 0 : 1]$. The fixed points of $\hat{g}^n$ are e_1, e_2, e_3 for every $n \in \mathbb{Z}$, then $L_0(\hat{g}) = \{e_1, e_2, e_3\}$.

One has that $\hat{g}^n(\cdot) \to e_3$ as $n \to \infty$, uniformly on compact subsets of $\mathbb{P}^2_{\mathbb{C}} \setminus \overleftrightarrow{e_1, e_2}$; and $\hat{g}^{-n}(\cdot) \to e_1$ as $n \to \infty$, uniformly on compact subsets of $\mathbb{P}^2_{\mathbb{C}} \setminus \overleftrightarrow{e_2, e_3}$. Now applying Corollary 3.3.5 we deduce that $\mathrm{Eq}(\langle \hat{g} \rangle) = \mathbb{P}^2_{\mathbb{C}} \setminus (\overleftrightarrow{e_1, e_2} \cup \overleftrightarrow{e_2, e_3})$. On the other hand observe that when we restrict $\hat{g}$ to the line $\overleftrightarrow{e_1, e_2}$ we obtain a loxodromic transformation such that e_2 is an attracting point and e_1 is a repelling point. When we restrict $\hat{g}$ to the line $\overleftrightarrow{e_2, e_3}$ we obtain a loxodromic transformation such that e_3 is an attracting point and e_2 is a repelling point. Therefore $L_1(\hat{g}) = \{e_1, e_2, e_3\}$.

Proposition 3.3.6 implies $L_2(\hat{g}) \subset \overleftrightarrow{e_1, e_2} \cup \overleftrightarrow{e_2, e_3}$. To prove that $\overleftrightarrow{e_1, e_2} \cup \overleftrightarrow{e_2, e_3} \subset L_2(\hat{g})$ we take the compact set

$$K = \{[z_1 : z_2 : z_3] \mid |z_2|^2 + |z_3|^2 = |z_1|^2\} \subset \mathbb{P}^2_{\mathbb{C}} \setminus \{e_1, e_2, e_3\}.$$

If $[0 : z_2 : z_3] \in \overleftrightarrow{e_2, e_3} \setminus \{e_3\}$, we take the sequence

$$k_n = [(|\lambda_3^n z_2|^2 + |\lambda_2^n z_3|^2)^{1/2} : \lambda_3^n z_2 : \lambda_2^n z_3] \in K, \quad n \in \mathbb{N},$$

and we see that

$$\hat{g}^n(k^n) = [\left|\frac{\lambda_1^n}{\lambda_2^n}\right|^2 (|z_2|^2 + \left|\frac{\lambda_2^n}{\lambda_3^n}\right|^2 |z_3|^2) : z_2 : z_3] \to [0 : z_2 : z_3]$$

as $n \to \infty$. Then the point $[0 : z_2 : z_3]$ is a cluster point of the family of compact sets $\hat{g}^n(K)$. This implies that $[0 : z_2 : z_3] \in L_2(\hat{g})$, and clearly $e_3 \in L_2(\hat{g})$. Then $\overleftrightarrow{e_2, e_3} \subset L_2(\hat{g})$. An analogous reasoning applied to $\hat{g}^{-1}$ implies $\overleftrightarrow{e_1, e_2} \subset L_2(\hat{g})$. Therefore $L_2(\hat{g}) = \overleftrightarrow{e_1, e_2} \cup \overleftrightarrow{e_2, e_3}$. $\qquad\square$

4.3   The classification theorems

We summarise in the classification theorems below, the information obtained in the previous section. We prove as well the algebraic classification of the elements in $\mathrm{PSL}(3, \mathbb{C})$ according to their trace.

Theorem 4.3.1. (i) *Every element in* $\mathrm{PSL}(3, \mathbb{C}) \setminus \{\mathrm{Id}\}$ *is of one and only one of the classes we have defined: elliptic, parabolic or loxodromic. Moreover:*

(ii) *An elliptic transformation belongs to one and only one of the following classes: regular (it has a lift whose eigenvalues are pairwise different) or conjugate to a complex reflection (it has a lift such that two eigenvalues are repeated).*

(iii) *A parabolic transformation belongs to one and only one of the following classes:* unipotent *(it has a lift whose eigenvalues are equal to one), or* ellipto-parabolic *(it is not unipotent).*

(iv) *A loxodromic element belongs to one and only one of the following four classes: loxo-parabolic, homothety, screw or strongly loxodromic.*

Proof. Let $\hat{g}$ be an element in $\mathrm{PSL}(3, \mathbb{C}) \setminus \{\mathrm{Id}\}$, and $g \in \mathrm{SL}(3, \mathbb{C})$ a lift with eigenvalues λ_i, $i = 1, 2, 3$. We have the following cases:

1) g has repeated eigenvalues.

    a) $|\lambda_i| = 1$, $i = 1, 2, 3$.
- $\hat{g}$ is conjugate to a complex reflection if and only if g is diagonalizable (Corollary 4.2.4).

- $\hat{g}$ is parabolic unipotent or ellipto-parabolic if and only if g is not diagonalizable. In this case, whether $\hat{g}$ is parabolic unipotent or ellipto-parabolic depends on whether $\lambda_i \in C_3$ for all $i = 1, 2, 3$, or not. As before, $C_3 = \{1, \omega, \omega^2\}$ is the set of cubic roots of unity.

    b) If $|\lambda_i| \neq 1$ for some $i = 1, 2, 3$, then $\hat{g}$ is loxodromic (Proposition 4.2.20), and:
- $\hat{g}$ is a complex homothety if and only if g is diagonalizable.

- $\hat{g}$ is loxo-parabolic if and only if g is not diagonalizable.

2) g has no repeated eigenvalues.

    a) $\hat{g}$ is elliptic if and only if the eigenvalues have the same module (equal to 1) (Corollary 4.2.4).

    b) $\hat{g}$ is a screw if and only if there are precisely two eigenvalues having the same module.

    c) $\hat{g}$ is strongly loxodromic if and only if $|\lambda_i| \neq |\lambda_j|$ when $i \neq j$ (Proposition 4.2.27). $\qquad\square$

To state the classification theorem according to trace (Theorem 4.3.3) we have to introduce new terminology: a *regular loxodromic transformation* is a loxodromic transformation which has a lift such that all its eigenvalues are pairwise different. Notice that every regular loxodromic transformation is strongly loxodromic or a screw, and conversely.

Having these remarks in mind, it is easy to see that all the classes of elements cited in Theorem 4.3.3 form a partition of $\mathrm{PSL}(3, \mathbb{C})$. We remark that every elliptic, parabolic or loxodromic element in $\mathrm{PU}(2, 1)$ is in one of the corresponding classes in $\mathrm{PSL}(3, \mathbb{C})$.

Lemma 4.3.2. *For each element $g \in \mathrm{SL}(3, \mathbb{C})$ we denote by $\tau(g)$ its trace and by $\overline{g}$ its conjugate, i.e., the matrix whose entries are the complex conjugate of the entries in g. Let $g \in \mathrm{SL}(3, \mathbb{C})$. Then the set of eigenvalues of g is invariant under inversion on the unit circle if and only if $\tau(g^{-1}) = \overline{\tau(g)}$.*

Proof. Let λ_i, $i = 1, 2, 3$ be the eigenvalues of g. If the set $\{\lambda_1, \lambda_2, \lambda_3\}$ is invariant under inversion on the unitary circle, then $\tau(g^{-1}) = \overline{\tau(g)}$. To prove the converse it suffices to show that g and $\overline{(g)^{-1}}$ have the same characteristic polynomial. The characteristic polynomial of g is $\chi_g(t) = t^3 - xt^2 + yt - 1$, where $x = \tau(g)$ and $y = \tau(g^{-1})$, but the hypothesis implies that $\tau(g) = \tau(\overline{g^{-1}})$ and $\tau(g^{-1}) = \overline{\tau(g)} = \tau((\overline{g^{-1}})^{-1})$. Therefore $\chi_g = \chi_{\overline{(g)^{-1}}}$ $\qquad\square$

The following theorem extends in a natural way the theorem of classification of elements of $SU(2,1)$ according to trace (Theorem 2.4.9).

Theorem 4.3.3. *Let $F(x, y)$ be the complex polynomial*

$$F(x, y) = x^2 y^2 - 4(x^3 + y^3) + 18xy - 27\,,$$

and $g \in SL(3, \mathbb{C})$. Assume that $\hat{g}$ is the transformation of $PSL(3, \mathbb{C})$ induced by g. Then the transformation $\hat{g}$ is:

(i) *Regular elliptic if and only if $\overline{\tau(g)} = \tau(g^{-1})$ and $F(\tau(g), \overline{\tau(g)}) < 0$.*

(ii) *Complex hyperbolic and loxodromic if and only if $\overline{\tau(g)} = \tau(g^{-1})$ and $F(\tau(g), \overline{\tau(g)}) > 0$.*

(iii) *Ellipto-parabolic if and only if $\overline{\tau(g)} = \tau(g^{-1})$, $F(\tau(g), \overline{\tau(g)}) = 0$, $\tau(g) \notin 3C_3$, and $\hat{g}$ is not elliptic.*

(iv) *Conjugate to a complex reflection if and only if $\hat{g}$ is elliptic, $\tau(g) \notin 3C_3$, $\overline{\tau(g)} = \tau(g^{-1})$ and $F(\tau(g), \overline{\tau(g)}) = 0$.*

(v) *Unipotent parabolic transformation if and only if $\tau(g) \in 3C_3$, $\tau(g^{-1}) = \overline{\tau(g)}$ and $\hat{g}$ is not the identity element.*

(vi) *Regular loxodromic but not complex hyperbolic if and only if $\overline{\tau(g)} \neq \tau(g^{-1})$ and $F(\tau(g), \tau(g^{-1})) \neq 0$.*

(vii) *A complex homothety if and only if g is diagonalizable, $\overline{\tau(g)} \neq \tau(g^{-1})$ and $F(\tau(g), \tau(g^{-1})) = 0$.*

(viii) *Loxo-parabolic if and only if $\overline{\tau(g)} \neq \tau(g^{-1})$, $F(\tau(g), \tau(g^{-1})) = 0$, and $\tau(g)$ is not diagonalizable.*

Proof. The characteristic polynomial of g has the form $\chi_g(t) = t^3 - \tau(g)t^2 + \tau(g^{-1})t - 1$, so g has repeated eigenvalues if and only if the discriminant of χ_g, i.e., the complex number $\tilde{f}(\tau(g), \tau(g^{-1}))$, is equal to 0. However $F = -\tilde{f}$, therefore g has repeated eigenvalues if and only if $F(\tau(g), \tau(g^{-1})) = 0$.

(i) Assume $\hat{g}$ is regular elliptic, then $\hat{g}$ is conjugate to a regular elliptic element in $PU(2, 1)$. Therefore every eigenvalue of g has module 1, $\tau(g^{-1}) = \overline{\tau(g)}$ and, by Theorem 2.4.9, $F(\tau(g), \overline{\tau(g)}) < 0$. Conversely, given that $\tau(g^{-1}) = \overline{\tau(g)}$, the set of eigenvalues of g is invariant under inversion on the unitary circle. Given that $F(\tau(g), \overline{\tau(g)}) < 0$, Corollary 2.4.13 implies that the eigenvalues of g are pairwise different and have module 1. Therefore $\hat{g}$ is regular elliptic.

(ii) Assume $\hat{g}$ is complex hyperbolic and loxodromic, then it is conjugate to a loxodromic element in $\mathrm{PU}(2,1)$. Therefore $\tau(g^{-1}) = \overline{\tau(g)}$ and, by Theorem 2.4.9, $F(\tau(g), \overline{\tau(g)}) > 0$.

Conversely, given that $\tau(g^{-1}) = \overline{\tau(g)}$ and $F(\tau(g), \overline{\tau(g)}) > 0$, then the set of eigenvalues of g is invariant under inversion on the unitary circle and, by Corollary 2.4.13, only one of them has module 1. It follows that g is conjugate to a transformation of the form

$$A = \begin{pmatrix} re^{i\theta_0} & & \\ & e^{-2i\theta_0} & \\ & & \frac{1}{r}e^{i\theta_0} \end{pmatrix},$$

where, $0 < r < 1$. Let $u \in \mathbb{R}^+$ be such that $r = e^{-u}$. It is possible to find a matrix B such that BAB^{-1} has the form

$$\begin{pmatrix} \cosh(u)e^{i\theta_0} & 0 & \sinh(u)e^{i\theta_0} \\ 0 & e^{-2i\theta_0} & 0 \\ \sinh(u)e^{i\theta_0} & 0 & \cosh(u)e^{i\theta_0} \end{pmatrix}.$$

We conclude that $\hat{g}$ is complex hyperbolic.

(iii) Assume $\hat{g}$ is ellipto-parabolic, then every eigenvalue of g has module 1, so $\tau(g^{-1}) = \overline{\tau(g)}$. Moreover, $F(\tau(g), \overline{\tau(g)}) = 0$, because g has a repeated eigenvalue. Clearly $\hat{g}$ is not elliptic. Finally, $\tau(g) \notin 3C_3$ (otherwise $\hat{g}$ would be unipotent).

Conversely, if $\tau(g^{-1}) = \overline{\tau(g)}$, then the set of eigenvalues of g is invariant under inversion on the unitary circle. On the other hand, $F(\tau(g), \overline{\tau(g)}) = 0$ implies g has repeated eigenvalues, so every eigenvalue has module 1. We have that $\tau(g) \notin 3C_3$, it follows that g has precisely two different eigenvalues. By hypothesis, $\hat{g}$ is not elliptic, then g is not diagonalizable, therefore $\hat{g}$ is ellipto-parabolic.

(iv) Assume $\hat{g}$ is conjugate to a complex reflection, then it is elliptic and g has an eigenvalue of multiplicity 2. It follows that $\tau(g^{-1}) = \overline{\tau(g)}$, $F(\tau(g), \overline{\tau(g)}) = 0$ and $\tau(g) \notin 3C_3$.

Conversely, $\tau(g^{-1}) = \overline{\tau(g)}$ implies that the set of eigenvalues of g is invariant under inversion on the unitary circle. The equation $F(\tau(g), \overline{\tau(g)}) = 0$ implies that g has repeated eigenvalues, and it follows that every eigenvalue of g has module 1. Given that $\tau(g) \notin 3C_3$, then one of the eigenvalues of g has multiplicity 2. By hypothesis $\hat{g}$ is elliptic, therefore $\hat{g}$ is conjugate to a complex reflection.

(v) If $\hat{g}$ is parabolic unipotent, then clearly $\tau(g) \in 3C_3$ and $\tau(g^{-1}) = \overline{\tau(g)}$.

Conversely, if $\tau(g) \in 3C_3$ and $\tau(g^{-1}) = \overline{\tau(g)}$, then the characteristic polynomial of g has the form $t^3 - 3\omega t^2 + 3\omega^2 t - 1$, where $\omega^3 = 1$. Then every eigenvalue of g is equal to ω and given that $\hat{g}$ is not the identity we conclude that $\hat{g}$ is parabolic unipotent.

(vi) Assume $\hat{g}$ is regular loxodromic but not complex hyperbolic, then the eigenvalues of g have not the same module. If $\tau(g^{-1}) = \overline{\tau(g)}$, then g would be complex hyperbolic. It follows that $\tau(g^{-1}) \neq \overline{\tau(g)}$. Now, $F(\tau(g), \overline{\tau(g)}) \neq 0$ because $\hat{g}$ is regular.

Conversely, if $F(\tau(g), \overline{\tau(g)}) \neq 0$, then $\hat{g}$ is regular. Now the eigenvalues of g have not the same module (otherwise every eigenvalue has module 1 and therefore $\tau(g^{-1}) = \overline{\tau(g)}$). It follows that $\hat{g}$ is loxodromic. Finally, $\hat{g}$ is not complex hyperbolic because $\tau(g^{-1}) \neq \overline{\tau(g)}$.

(vii) Assume $\hat{g}$ is a complex homothety, then g is diagonalizable, it has an eigenvalue with multiplicity equal to 2, and the eigenvalues do not have the same module. Then $\tau(g^{-1}) \neq \overline{\tau(g)}$ (otherwise the set of eigenvalues is invariant under the inversion on the unitary circle, and in this case every eigenvalue has module 1), and clearly $F(\tau(g), \overline{\tau(g)}) = 0$.

Conversely, if $F(\tau(g), \overline{\tau(g)}) = 0$, then g has a repeated eigenvalue. The hypothesis $\tau(g^{-1}) \neq \overline{\tau(g)}$ implies that the eigenvalues have not the same module. Finally, g is diagonalizable, therefore $\hat{g}$ is a complex homothety.

(viii) Assume $\hat{g}$ is loxo-parabolic, then g has an eigenvalue with multiplicity 2, and the eigenvalues of g do not have the same module. Therefore $F(\tau(g), \overline{\tau(g)}) = 0$, $\tau(g^{-1}) \neq \overline{\tau(g)}$ and g is not diagonalizable.

Conversely, assume g is not diagonalizable and $\tau(g^{-1}) \neq \overline{\tau(g)}$, then g has one eigenvalue of module not equal to 1 (otherwise, $\tau(g^{-1}) = \overline{\tau(g)}$). Therefore $\hat{g}$ is loxo-parabolic. $\square$

The following table summarises the descriptions of the limit set in the sense of Kulkarni, the maximal regions of discontinuity and equicontinuity regions for cyclic groups, in terms of the normal Jordan form of the generator:

Normal Form of $\tilde{\gamma}$	Condition over the λ's	Transformation Type	Sub-type	$\Omega_{Kul}(<\gamma>)$	$Eq(<\gamma>)$	Maximal Regions of Discontinuity
$\begin{pmatrix} 1 & 1 & 0 \\ 0 & 1 & 1 \\ 0 & 0 & 1 \end{pmatrix}$	None	Parabolic	Unipotent	$\mathbb{P}^2_{\mathbb{C}} \setminus \{\hat{e}_1, \hat{e}_2\}$	$\Omega_{Kul}(<\gamma>)$	$\Omega_{Kul}(<\gamma>)$
$\begin{pmatrix} \lambda_1 & 0 & 0 \\ 0 & \lambda_2 & 0 \\ 0 & 0 & (\lambda_1\lambda_2)^{-1} \end{pmatrix}$	$\lambda_1^n = \lambda_2^n = \lambda_3^n = 1$ for some n	Elliptic	Regular if $\lambda_1\lambda_2^{-1} \neq 1$. Complex Reflection if $\lambda_1\lambda_2^{-1} = 1$.	$\mathbb{P}^2_{\mathbb{C}}$	$\mathbb{P}^2_{\mathbb{C}}$	$\mathbb{P}^2_{\mathbb{C}}$
	$|\lambda_1| = |\lambda_2| = 1$ and $\lambda_3^n \neq 1$ for all n	Elliptic	Regular if $\lambda_1\lambda_2^{-1} \neq 1$. Complex Reflection if $\lambda_1\lambda_2^{-1} = 1$. Irrational screw, in other case	$\mathbb{P}^2_{\mathbb{C}} \setminus (\{\hat{e}_1, \hat{e}_2\} \cup \{e_3\})$	$\Omega_{Kul}(<\gamma>)$	$\Omega_{Kul}(<\gamma>)$
	$|\lambda_3| \neq |\lambda_1| = |\lambda_2|$, and $|\lambda_2| \neq 1$.	Loxodromic	Rational screw if $\lambda_1\lambda_2^{-1} \neq 1$ is a root of unity. Complex Homothety if $\lambda_1\lambda_2^{-1} = 1$.	$\emptyset$	$\mathbb{P}^2_{\mathbb{C}}$	$\mathbb{P}^2_{\mathbb{C}}$
	$|\lambda_1| < |\lambda_2| < |\lambda_3|$	Loxodromic	Strongly Loxodromic	$\mathbb{P}^2_{\mathbb{C}} \setminus (\{\hat{e}_1, \hat{e}_2\} \cup \{\hat{e}_1, \hat{e}_3\})$	$\Omega_{Kul}(<\gamma>)$	$\Omega_{Kul}(<\gamma>)$
$\begin{pmatrix} \lambda_1 & 1 & 0 \\ 0 & \lambda_1 & 0 \\ 0 & 0 & \lambda_1^{-2} \end{pmatrix}$	$|\lambda_1| = 1$	Parabolic	Rational Ellipto-Parabolic, if $\lambda \neq 1$ is a root of unity. Irrational Ellipto-Parabolic, in other case	$\mathbb{P}^2_{\mathbb{C}} \setminus \{\hat{e}_1, \hat{e}_3\}$	$\Omega_{Kul}(<\gamma>)$	$\Omega_{Kul}(<\gamma>)$
	$|\lambda_1| \neq 1$	Loxodromic	Loxo-parabolic	$\mathbb{P}^2_{\mathbb{C}} \setminus (\{\hat{e}_1, \hat{e}_3\} \cup \{\hat{e}_1, \hat{e}_3\})$	$\Omega_{Kul}(<\gamma>)$	$\mathbb{P}^2_{\mathbb{C}} \setminus (\{\hat{e}_1, \hat{e}_2\} \cup \{e_3\})$, $\mathbb{P}^2_{\mathbb{C}} \setminus (\{\hat{e}_3, \hat{e}_2\} \cup \{e_1\})$

From the previous discussion we also obtain information about the set of invariant lines for each cyclic group. This information is given in the table below and it will be used later in the text:

Corollary 4.3.4 (See [41]). *Let* $\gamma \in PSL_3(\mathbb{C}) - \{\mathrm{Id}\}$ *and* $\tilde{\gamma}$ *be a lift of* γ. *Consider the normal Jordan form of* $\tilde{\gamma}$. *Then the set of* γ-*invariant lines is given by:*

Normal Form	Condition over the λ's	Invariant Lines
$\begin{pmatrix} \lambda_1 & 0 & 0 \\ 0 & \lambda_2 & 0 \\ 0 & 0 & \lambda_3 \\ \lambda_1\lambda_2\lambda_3 = 1 \end{pmatrix}$	$\lambda_1 \neq \lambda_2 \neq \lambda_3 \neq \lambda_1$ $\lambda_1 = \lambda_2 \neq \lambda_3$	$\{\overleftrightarrow{e_1, e_2},\ \overleftrightarrow{e_1, e_3}.\ \overleftrightarrow{e_3, e_2}\}$ $\{\overleftrightarrow{e_1, e_2}\} \cup \{$*The set of projective lines through* $e_3\}$
$\begin{pmatrix} \lambda_1 & 1 & 0 \\ 0 & \lambda_1 & 0 \\ 0 & 0 & \lambda_1^{-2} \end{pmatrix}$	$\lambda_1^3 \neq 1$ $\lambda_1 = 1$	$\{\overleftrightarrow{e_1, e_3},\ \overleftrightarrow{e_1, e_2}\}$ $\{$*The set of projective lines through* $e_1\}$
$\begin{pmatrix} 1 & 1 & 0 \\ 0 & 1 & 1 \\ 0 & 0 & 1 \end{pmatrix}$		$\{\overleftrightarrow{e_1, e_2}\}$

Remark 4.3.5. It is natural to search for a classification of the elements in $\mathrm{PSL}(n+1, \mathbb{C})$ generalizing the results described in this chapter for the case $n = 2$. There are several articles in this direction by K. Gongopadhyay, J. R. Parker and others, classifying the transformations which are isometries of either complex or quaternionic hyperbolic spaces. See for instance [170], [43], [73] and [74], where the authors give algebraic characterizations of the isometries in complex and quaternionic projective spaces. Notice too that all these transformations are isometries of spaces which are Gromov-hyperbolic, so their study actually fits within that general framework (see for instance [32]).

A classification of the elements in $\mathrm{PSL}(n + 1, \mathbb{C})$ in general has been done recently in [39], generalizing the results explained in this chapter. This is in terms of the dynamics and the eigenvalues of the corresponding transformations. They describe in each case the corresponding Kulkarni limit set, as well as the regions of discontinuity, of equicontinuity, and the maximal region where the action is properly discontinuous. It remains to give an algebraic characterization of the elements in $\mathrm{PSL}(n+1, \mathbb{C})$, extending to this setting the known results for complex and quaternionic hyperbolic isometries.

Chapter 5

Kleinian Groups with a Control Group

Nowadays the literature and the knowledge about subgroups of $\mathrm{PSL}(2,\mathbb{C})$ is vast. Thence, when going up into higher dimensions, a natural step is to consider Kleinian subgroups of $\mathrm{PSL}(3,\mathbb{C})$ whose geometry and dynamics are "governed" by a subgroup of $\mathrm{PSL}(2,\mathbb{C})$. That is the subject we address in this chapter. The corresponding subgroup in $\mathrm{PSL}(2,\mathbb{C})$ is the *control group*. These groups play a significant role in the classification theorems we give in Chapter 8. The simplest of these are the groups constructed by suspension, already mentioned in Chapter 3. That construction extracts discrete subgroups of $\mathrm{PSL}(n+1,\mathbb{C})$ from discrete subgroups of $\mathrm{PSL}(n,\mathbb{C})$. More generally, if a subgroup $\Gamma \subset \mathrm{PSL}(n+1,\mathbb{C})$ acts on $\mathbb{P}^n_{\mathbb{C}}$ with a fixed point p, then a choice of a projective hyperplane $\mathcal{L} \subset \mathbb{P}^n_{\mathbb{C}} \setminus \{p\}$ defines canonically a holomorphic fibre bundle $\pi_{p,\mathcal{L}} : \mathbb{P}^n_{\mathbb{C}} \setminus \{p\} \to \mathcal{L}$, that gives rise to a group homomorphism

$$\Pi := \Pi_{p,\mathcal{L}} : \mathrm{PSL}(n+1,\mathbb{C}) \to \mathrm{PSL}(n,\mathbb{C}),$$

which is independent of the choice of $\mathcal{L}$ up to conjugation in $\mathrm{PSL}(n+1,\mathbb{C})$. Then $\Pi(\Gamma) \subset \mathrm{PSL}(n,\mathbb{C})$ is the control group of Γ. Under appropriate conditions, the control group indeed dictates much of the dynamics and the geometry of the Γ-action on $\mathbb{P}^n_{\mathbb{C}}$. If, for instance, Γ is the suspension of a Kleinian subgroup $\Sigma \subset \mathrm{PSL}(n,\mathbb{C})$, then Σ itself is the control group. In the general setting of discrete subgroups of $\mathrm{PSL}(3,\mathbb{C})$ with a control group in $\mathrm{PSL}(2,\mathbb{C})$, one can have that the control group $\Pi(\Gamma)$ is nondiscrete, even if $\Gamma \subset \mathrm{PSL}(n+1,\mathbb{C})$ is discrete.

Hence we begin this chapter, which is based on [41], by studying nondiscrete subgroups of $\mathrm{PSL}(2,\mathbb{C})$. We look at their limit set in the sense of [76], which turns out to be the complement of the equicontinuity region in the Riemann sphere (notice that for nondiscrete groups the region of discontinuity is empty by definition). Then we classify the nondiscrete groups whose region of equicontinuity misses at most two points in the Riemann sphere: the elementary groups, as in the discrete case. This is used in the sequel.

Next, we study the suspension constructions as given in [201] and generalised in [160], [41]. Using the corresponding control group we determine the Kulkarni limit set of the suspension group, and we show that this coincides with the complement of the equicontinuity region. We then look at subgroups $\Gamma \subset \mathrm{PSL}(3, \mathbb{C})$ having a control group in $\mathrm{PSL}(2, \mathbb{C})$. We show that under a certain hypothesis, the equicontinuity region in $\mathbb{P}^1_{\mathbb{C}}$ of the control group determines the equicontinuity region $\mathrm{Eq}\,(\Gamma)$ in $\mathbb{P}^2_{\mathbb{C}}$, as well as its Kulkarni limit set and the corresponding region of discontinuity. These results are used in the following chapter.

5.1 $\mathrm{PSL}(2, \mathbb{C})$ revisited: nondiscrete subgroups

We recall (see Definition 1.2.20) that given a subgroup $\Sigma \subset \mathrm{PSL}(2, \mathbb{C})$, its *equicontinuity region* $\mathrm{Eq}\,(\Sigma)$ in $\mathbb{P}^1_{\mathbb{C}}$ is the set of points having an open neighbourhood where Σ forms a normal family. In Theorem 1.2.21 we proved that for discrete subgroups of $\mathrm{PSL}(2, \mathbb{C})$, the equicontinuity and discontinuity regions coincide. For nondiscrete subgroups of $\mathrm{PSL}(2, \mathbb{C})$ the concept of "discontinuity" is not interesting, since the discontinuity region is always empty, by definition. Now it is equicontinuity that plays a significant role.

We consider the usual identification of $\mathrm{PSL}(2, \mathbb{C})$ with the isometries group of the hyperbolic 3-space $\mathbb{H}^3_{\mathbb{R}}$, and the corresponding identification of $\mathbb{P}^1_{\mathbb{C}}$ with the Riemann sphere regarded as the sphere at infinity $\mathbb{S}^2_{\infty} := \partial \mathbb{H}^3_{\mathbb{R}}$ (see Chapter 1).

Throughout this chapter Σ denotes a possibly nondiscrete subgroup of $\mathrm{PSL}(2, \mathbb{C})$.

Definition 5.1.1. The *limit set of* Σ *in the sense of Greenberg* (see [76]), denoted by $\Lambda_{Gr}(\Sigma)$, is the intersection with $\mathbb{S}^2_{\infty}$ of the set of accumulation points of the orbits of points in $\mathbb{H}^3$.

Of course this is the usual limit set when the group is discrete.

Definition 5.1.2. The (nondiscrete) group Σ is *elementary* if its equicontinuity set omits at most 2 points in $\mathbb{P}^1_{\mathbb{C}}$.

5.1.1   Main theorems for nondiscrete subgroups of $\mathrm{PSL}(2, \mathbb{C})$

We prove in the sequel that as in the case of discrete groups, if Σ is elementary, then its limit set contains at most two points. Moreover, we will prove the two theorems below.

Theorem 5.1.3. *Let* $\Sigma \subset \mathrm{PSL}(2, \mathbb{C})$ *be a nondiscrete group. Then:*

(i) *The limit set* $\Lambda_{Gr}(\Sigma)$ *equals* $\mathbb{P}^1_{\mathbb{C}} \setminus \mathrm{Eq}\,(\Sigma)$, *the complement of the region of equicontinuity in* $\mathbb{P}^1_{\mathbb{C}}$.

(ii) *If Σ is nonelementary, then the limit set $\Lambda_{Gr}(\Sigma)$ is independent of the choice of orbit used to define it.*

(iii) *If Σ is nonelementary and $\mathrm{Eq}(\Sigma)$ is nonempty, then up to conjugation by an element of $\mathrm{M\ddot{o}b}(\mathbb{C})$, one has that $\Lambda_{Gr}(\Sigma)$ is a real projective line: $\Lambda_{Gr}(\Sigma) = \mathbb{R} \cup \{\infty\}$.*

In the following theorem, we let $\mathrm{Epa}(\mathbb{C})$ be the group generated by all the elliptic and parabolic elements which leave invariant $\mathbb{C}$ (see Example 5.2.4 below), and let Dih_∞ the infinite dihedral group (see Example 5.2.3).

Theorem 5.1.4. *Let $\Sigma \subset \mathrm{PSL}(2, \mathbb{C})$ be a nondiscrete elementary group. Then:*

(i) *Its limit set consists of at most two points.*

(ii) *Its limit set contains exactly two points if and only if Σ is conjugate to a subgroup Σ_* of $\mathrm{M\ddot{o}b}(\mathbb{C}^*)$ such that Σ_* contains a loxodromic element.*

(iii) *Its limit set contains exactly one point if and only if Σ is conjugate to a subgroup Σ_* of $\mathrm{Epa}(\mathbb{C})$ such that $\overline{\Sigma_*}$ contains a parabolic element.*

(iv) *Its limit set is empty if and only if Σ is either conjugate to a subgroup Σ_* of Dih_∞ or $S0(3)$.*

Theorem 5.1.4 will follow from Theorems 5.3.7 and 5.1.3.

The main tools for proving these theorems are the following two results due to Greenberg (see Theorem 1 and Proposition 12 in [76]), that we state without a proof.

Theorem 5.1.5. *Let G be a connected Lie subgroup of $\mathrm{PSL}(2, \mathbb{C})$. Then one of the following assertions applies:*

(i) *The elements of G have a common fixed point in $\mathbb{H}^3$, and G is conjugate to a Lie subgroup of $O(3)$.*

(ii) *The elements of G have a common fixed point in $\mathbb{P}^1_\mathbb{C}$.*

(iii) *There exists a hyperbolic geodesic line $\ell \subset \mathbb{H}^3$ such that ℓ is G-invariant.*

(iv) *There exists a hyperbolic plane $L \subset \mathbb{H}^3$ such that L is G-invariant.*

(v) *$G = \mathrm{PSL}(2, \mathbb{C})$.*

Theorem 5.1.6. *Let $\Sigma \subset \mathrm{PSL}(2, \mathbb{C})$ be a subgroup with $\mathrm{card}(\Lambda_{Gr}(\Sigma)) \geq 2$. Then $\Lambda_{Gr}(\Sigma)$ is the closure of the loxodromic fixed points.*

The following proposition is used in the sequel.

Proposition 5.1.7. *Let Σ be a subgroup of $\mathrm{PSL}(2, \mathbb{C})$, then:*

(i) *If $\gamma \in \Sigma$ is not elliptic, then $\mathrm{Fix}(\gamma) \subset \mathbb{P}^1_\mathbb{C} \setminus \mathrm{Eq}(\Sigma)$.*

(ii) *The set $\mathrm{Eq}(\Sigma)$ is open and Σ-invariant.*

(iii) *If $\overline{\Sigma}$ is the topological closure of Σ in $\mathrm{PSL}(2,\mathbb{C})$, then $\mathrm{Eq}\,(\Sigma) = \mathrm{Eq}\,(\overline{\Sigma})$.*

Proof. Statement (i) is immediate from Theorem 1.2.21. Statement (ii) is obvious. For statement (iii) observe that we have trivially $\mathrm{Eq}\,(\overline{\Sigma}) \subset \mathrm{Eq}\,(\Sigma)$. Now take $(\gamma_m) \subset \overline{\Sigma}$, let $U \subset \mathrm{Eq}\,(\Sigma)$ be an open set whose closure is compact and is fully contained in $\mathrm{Eq}\,(\Sigma)$, and $\tau_m \in \Sigma$ such that

$$d_\infty(\tau_m|_{\overline{U}}, \gamma_m|_{\overline{U}})) < 2^{-m}$$

where d_∞ denotes the metric of the uniform convergence.

By definition of $\mathrm{Eq}\,(\Sigma)$ there is a continuous function $\tau : \overline{U} \to \mathbb{P}^1_{\mathbb{C}}$ and a subsequence of (τ_m), still denoted by (τ_m), such that

$$d_\infty(\tau_m|_{\overline{U}}, \tau|_{\overline{U}})) < 2^{-m}.$$

Now the assertion follows from the following inequality:

$$d_\infty(\tau|_{\overline{U}}, \gamma_m|_{\overline{U}}) \leq d_\infty(\tau|_{\overline{U}}, \tau_m|_{\overline{U}}) + d_\infty(\tau_m|_{\overline{U}}, \gamma_m|_{\overline{U}}) \leq 2^{1-m}. \qquad \square$$

5.2   Some basic examples

We now give some examples that play a significant role in the sequel because we will show that up to conjugation, every nondiscrete subgroup of $\mathrm{PSL}(2,\mathbb{C})$ with nonempty region of equicontinuity is a subgroup of one of the groups in Examples 5.2.1 to 5.2.5.

Example 5.2.1. Denote by $\widehat{\mathbb{R}}$ the subset $\mathbb{R} \cup \infty$ of $\widehat{\mathbb{C}} = \mathbb{C} \cup \infty \cong \mathbb{P}^1_{\mathbb{C}}$. Notice that this is a copy of $\mathbb{P}^1_{\mathbb{R}}$. Let $\mathrm{M\ddot{o}b}(\widehat{\mathbb{R}})$ be the subgroup of $\mathrm{PSL}(2,\mathbb{C})$ generated by the elements in $\mathrm{PSL}(2,\mathbb{R})$ and a new generator $z \mapsto -\bar{z}$, i.e., $\mathrm{M\ddot{o}b}(\widehat{\mathbb{R}}) = \langle \mathrm{PSL}(2,\mathbb{R}), -\bar{z} \rangle$. Then $\Lambda_{Gr}(\mathrm{M\ddot{o}b}(\widehat{\mathbb{R}})) = \widehat{\mathbb{R}}$ and $\mathrm{Eq}\,(\mathrm{M\ddot{o}b}(\widehat{\mathbb{R}})) = \mathbb{P}^1_{\mathbb{C}} \setminus \widehat{\mathbb{R}} = \mathbb{C} \setminus \mathbb{R}$.

The following are examples of elementary groups.

Example 5.2.2. Set $\mathrm{Rot}_\infty = \{T(z) = az : a \in \mathbb{S}^1\} \cong \mathbb{S}^1$, the group of rotations, and $\mathrm{Dih}_\infty = \langle \mathrm{Rot}_\infty, 1/z \rangle$, the infinite dihedral group. Then the limit set Λ_{Gr} is empty in both cases, and we have $\mathrm{Eq}\,(\mathrm{Rot}_\infty) = \mathrm{Eq}\,(\mathrm{Dih}_\infty) = \mathbb{P}^1_{\mathbb{C}} \setminus \Lambda_{Gr}(\mathrm{Dih}_\infty) = \mathbb{P}^1_{\mathbb{C}}$.

Example 5.2.3. Consider now the group of all Möbius transformations that leave invariant $\mathbb{C}^*$ and denote it by $\mathrm{M\ddot{o}b}(\mathbb{C}^*)$. That is,

$$\mathrm{M\ddot{o}b}(\mathbb{C}^*) = \{T(z) = az,\ az^{-1} : a \in \mathbb{C}^*\}.$$

Then $\Lambda_{Gr}(\mathrm{M\ddot{o}b}(\mathbb{C}^*))$ consists of the points $\{0, \infty\}$ and we have:

$$\mathrm{Eq}\,(\mathrm{M\ddot{o}b}(\mathbb{C}^*)) = \mathbb{P}^1_{\mathbb{C}} \setminus \Lambda_{Gr}(\mathrm{M\ddot{o}b}(\mathbb{C}^*)) = \mathbb{C}^*.$$

Example 5.2.4. The following is an example where the limit set consists of a single point. Consider the group generated by the set of all elliptic or parabolic elements in the affine group and denote it by $\mathrm{Epa}\,(\mathbb{C})$. That is

$$\mathrm{Epa}\,(\mathbb{C}) = \{\gamma(z) = az + b : |a| = 1 \text{ and } b \in \mathbb{C}\}.$$

Then $\Lambda_{Gr}(\mathrm{M\ddot{o}b}(\mathbb{C}^*)) = \infty$ and $\mathrm{Eq}\,(\mathrm{Epa}\,(\mathbb{C})) = \mathbb{P}^1_{\mathbb{C}} \setminus \Lambda_{Gr}(\mathrm{M\ddot{o}b}(\mathbb{C}^*)) = \mathbb{C}$.

Example 5.2.5. Consider now the group G defined by

$$G \cong \left\{ \begin{pmatrix} a & -\bar{c} \\ c & \bar{a} \end{pmatrix} \in \mathrm{PSL}(2,\mathbb{C}) : |a|^2 + |c|^2 = 1 \right\}.$$

In other words, G is the projectivisation of the group $\mathrm{SU}(2)$, which is isomorphic to the sphere $\mathbb{S}^3$ (regarded as a Lie group). Hence G is isomorphic to the group of rotations $\mathrm{SO}(3)$, so as a manifold it is diffeomorphic to $\mathbb{P}^3_{\mathbb{R}}$. In this case the limit set is empty, since no orbit of a point in the ball converges to the sphere at infinity, and one has $\mathrm{Eq}\,(\mathrm{SO}(3)) = \mathbb{P}^1_{\mathbb{C}}$.

5.3 Elementary groups

In this section we classify the *elementary* groups: those whose equicontinuity region misses at most two points in $\mathbb{P}^1_{\mathbb{C}}$.

5.3.1 Groups whose equicontinuity region is the whole sphere

We consider now nondiscrete subgroups of $\mathrm{PSL}(2,\mathbb{C})$ whose equicontinuity set is the whole Riemann sphere. For this we recall the following well-known result that essentially goes back to A. Cayley in the 19th Century (see for instance [144]).

Theorem 5.3.1. *The finite groups of* $\mathrm{PSL}(2,\mathbb{C})$ *are the cyclic, dihedral, tetrahedral, octahedral and icosahedral groups, up to conjugation.*

So if we start with a group whose equicontinuty region is the whole Riemann sphere, by Proposition 5.1.7 we deduce that Σ is either finite or a purely elliptic nondiscrete group. Thus our first problem is finding conditions under which the composition of a pair of elliptic transformations is elliptic. A way to search for the solution of this problem is by looking at the fixed points (see [144, p. 11, 12 and 19]):

Lemma 5.3.2. *Let $f, g \in \mathrm{PSL}(2,\mathbb{C})$ be two nontrivial elements, then:*

(i) *They commute if and only if either they have exactly the same fixed point set, or else each element is elliptic of order 2, and each of them interchanges the fixed points of the other.*

(ii) *Assume f has exactly two fixed points. If f, g share exactly one fixed point, then $fgf^{-1}g^{-1}$ is parabolic. If f is loxodromic, then there is a sequence $(g_m)_{m \in \mathbb{N}}$ contained in the group $\langle f, g \rangle$ generated by f, g, of distinct parabolic elements such that $g_m \xrightarrow[m \to \infty]{} \mathrm{Id}$ uniformly on $\mathbb{P}^1_{\mathbb{C}}$.*

Hence the initial problem becomes that of finding conditions so that two elliptic transformations whose fixed points sets are pairwise disjoint generate a purely elliptic group. Such a condition, as we see below, is given in terms of the cross-ratios of the fixed points sets of the elliptic transformations. First we need the following two technical lemmas from [41]:

Lemma 5.3.3. *Let $\gamma = \begin{pmatrix} a & b \\ c & d \end{pmatrix} \in \mathrm{PSL}(2, \mathbb{C})$ be an elliptic element with $\mathrm{Fix}(\gamma) = \{1, p\} \subset \overline{\mathbb{D}}$. Then:*

(i) *$a = \bar{d}$ if and only if $p \in \mathbb{R}$.*

(ii) *If $p \in \mathbb{R}$, then $|a| = 1$ if and only if $p = 0$.*

(iii) *If $p \in \mathbb{R}$, then $|a| < 1$ if and only if $p < 0$.*

Lemma 5.3.4. *Let $\gamma_1, \gamma_2 \in \mathrm{PSL}(2, \mathbb{C})$ be elliptic elements with infinite order, whose fixed points $\mathrm{Fix}(\gamma_1) = \{z_1, z_2\}$, $\mathrm{Fix}(\gamma_2) = \{w_1, w_2\}$, are such that their cross-ratio satisfies $[z_1; z_2; w_1; w_2] \in \mathbb{C} \setminus (\mathbb{R}_- \cup \{0\})$. Then $\langle \gamma_1, \gamma_2 \rangle$ contains a loxodromic element.*

One can also ask if the previous lemma holds when at least one of the elements appearing in the hypothesis has finite order, however for our purposes it is enough to consider this case since we always have two distinct elements of infinite order.

We refer to [41] for the proof of these lemmas. Let us now introduce another interesting family of groups that we use in the sequel.

Definition 5.3.5. For each real number $p < 0$, the *Chinese Rings group* at p is defined as

$$\mathrm{Cr}\,(p) = \overline{\langle \mathrm{Rot}\,_\infty, \tau_p^{-1} \mathrm{Rot}\,_\infty \tau_p \rangle},$$

where $\mathrm{Rot}\,_\infty$ is the group of rotations in Example 5.2.2, and $\tau_p(z)$ is the Möbius transformation

$$\tau_p(z) = \frac{z - p}{z - 1}.$$

In other words, this group is generated by $\mathrm{Rot}\,_\infty \cong \mathrm{SO}(2)$ and its conjugate $\tau_p^{-1} \mathrm{Rot}\,_\infty \tau_p$. Hence the generators are all elliptic. Notice also that the fixed points of the first group are $\{0, \infty\}$ while those of the second group are 1 and p, which is negative. Hence these four points are contained in a real projective line and they are intercalated. We will see that this property is shared by every pair of noncommuting elements in the corresponding Chinese Rings group. Notice also that τ_p is an involution, so it coincides with τ_p^{-1}.

The Chinese Rings groups have the following properties (see [41] for the proof):

Proposition 5.3.6. (i) *For each* $\gamma \in \mathrm{Cr}\,(p)$, *there is* $\lambda = e^{\pi i \vartheta}$ *such that*

$$\gamma(z) = \frac{\frac{p\bar{\lambda}-\lambda}{p-1}z + \frac{p(\lambda-\bar{\lambda})}{p-1}}{\frac{\bar{\lambda}-\lambda}{p-1}z + \frac{p\lambda-\bar{\lambda}}{p-1}}. \tag{5.3.6}$$

(ii) *For each* $z \in \mathbb{P}^1_{\mathbb{C}}$, *its* $\mathrm{Cr}\,(-1)$*-orbit is all of* $\mathbb{P}^1_{\mathbb{C}}$.

(iii) *One has* $\mathrm{Cr}\,(-1) = \mathrm{SO}(3)$.

(iv) *For each* $p < 0$ *there are* $z_p \in \mathbb{C}$ *and* $\gamma_p \in \mathrm{Cr}\,(p)$ *such that* $\overline{\langle \mathrm{Rot}\,_\infty, \gamma_p \rangle} = \mathrm{Cr}\,(p)$ *and* $\mathrm{Fix}(\gamma_p) = \{z_p, -z_p\}$.

(v) *For each* $p < 0$ *one has that* $\mathrm{Cr}\,(p)$ *is conjugate to* $\mathrm{SO}(3)$.

Now we have:

Theorem 5.3.7. *Let* Σ *be an infinite subgroup of* $\mathrm{PSL}(2, \mathbb{C})$ *whose equicontinuity region is the whole sphere. Then* Σ *is conjugate to a subgroup of either* Dih_∞ *or* $\mathrm{SO}(3)$.

Proof. Assume that $Cr(p)$ is not conjugate to a subgroup of Dih_∞. Then we claim that Σ contains an elliptic element γ_1 with infinite order. If this is not the case we have by Proposition 5.1.7 that Σ is a purely elliptic group, where each element has finite order. Now, by Selberg's lemma and the classification of the finite groups of $\mathrm{PSL}(2, \mathbb{C})$, Theorem 5.3.1, we deduce that the collection

$$\mathcal{B} = \{\langle A \rangle \mid A \text{ is a nonempty finite subset of } \Sigma\},$$

is an infinite set where each element is either a cyclic or a dihedral group. From this and (i) of Lemma 5.3.2 we deduce that Σ is conjugate to a subgroup of Dih_∞, which is a contradiction.

Now, by Lemma 5.3.4 and statement (i) in Lemma 5.3.2, we deduce that if $\mathrm{Fix}(\gamma) = \mathrm{Fix}(\gamma_1)$ for each element $\gamma \in \Sigma$ with $o(\gamma_1) = \infty$, then $\gamma(\mathrm{Fix}(\gamma_1)) = \mathrm{Fix}(\gamma_1)$ for each $\gamma \in \Sigma$ with $o(\gamma) < \infty$. Hence Σ is conjugate to a subgroup of Dih_∞. Thus we may assume there is an element γ_2 with infinite order and such that $\mathrm{Fix}(\gamma_1) \neq \mathrm{Fix}(\gamma_2)$. Then by Proposition 5.1.7 and Lemmas 5.3.3 and 5.3.4, we can assume that $\mathrm{Fix}(\gamma_2) = \{0, \infty\}$ and $\mathrm{Fix}(\gamma_2) = \{1, p\}$ where $p < 0$. Thence $\langle \gamma_1, \gamma_2 \rangle = \mathrm{Cr}\,(p)$. Therefore by property (v) of Example 5.2.5 we can assume that $p = -1$. Finally, if $\gamma_3 \in \Sigma$ is another element, by Proposition 5.1.7, the proof of property (ii) of $\mathrm{Cr}\,$, the property (iii) of Example 5.2.5 and statement (ii) of Lemma 5.3.2, we deduce that there is $z \in \mathbb{C}$ such that $\mathrm{Fix}(\gamma_3) = \{z, \bar{z}^{-1}\}$. Hence the proof of property (ii) of Example 5.2.5 says that $\gamma_3 \in \mathrm{Cr}\,$. Thence Σ is conjugate to a subgroup of $\mathrm{Cr}\,$. $\square$

From the proof of Theorem 5.3.7 one gets:

Corollary 5.3.8. *Let* $\Sigma \subset \mathrm{PSL}(2, \mathbb{C})$ *be an infinite closed group. Then* Σ *is purely elliptic if and only if* $\mathrm{Eq}\,(\Sigma) = \mathbb{P}^1_{\mathbb{C}}$.

5.3.2 Classification of elementary groups

To finish this section we give the classification of the elementary nondiscrete groups in $PSL(2, \mathbb{C})$. Recall that elementary means that its limit set consists of at most two points, or equivalently that its equicontinuity region misses at most two points. The proof of Theorem 5.3.9 follows easily from Proposition 5.1.7 and Corollary 5.3.8.

Theorem 5.3.9. *Let* $\Sigma \subset PSL(2, \mathbb{C})$ *be an elementary nondiscrete subgroup. Then up to conjugation by a projective transformation one has:*

(i) *The set* $Eq(\Sigma)$ *is* $\mathbb{P}^1_{\mathbb{C}}$ *if and only if* Σ *is purely elliptic (contained in* Dih_∞ *or* $SO(3)$*).*

(ii) *The set* $Eq(\Sigma)$ *is* $\mathbb{C}$ *if and only if* Σ *is conjugate to a subgroup* Σ_* *of* $Epa(\mathbb{C})$ *such that* $\overline{\Sigma_*}$ *contains a parabolic element, where* $Epa(\mathbb{C})$ *is as in Example 5.2.4.*

(iii) *The set* $Eq(\Sigma)$ *is* $\mathbb{C}^*$ *if and only if* Σ *is conjugate to a subgroup* Σ_* *of* $M\ddot{o}b(\mathbb{C}^*)$ *such that* Σ_* *contains a loxodromic element.*

Proof. The proof of (i) follows from Theorem 5.3.7. Let us show (ii). If $\Lambda_{Gr}(\Sigma)$ has exactly one point, then after conjugating with a Möbius transformation we can assume that $\Lambda_{Gr}(\Sigma) = \{\infty\}$. Since loxodromic elements contribute with two points to the limit set we conclude that Σ has only parabolic or elliptic elements, each one leaving ∞ fixed. By the definition of $Epa(\mathbb{C})$ this concludes the proof.

Let us show (iii). If $\Lambda_{Gr}(\Sigma)$ has exactly two points, then after conjugating with a Möbius transformation we can assume that it has exactly one fixed point p and any other point has infinity orbit whose closure is p. Then Σ has only loxodromic and parabolic points where Λ_{Gr} is the fixed set of each loxodromic element, and elliptic elements act as permutation group of $\{0, \infty\}$. By the definition of $M\ddot{o}b(\mathbb{C}^*)$ this concludes the proof. □

5.4 Consequences of the classification theorem

In this section we deduce several corollaries of the previous discussion, and we prove Theorem 5.1.3.

Corollary 5.4.1. *Let* $\Sigma \subset PSL(2, \mathbb{C})$ *be nondiscrete and assume* $\mathcal{H}$ *is an infinite normal subgroup of* Σ *such that* $card(\Lambda_{Gr}(\mathcal{H})) = 0, 2$. *Then* Σ *is either elementary or conjugate to a subgroup of* Cr.

Proof. By the proof of Theorem 5.3.7 we deduce that there is an $\mathcal{H}$-invariant set $\mathcal{P}$ with $card(\mathcal{P}) = 2$ and with the following property: If $\mathcal{R}$ is another finite $\mathcal{H}$-invariant set, then $\mathcal{R} \subset \mathcal{P}$. Since $\mathcal{H}$ is a normal subgroup it follows that $\mathcal{H}g(\mathcal{P}) = g(\mathcal{P})$ for all $g \in \Sigma$, which implies $\mathcal{P} = g(\mathcal{P})$. This implies that Σ is conjugate to a subgroup of either $M\ddot{o}b(\mathbb{C}^*)$ or Dih_∞ and the result follows from Theorem 5.3.9. □

We remark that the statement above is false if $\mathrm{card}(\Lambda_{Gr}(\mathcal{H})) = 1$. In fact consider for instance the case when $\Sigma \subset \mathrm{PSL}(2,\mathbb{R})$ is the isotropy group of ∞ and $\mathcal{H}$ is the subset of Σ of parabolic elements. Then $\mathcal{H}$ is normal in Σ and Σ is neither elementary nor a subgroup of Cr.

Now we have:

Proposition 5.4.2. *Let $\gamma_1, \gamma_2 \in \mathrm{PSL}(2,\mathbb{C})$ be parabolic elements with distinct fixed points, i.e., $\mathrm{Fix}(\gamma_1) \cap \mathrm{Fix}(\gamma_2) = \emptyset$. Then $\langle \gamma_1, \gamma_2 \rangle$ contains a loxodromic element.*

Proof. Without loss of generality we can assume that 0 is the fixed point of Γ_1 and ∞ is that of Γ_2. Then these maps are of the form: $\gamma_1(z) = z + \alpha$ and $\gamma_2(z) = \frac{z}{\beta z + 1}$ for some $\alpha, \beta \in \mathbb{C}^*$. Consider the composition $\gamma_2^m \gamma_1$. Its trace satisfies

$$Tr^2(\gamma_2^m \gamma_1) = (m\alpha\beta + 2)^2,$$

so it tends to ∞ as $m \to \infty$. Then $\gamma_2^m \gamma_1$ is loxodromic for m large. $\square$

Corollary 5.4.3. *Let $\Sigma \subset \mathrm{PSL}(2,\mathbb{C})$ be a nonelementary subgroup. Then Σ contains a loxodromic element.*

Proof. Assume that Σ does not contain loxodromic elements. Then by Corollary 5.3.8 and Proposition 5.1.7 we deduce that $\overline{\Sigma}$ contains a parabolic element γ. Assume that ∞ is the unique fixed point of γ. Then by Proposition 5.4.2 we conclude that $\overline{\Sigma}\infty = \infty$. Hence every element in Σ has the form $az + b$ with $|a| = 1$. Hence $\Sigma \subset \mathrm{Epa}(\mathbb{C})$, which is a contradiction with Theorem 5.3.9, by (ii). $\square$

Using Corollary 5.4.3 and standard arguments, see [144], we can show the following corollaries. The first of these proves statement (ii) and part of statement (i) in Theorem 5.1.3.

Corollary 5.4.4. *Let $\Sigma \subset \mathrm{PSL}(2,\mathbb{C})$ be a nonelementary group. Then $\mathbb{P}^1_{\mathbb{C}} \setminus \mathrm{Eq}(\Sigma)$ is the closure of the set of fixed points of loxodromic elements.*

Corollary 5.4.5. *Let $\Sigma \subset \mathrm{PSL}(2,\mathbb{C})$ be a nonelementary subgroup and define $Ex(\Sigma) = \{z \in \mathbb{P}^1_{\mathbb{C}} \setminus \mathrm{Eq}(\Sigma) : \overline{\Sigma z} \neq \mathbb{P}^1_{\mathbb{C}} \setminus \mathrm{Eq}(\Sigma)\}$. Then $\mathrm{card}(Ex(\Sigma)) = 0, 1$. Moreover, Σ is nondiscrete if $Ex(\Sigma) \neq \emptyset$.*

Corollary 5.4.6. *Let $\Sigma \subset \mathrm{PSL}(2,\mathbb{C})$ be a subgroup and $\mathcal{C} \neq Ex(\Sigma)$ a closed Σ-invariant set. Then $\Lambda_{Gr}(\Sigma) \subset \mathcal{C}$.*

As a corollary we obtain the proof of the rest of statement (i) in Theorem 5.1.3: The limit set is the complement of the region of equicontinuity. More precisely, if $\Sigma \subset \mathrm{PSL}(2,\mathbb{C})$ is nonelementary, then

$$\mathrm{Eq}(\Sigma) = \mathbb{P}^1_{\mathbb{C}} \setminus \Lambda_{Gr}(\Sigma).$$

To prove the second statement in Theorem 5.1.3 we need the lemmas below.

Lemma 5.4.7. *Let $\Sigma \subset \mathrm{PSL}(2,\mathbb{C})$ be a purely parabolic closed Lie group with $\dim_{\mathbb{R}}(\Sigma) = 1$ and such that ∞ is a fixed point of every element in Σ. Let γ_1, γ_2 be loxodromic elements in $\mathrm{PSL}(2,\mathbb{C})$ such that $\gamma_1(\infty) = \gamma_2(\infty) = \infty$ and $\mathrm{Fix}(\gamma_2)$ is not in the orbit $\Sigma(\mathrm{Fix}(\gamma_1))$. Then*

$$\Sigma_0 = \{\gamma \in \langle \Sigma, \gamma_1, \gamma_2 \rangle : Tr^2(\gamma) = 4\}$$

is a Lie group with $\dim_{\mathbb{R}}(\Sigma) = 2$.

Proof. Assume on the contrary that $\dim_{\mathbb{R}}(\Sigma_0) \leq 1$. Since $\Sigma \subset \Sigma_0$ we have that $\dim_{\mathbb{R}}(\Sigma_0)$. Since $\gamma_i(\infty) = \infty$, for $i = 1, 2$, we can assume that the γ_i are given by:

$$\gamma_1 = \begin{pmatrix} t & 0 \\ 0 & t^{-1} \end{pmatrix} ; \; \gamma_2 = \begin{pmatrix} a & b \\ 0 & a^{-1} \end{pmatrix} .$$

Then $\gamma_2 \Sigma \gamma_2^{-1} = \{z + t^2 r : r \in \mathbb{R}\}$, $\gamma_1 \Sigma \gamma_1^{-1} = \{z + a^2 r : r \in \mathbb{R}\}$. This implies that $a^2, t^2 \in \mathbb{R}$. On the other hand, observe that for all $n \in \mathbb{Z}$ we have

$$\gamma_1^n \gamma_2 \gamma_1^{-n} \gamma_2 = \begin{pmatrix} 1 & ab(1 - t^{2n}) \\ 0 & 1 \end{pmatrix} .$$

Hence $\{z + abr : r \in \mathbb{R}\} \subset \langle \Sigma, \gamma_1, \gamma_2 \rangle$ and therefore $ab \in \mathbb{R}$, which is a contradiction since $ab(1 - a^2)^{-1} \in \mathrm{Fix}(\gamma_2)$. $\square$

Lemma 5.4.8. *Let $\Sigma \subset \mathrm{PSL}(2,\mathbb{C})$ be a connected Lie group, $g \in \Sigma$ a loxodromic element and U be a neighbourhood of g in Σ such that $hgh^{-1}g^{-1} = \mathrm{Id}$ for each $h \in U$. Then $hgh^{-1}g^{-1} = \mathrm{Id}$ for each $h \in \Sigma$.*

Proof. Let $\tau \in \Sigma$ be such that $\tau g \tau^{-1} g^{-1} = \mathrm{Id}$. Observe that $\tau g^{-1} U = W$ is an open neighbourhood of τ and for each $\kappa \in W$ there is $u \in U$ such that $\kappa = \tau g^{-1} u$. Therefore $\kappa g \kappa^{-1} g^{-1} = \tau g^{-1} u g u^{-1} g \tau^{-1} g^{-1} = \mathrm{Id}$. That is, the set $\mathrm{Comm}(g) = \{h \in \Sigma : hgh^{-1}g^{-1} = \mathrm{Id}\}$ is open. Now let $(h_n)_{n \in \mathbb{N}} \subset \mathrm{Comm}(g)$ be a sequence and $h_0 \in \Sigma$ such that $h_n \xrightarrow[n \to \infty]{} h_0$. Then $\mathrm{Id} = h_n g h_n^{-1} g^{-1} \xrightarrow[n \to \infty]{} h_0 g h_0^{-1} g^{-1}$. Therefore $\mathrm{Comm}(g)$ is closed, concluding the proof. $\square$

Now we have:

Proof of Theorem 5.1.3. Since $\mathrm{PSL}(2,\mathbb{C})$ is a Lie group we deduce that $\overline{\Sigma}$ is a Lie group (see [229]). Thus, if $\mathcal{H}$ is the connected component of the identity in $\overline{\Sigma}$, we conclude that $\mathcal{H}$ is a connected and normal subgroup of $\overline{\Sigma}$. By Theorem 5.1.5, the property (ii) of the group Cr, and Corollary 5.4.4, we have one of the following cases:

Case 1. There is exactly one point $p \in \mathbb{P}^1_{\mathbb{C}}$ such that $\mathcal{H}p = p$. Since $\mathcal{H}$ is normal this implies $\Sigma p = p$. And by Lemma 5.4.7 we conclude that there is a purely elliptic Lie group $\mathcal{K} \subset \mathcal{H}$ with $\dim_{\mathbb{R}}(\mathcal{K}) = 1$. Let $\gamma_0 \in \Sigma$ be a loxodromic element. Then by Lemma 5.4.7 we deduce that for each $\gamma \in \Sigma$ one has that $\mathrm{Fix}(\gamma) \subset \mathcal{K}(\mathrm{Fix}(\gamma_0))$.

Moreover by Corollary 5.4.6 we conclude that $\mathcal{K}(\text{Fix}(\gamma_0)) = \Lambda_{Gr}(\Sigma)$. The result now follows because $\mathcal{K}(\text{Fix}(\gamma_0))$ is a circle in $\mathbb{P}^1_{\mathbb{C}}$.

Case 2. $\text{card}(\Lambda_{Gr}(\mathcal{H})) \geq 2$ and there is a circle $\mathcal{C}$ such that $\mathcal{H}\mathcal{C} = \mathcal{C}$. In this case we can assume that $\mathcal{C} = \widehat{\mathbb{R}}$. Thus for each loxodromic element $g \in \mathcal{H}$ one has $\text{Fix}(g) \subset \widehat{\mathbb{R}}$. Let $g \in \mathcal{H}$ be a loxodromic element such that $\text{Fix}(g) \in \mathbb{R}$ and let p_1, p_2 be the fixed points of g. Then there exist a neighbourhood $W \subset \mathbb{R}^{\dim_{\mathbb{R}}(\mathcal{H})}$ of 0 and real analytic maps $a, b, c, d : W \to \mathbb{C}$ such that the map $\phi : W \to \mathcal{H}$ defined by:

$$\phi(w)(z) = \frac{a(w)z + b(w)}{c(w) + d(w)}$$

is a local chart for $\mathcal{H}$ at g, i.e., a diffeomorphism over its image, such that $\phi(0) = g$. Set $F(w, z) = \phi(w)(z) - z$, so $\partial_z F(0, p_i) = g'(p_i) - 1 \neq 0$. Then by the implicit function theorem there is a neighbourhood W_1 of 0, and continuous functions $g_i : W \to \mathbb{C}$ such that $F(w, g_i(w)) = 0$; one has $\{g_1(w), g_2(w)\} = \text{Fix}(\phi(w))$. Thus by Lemma 5.4.8 we can assume that g_1 is nonconstant and $\phi(W_1)$ contains only loxodromic elements. Hence $g_1(W_1) \subset \Lambda_{Gr}(\Sigma) \subset \mathbb{R}$. That is, $\Lambda_{Gr}(\Sigma)$ contains an open interval, which implies that $\Lambda_{Gr}(\Sigma) = \widehat{\mathbb{R}}$. $\qquad\square$

We finish this section with an easy consequence of Theorem 5.1.3. This result says that if a sequence of Möbius transformations converges uniformly on compact sets of $\text{Eq}(\Sigma)$ to a certain endomorphism g, then g is either constant or an element in $\text{PSL}(2, \mathbb{C})$ and the convergence is uniform on all of $\mathbb{P}^1_{\mathbb{C}}$.

Corollary 5.4.9. *Let $\Sigma \subset \text{PSL}(2, \mathbb{C})$ be a subgroup such that $\text{Eq}(\Sigma) \neq \emptyset$. If $(g_n)_{n \in \mathbb{N}} \subset \Sigma$ and $g_n \xrightarrow[n \to \infty]{} g$ uniformly on compact sets of $\text{Eq}(\Sigma)$, where $g : \text{Eq}(\Gamma) \to \mathbb{P}^1_{\mathbb{C}}$, then g is either a constant function $c \in \Lambda_{Gr}(\Sigma)$ or $g \in \text{PSL}(2, \mathbb{C})$ with $g_n \xrightarrow[n \to \infty]{} g$ uniformly on $\mathbb{P}^1_{\mathbb{C}}$.*

5.5 Controllable and control groups: definitions

5.5.1 Suspensions

We recall the suspension construction introduced in Chapter 3; this appears also in Chapter 9 for constructing kissing-Schottky groups in $\mathbb{P}^2_{\mathbb{C}}$.

Consider the covering map $[[\ ]]_n : \text{SL}(n, \mathbb{C}) \to \text{PSL}(n, \mathbb{C})$, let Γ be a subgroup of $\text{PSL}(n, \mathbb{C})$ and let $\widehat{\Gamma} \subset \text{SL}(n, \mathbb{C})$ be $[[\ ]]_n^{-1}(\Gamma)$. Think of $\mathbb{P}^n_{\mathbb{C}}$ as being $\mathbb{C}^n \cup \mathbb{P}^{n-1}_{\mathbb{C}}$ and let $\widetilde{\Gamma}$ be the subgroup of $\text{PSL}(n + 1, \mathbb{C})$ that acts as $\widehat{\Gamma}$ on $\mathbb{C}^n$, so its action on $\mathbb{P}^{n-1}_{\mathbb{C}}$, the hyperplane at infinity, coincides with that of Γ. We call $\widetilde{\Gamma}$ the *full suspension group* of Γ.

If we can actually lift $\Gamma \subset \text{PSL}(n, \mathbb{C})$ to a subgroup $\widehat{\Gamma} \subset \text{SL}(n, \mathbb{C})$ that intersects the kernel of $[[\ ]]_n$ only at the identity, so that $\widehat{\Gamma}$ is isomorphic to Γ, then we can define $\widetilde{\Gamma}$ to be the subgroup of $\text{PSL}(n + 1, \mathbb{C})$ that acts as $\widehat{\Gamma}$ on $\mathbb{C}^n$

and as Γ on $\mathbb{P}_{\mathbb{C}}^{n-1}$, the hyperplane at infinity. We now call $\widetilde{\Gamma}$ a *(simple) suspension group* of Γ.

From now on we restrict the discussion to subgroups of $\mathrm{PSL}(2, \mathbb{C})$, though parts of what we say actually extend to higher dimensions. Notice that given $g \in \mathrm{PSL}(2, \mathbb{C})$, its liftings to $\mathrm{SL}(2, \mathbb{C})$ are $\pm g$, so the corresponding elements in $\mathrm{PSL}(3, \mathbb{C})$ induced by the suspension construction are

$$\begin{pmatrix} g & 0 \\ 0 & 1 \end{pmatrix}; \quad \begin{pmatrix} -g & 0 \\ 0 & 1 \end{pmatrix}.$$

Each of these two elements gives a simple suspension of the cyclic group generated by g. In this case the covering $\mathrm{SL}(2, \mathbb{C}) \to \mathrm{PSL}(2, \mathbb{C})$ is two-to-one and the full suspension group can be also called a double suspension.

Now, given $\Gamma \subset \mathrm{PSL}(2, \mathbb{C})$ a discrete group with a nonempty discontinuity region, notice that there is an inclusion $\iota : \mathrm{SL}(2, \mathbb{C}) \to \mathrm{SL}(3, \mathbb{C})$ given by

$$\iota(h) = \begin{pmatrix} h & 0 \\ 0 & 1 \end{pmatrix}.$$

Then the full suspension can be regarded as

$$\widehat{\Gamma} = \{\iota(\pm\gamma) : \gamma \in \Gamma\}.$$

Also in the case where Γ can be lifted to a group $\hat{\Gamma} \subset \mathrm{SL}(2, \mathbb{C})$ which is isomorphic to Γ, the corresponding simple suspension is given by

$$\tilde{\Gamma} = \{\iota(\pm\gamma) : \gamma \in \hat{\Gamma}\}.$$

Observe that in all cases the line $\overleftrightarrow{e_1, e_2} \subset \mathbb{P}_{\mathbb{C}}^2$ is $\widehat{\Gamma}$-invariant, where e_1, e_2 represent here the points in $\mathbb{P}_{\mathbb{C}}^2$ that correspond to the points $(1, 0, 0)$ and $(0, 1, 0)$ in $\mathbb{C}^3$. The actions of the full and the simple suspensions on this line coincide with the action of Γ on $\mathbb{P}_{\mathbb{C}}^1$.

Now consider the following generalisation of the suspension construction. Let $\Gamma \subset \mathrm{PSL}(2, \mathbb{C})$ be a Kleinian group and $G \subset \mathbb{C}^*$ a discrete subgroup. *The suspension of Γ extended by the group G*, denoted by $\mathrm{Susp}(\Gamma, G)$, is the group generated by the full suspension $\widehat{\Gamma} \subset \mathrm{PSL}(3, \mathbb{C})$ and all the elements in $\mathrm{PSL}(3, \mathbb{C})$ represented by diagonal matrices having in its diagonal the vector (g, g, g^{-2}), for each $g \in G$. That is:

$$\mathrm{Susp}(\Gamma, G) = \left\langle \{\iota(h) : h \in [[\ ]]_2^{-1}(\Gamma)\}, \left\{ \begin{pmatrix} g & 0 & 0 \\ 0 & g & 0 \\ 0 & 0 & g^{-2} \end{pmatrix} : g \in G \right\} \right\rangle.$$

Observe that if G is trivial or $G = \{\pm 1\} = \mathbb{Z}_2$, then $\mathrm{Susp}(\Gamma, \mathbb{Z}_2)$ coincides with the double suspension of Γ.

The following example gives a taste of the groups one gets through the preceding definition.

Example 5.5.1. Consider the following Möbius transformations:

$$a(z) = \frac{(s+t)z - 2st}{-2z + (s+t)}, \quad b(z) = \frac{(s+t)z + 2}{2stz + (s+t)},$$

where $0 < s < t < 1$, see see [158]. Then we can check that b maps the outside of the circle C_B into the inside of C_b and a maps the outside of C_A into the inside of C_a. That is $\Gamma_{s,t} = \langle a, b \rangle$ is a Schottky group whose limit set is a Cantor set contained in the real line.

Let $G = \{2^n : \in \mathbb{Z}\}$, then the suspension of Γ extended by the group G is the group generated by

$$A = \begin{pmatrix} \pm\frac{s+t}{s-t} & \mp\frac{2st}{s-t} & 0 \\ \mp\frac{2}{s-t} & \pm\frac{s+t}{s-t} & 0 \\ 0 & 0 & 1 \end{pmatrix}, \ B = \begin{pmatrix} \pm\frac{s+t}{t-s} & \pm\frac{2}{t-s} & 0 \\ \pm\frac{2st}{t-s} & \pm\frac{s+t}{t-s} & 0 \\ 0 & 0 & 1 \end{pmatrix}, \ C = \begin{pmatrix} 2 & 0 & 0 \\ 0 & 2 & 0 \\ 0 & 0 & 2^{-1} \end{pmatrix}.$$

That is $\text{Susp}(\Gamma, G) = \langle [[A]]_3, [[B]]_3, [[C]]_3 \rangle$, where $[[\ ]]$ is the projection $\text{SL}(3, \mathbb{C}) \to \text{PSL}(3, \mathbb{C})$. Observe that since every point in the line $\overleftrightarrow{e_1, e_2}$ has infinite isotropy group, we have that $\text{Susp}(\Gamma, G)$ is not topologically conjugate to either a full or a simple suspension. In fact the same argument can be used to show that a suspension extended by an infinite group can never be topologically conjugate to either a full or a simple suspension group. Moreover, this argument, combined with Theorem 7.3.1 yields that a suspension extended by an infinite group can not be topologically conjugate to a complex hyperbolic group.

5.6 Controllable groups

We see that in all suspension constructions we have a fixed point and an action of the group on an invariant line. As we will see, this allows us to describe (or "control") the geometry and dynamics of the suspension group in terms of a lower-dimensional model. This motivates the following definition:

Definition 5.6.1. Let $\Gamma \subset \text{PSL}(3, \mathbb{C})$ be a (discrete or not) subgroup. We say Γ is a *controllable group* if there is a line ℓ and a point $p \notin \ell$ that are invariant under the action of Γ. The group $\Gamma|_\ell = G$ is called the *control group* and $K = \{h \in \Gamma : h(x) = x \text{ for all } x \in \ell\}$ is the *kernel* of Γ. We call the line ℓ the *horizon* (a name suggested to us by A. Verjovsky).

One can easily check:

Proposition 5.6.2. *Let Γ be a suspension of a group Σ (full, simple or extended), then:*

(i) *The control group of Γ is Σ.*

(ii) *If Γ is a simple suspension, then $\text{Ker}(\Gamma)$ is trivial.*

(iii) *If* Γ *is a full suspension, then* $\mathrm{Ker}(\Gamma) = \mathbb{Z}_2$.

(iv) *If* Γ *is a suspension extended by a group* G, *then*

$$
\mathrm{Ker}(\Gamma) = \left\langle \left\{ \begin{pmatrix} g & 0 & 0 \\ 0 & g & 0 \\ 0 & 0 & g^{-2} \end{pmatrix} : g \in G \right\}, \begin{pmatrix} -1 & 0 & 0 \\ 0 & -1 & 0 \\ 0 & 0 & 1 \end{pmatrix} \right\rangle.
$$

Now let us construct an example of a controllable group whose control group does not contain subgroups of finite index conjugate to a subgroup of a suspension (simple, full or extended by a group). This implies that the class of controllable groups is larger than the class of suspension groups. First we need the following definition.

Definition-Proposition 5.6.3. Let $\gamma = [[\tilde{\gamma}]]_3 \in \mathrm{PSL}(3, \mathbb{C})$, with $\tilde{\gamma} = (\gamma_{ij})_{i,j=1,3}$, be an element with a fixed point at e_3, i.e., $\gamma(e_3) = e_3$. Let us set

$$
\psi(\gamma) = \gamma_{33}^3.
$$

If $\Gamma \subset \mathrm{PSL}(3, \mathbb{C})$ is a group, $p \in \mathbb{P}_{\mathbb{C}}^2$ is fixed by Γ and $\tau \in \mathrm{PSL}(3, \mathbb{C})$ verifies $\tau(p) = e_3$, then $\Psi_p : \Gamma \to \mathbb{C}$ defined by $\Psi_p(\gamma) = \psi(\tau\gamma\tau^{-1})$ is a group morphism.

Proof. Let $\gamma = [[\tilde{\gamma}]]_3$ be as above, $\tilde{\gamma} = (\gamma_{ij})_{i,j=1,3}$. Then $\psi(\gamma)$ is the eigenvalue of $\tilde{\gamma}$ corresponding to e_3, to the cubic exponent. It follows that $\Psi(\gamma)$ does not depend on the choice of the lift $\tilde{\gamma}$.

Now let $\tau = [[\tilde{\tau}]]_3 \in \mathrm{PSL}(3, \mathbb{C})$, with $\tau = (\tau_{ij})_{i,j=1,3}$, be another element which leaves invariant e_3. Then $\tilde{\tau}(\tilde{\gamma}(e_3)) = \tau_{33}\gamma_{33}e_3$. Hence Ψ is a group morphism. To finish the proof it is enough to recall that eigenvalues are invariant under conjugation. $\qquad \square$

Remark 5.6.4. If Γ is a suspension (full or simple), then $\Psi_{e_3}(\Gamma)$ is trivial and if Γ is a suspension extended by a group G, then $\Psi_{e_3}(\Gamma) = \{g^6 : g \in G\}$. That is, the image of Ψ_{e_3} is a discrete subgroup of $\mathbb{C}^*$.

Example 5.6.5. Let $\Gamma_{\varepsilon,\lambda,s,t}$ be the group generated by the transformations

$$
A_{\lambda,s,t} = \begin{pmatrix} \lambda\varepsilon\frac{s+t}{t-s} & \lambda\varepsilon\frac{2st}{t-s} & 0 \\ \lambda\varepsilon\frac{2}{t-s} & \lambda\varepsilon\frac{s+t}{t-s} & 0 \\ 0 & 0 & (\lambda\varepsilon)^{-2} \end{pmatrix}, \quad B_{\varepsilon,s,t} = \begin{pmatrix} \varepsilon\frac{s+t}{t-s} & \varepsilon\frac{2}{t-s} & 0 \\ \varepsilon\frac{2st}{t-s} & \varepsilon\frac{s+t}{t-s} & 0 \\ 0 & 0 & \varepsilon^{-2} \end{pmatrix},
$$

where $\lambda = e^{2\pi i\theta}$, with $\theta \in \mathbb{R} - \mathbb{Q}$. Observe that $\Gamma_{\varepsilon,\lambda,s,t}$ leaves invariant e_3 and $\ell = \overleftrightarrow{e_1, e_2}$, hence $\Gamma_{\varepsilon,\lambda,s,t}$ is a controllable group. Moreover, the control group of $\Gamma_{\varepsilon,\lambda,s,t}$ is $\Gamma_{s,t}$, where $\Gamma_{s,t}$ is taken as in Example 5.5.1, and $\mathrm{Ker}(\Gamma_{\varepsilon,\lambda,s,t})$ is trivial. Furthermore, since $\Psi_{e_3}(A_{\lambda,s,t}B_{\varepsilon,s,t}) = \lambda^{-2}$, we conclude that Γ_ε is not conjugate in $\mathrm{PSL}(3, \mathbb{C})$ to a subgroup of a suspension, neither full nor simple nor extended by a group.

5.7 Groups with control

We now introduce a third class of groups, which essentially includes the previous two classes: the suspensions and the controllable groups. We call these semi-controllable groups.

Consider $\Gamma \subset \mathrm{PSL}(3, \mathbb{C})$ a (discrete or not) subgroup which acts on $\mathbb{P}_{\mathbb{C}}^2$ with a point p which is fixed by all of Γ. Choose an arbitrary line ℓ in $\mathbb{P}_{\mathbb{C}}^2 \setminus \{p\}$, and notice that we have a canonical projection

$$\pi = \pi_{p,\ell} : \mathbb{P}_{\mathbb{C}}^2 \setminus \{p\} \longrightarrow \ell,$$

given by $\pi(x) = \overleftrightarrow{x, p} \cap \ell$. It is clear that this map is holomorphic and it allows us to define a group homomorphism,

$$\Pi = \Pi_{p,\ell} : \Gamma \longrightarrow \mathrm{Bihol}(\ell) \cong \mathrm{PSL}(2, \mathbb{C}),$$

by $\Pi(g)(x) = \pi(g(x))$. If we choose another line, say ℓ', one gets similarly a projection $\pi' = \pi_{p,\ell'} : \mathbb{P}_{\mathbb{C}}^2 \setminus \{p\} \to \ell'$, and a group homomorphism $\Pi' = \Pi_{p,\ell'} : \Gamma \to \mathrm{PSL}(2, \mathbb{C})$. It is an exercise to see that Π and Π' are equivalent in the sense that there is a biholomorphism $h : \ell \to \ell'$ inducing an automorphism H of $\mathrm{PSL}(2, \mathbb{C})$ such that $H \circ \Pi = \Pi'$. As before, the line ℓ is called *the horizon*.

This leads to the following definition:

Definition 5.7.1. Let $\Gamma \subset \mathrm{PSL}(3, \mathbb{C})$ be a discrete group. We say that Γ is *weakly semi-controllable* if it acts with a fixed point in $\mathbb{P}_{\mathbb{C}}^2$. In this case a choice of a horizon ℓ determines a *control group* $\Pi(\Gamma) \subset \mathrm{PSL}(2, \mathbb{C})$, which is well-defined and independent of ℓ up to an automorphism of $\mathrm{PSL}(2, \mathbb{C})$.

It is clear that all suspension and all controllable groups are weakly semi-controllable, but this latter class is larger, as illustrated by the examples below. These are weakly semi-controllable but not controllable.

Example 5.7.2 (The fundamental groups of Inoue surfaces). These were introduced in Chapter 3, page 90. We know that there are three families of such surfaces: The S_M, S_N^+ and S_N^- families, all of these generating complex affine groups Γ that act properly discontinuously on $\mathbb{H} \times \mathbb{C}$ with compact quotient $(\mathbb{H} \times \mathbb{C})/\Gamma$.

These are all weakly semi-controllable groups, but one can prove (using the results in Chapter 8), that they are not even topologically conjugate to controllable groups.

Generally speaking, the control group $\Pi(\Gamma)$ of a weakly semi-controllable group allows us to get some information about the group Γ itself. Yet, the condition for being weakly semi-controllable (just having a fixed point) is too mild and one can not expect to gain too much information from this. The previous example leads to the following definition.

Definition 5.7.3. Let $\Gamma \subset \mathrm{PSL}(3,\mathbb{C})$ be a discrete group. We say that Γ is *semi-controllable* if there is an open set Ω in $\mathbb{P}^2_{\mathbb{C}}$ which is the largest open set on which Γ acts discontinuously, such that $\mathbb{P}^2_{\mathbb{C}} \setminus \Omega$ contains at least two lines and any three lines in $\mathbb{P}^2_{\mathbb{C}} \setminus \Omega$ are concurrent.

It is clear that every semi-controllable group has a fixed point and therefore it is weakly semi-controllable.

We refer to [41] for the proof of the following proposition:

Proposition 5.7.4. *Every nonelementary controllable group is semi-controllable.*

Notice this includes the suspensions of nonelementary Kleinian subgroups of $\mathrm{PSL}(2,\mathbb{C})$.

These concepts, and the results we prove in the rest of this chapter, will be used later in Chapter 8.

5.8   On the limit set

5.8.1   The limit set for suspensions extended by a group

We want to describe the Kulkarni limit set for the suspension groups. Recall that an element $\gamma \in \mathrm{PSL}(2,\mathbb{C})$ is *parabolic* if it has only one fixed point in $\mathbb{P}^1_{\mathbb{C}}$, and in that case it is conjugate to a translation of $\mathbb{C}$ with a fixed point at ∞.

We refer to [41] for the proof of the following theorem.

Theorem 5.8.1. *Let $\Sigma \subset \mathrm{PSL}(2,\mathbb{C})$ be a nonelementary Kleinian group with limit set $\Lambda(\Sigma) \subset \overleftrightarrow{e_1, e_2} \cong \mathbb{P}^1_{\mathbb{C}}$. Let G be a multiplicative subgroup of $\mathbb{C}^*$ and let Γ be the suspension of Σ extended by G. Let $\mathcal{C}$ be the complex cone with base $\Lambda(\Sigma)$ and vertex $e_3 = [0:0:1]$. Then:*

(i)
$$\Lambda_{\mathrm{Kul}}(\Gamma) = \left\{ \begin{array}{ll} \mathcal{C} & \text{if } G \text{ is finite,} \\ \mathcal{C} \cup \overleftrightarrow{e_1, e_2} & \text{if } G \text{ is infinite.} \end{array} \right.$$

(ii) *The set $\Omega_{\mathrm{Kul}}(\Gamma)$ is the largest open set where Γ acts discontinuously.*

(iii) *The Kulkarni region of discontinuity $\Omega_{\mathrm{Kul}}(\Gamma) := \mathbb{P}^2_{\mathbb{C}} \setminus \Lambda_{\mathrm{Kul}}(\Gamma)$ coincides with the region of equicontinuity,*

$$\Omega_{\mathrm{Kul}}(\Gamma) = \mathrm{Eq}\,(\Gamma).$$

5.8.2   A discontinuity region for some weakly semi-controllable groups

We now look at the largest class of weakly semi-controllable groups. The following result enables us to provide a region of discontinuity under certain conditions. Alas we are unable to give necessary conditions to insure that this open set coincides with the Kulkarni region of discontinuity, neither can we decide whether or not this region is maximal.

Theorem 5.8.2. *Let* $\Gamma \subset PSL(3, \mathbb{C})$ *be discrete and weakly semi-controllable, with* $p \in \mathbb{P}^2_{\mathbb{C}}$ *a* Γ*-invariant point and* ℓ *a complex line not containing* p*. Define* $\Pi = \Pi_{p,\ell} : \Gamma \longrightarrow \mathrm{Bihol}(\ell) \cong PSL(2, \mathbb{C})$ *as above. Consider the control group* $\Pi_{p,\ell}(\Gamma)$ *(which is independent of the choice of the line* ℓ *up to conjugation in* $PSL(2, \mathbb{C})$*).*

(i) *If* $\mathrm{Ker}(\Pi)$ *is finite and* $\Pi(\Gamma)$ *is discrete, then* Γ *acts properly discontinuously on*

$$\Omega = \Big(\bigcup_{z \in \Omega(\Pi(\Gamma))} \overleftrightarrow{z, p} \Big) - \{p\},$$

where $\Omega(\Pi(\Gamma)) \subset \ell \cong \mathbb{P}^1_{\mathbb{C}}$ *denotes the discontinuity set of* $\Pi(\Gamma)$*.*

(ii) *If* Γ *is controllable with an invariant line,* $\Pi(\Gamma)$ *is nondiscrete, then* Γ *acts properly discontinuously on*

$$\Omega = \bigcup_{z \in \mathrm{Eq}(\Pi(\Gamma))} \overleftrightarrow{z, p} - (\ell \cup \{p\}).$$

Proof. For (i), assume that Γ does not act discontinuously on Ω, so there is $K \subset \Omega$ a compact set such that $K(\Gamma) = \{g \in \Gamma : g(K) \cap K \neq \emptyset\}$ is infinite. Let $(g_n) \subset K(\Gamma)$ be a sequence of distinct elements. Since $\mathrm{Ker}(\Pi)$ is finite, there is a subsequence of (g_m), still denoted by (g_m), such that $\Pi(g_l) \neq \Pi(g_m)$, whenever $l \leq m$. Therefore $\{\Pi(h_n) : n \in \mathbb{N}\} \subset \{g \in \Pi(\Gamma) : g(\pi(K)) \cap \pi(K) \neq \emptyset\}$. This is a contradiction since $\Pi(\Gamma)$ acts discontinuously on $\Omega(\Pi(\Gamma))$.

Let us prove (ii). Take $p = e_3$, $l = \overleftrightarrow{e_1, e_2}$ and assume that the action is not properly discontinuous. Then there are $k = [z; h; w]$, $q \in \Omega_\Gamma$, $(k_n)_{n \in \mathbb{N}} \subset \Omega_\Gamma$ and $(g_n = (g_{ij}^{(n)})^3_{i,j=1})_{n \in \mathbb{N}} \subset \Gamma$, a sequence of distinct elements such that $k_n \xrightarrow[n \to \infty]{} k$ and $g_n(k_n) \xrightarrow[n \to \infty]{} q$. By Corollary 5.4.9 we can assume that there is a holomorphic map

$$f : \mathrm{Eq}(\Pi(\Gamma)) \longrightarrow \mathrm{Eq}(\Pi(\Gamma)),$$

such that $\Pi(g_n) \xrightarrow[n \to \infty]{} f$ uniformly on compact sets of $\mathrm{Eq}(\Pi(\Gamma))$. Moreover, either $f \in \mathrm{Bihol}(\ell)$ or f is a constant function $c \in \Lambda_{Gr}(\Pi(\Gamma))$. Since $\pi(g_n(k_n))$ converges to $\pi(q) \in \mathrm{Eq}(\Pi(\Gamma))$ as $\to \infty$, we conclude that f is not constant. Thus

$$f[z; w; 0] = [g_{11}z + g_{12}w; g_{21}z + g_{22}w; 0], \text{ with } g_{11}g_{22} - g_{12}g_{21} = 1.$$

Since $g_{13}^{(n)} = g_{23}^{(n)} = g_{31}^{(n)} = g_{32}^{(n)} = 0$, $g_{33}^{(n)}(g_{11}^{(n)} g_{22}^{(n)} - g_{12}^{(n)} g_{21}^{(n)}) = 1$ we can assume that $g_{ij}^{(n)} \sqrt{g_{33}^{(n)}} \xrightarrow[n \to \infty]{} g_{ij}$. Thus we can assume that there is $g \in \mathbb{C}^*$ such that $g_{33}^{(n)} \xrightarrow[n \to \infty]{} g \in \mathbb{C}^*$ (otherwise we may assume $g_{33}^{(n)} \xrightarrow[n \to \infty]{} 0$ or $g_{33}^{(n)} \xrightarrow[n \to \infty]{} \infty$;

this implies $g_n(k_n) \xrightarrow[n \to \infty]{} [g_{11}z + a_{12}h; g_{21}z + g_{22}h; 0]$ or $g_n(k_n) \xrightarrow[n \to \infty]{} [0; 0; 1]$, which is a contradiction). Thence,

$$g_n = \begin{pmatrix} g_{11}^{(n)} & g_{12}^{(n)} & 0 \\ g_{21}^{(n)} & g_{22}^{(n)} & 0 \\ 0 & 0 & g_{33}^{(n)} \end{pmatrix} \xrightarrow[n \to \infty]{} \begin{pmatrix} g_{11}\sqrt{g^{-1}} & g_{12}\sqrt{g^{-1}} & 0 \\ g_{21}\sqrt{g^{-1}} & g_{22}\sqrt{g^{-1}} & 0 \\ 0 & 0 & g \end{pmatrix} \in \mathrm{PSL}(3, \mathbb{C}).$$

This is a contradiction since Γ is discrete.   □

Remark 5.8.3. (See [14].) Let Γ be a semi-controllable group. Let Ω be its maximal region of discontinuity and set $\Lambda := \mathbb{P}_{\mathbb{C}}^2 \setminus \Omega$. Then the set Λ is Γ-invariant and contains infinitely many lines, all of them concurrent to a certain point, say p, which is therefore a fixed point of Γ.

Chapter 6

The Limit Set in Dimension 2

As we know already, there is no unique notion of "the limit set" for complex Kleinian groups in higher dimensions. There are instead several natural such notions, each with its own properties and characteristics, providing each a different kind of information about the geometry and dynamics of the group. The Kulkarni limit set has the property of "quasi-minimality", which is interesting for understanding the minimal invariant sets; and the action on its complement is properly discontinuous, which is useful for studying geometric properties of the group. Yet, this may not be the largest region where the action is properly discontinuous. There is also the region of equicontinuity, which provides a set where we can use the powerful tools of analysis to study the group action.

In this chapter we study, in the two-dimensional case, conditions under which all these notions coincide, as they do for one-dimensional Kleinian groups (see Theorem 6.3.6). This is interesting also from the viewpoint of having a Sullivan dictionary between Kleinian groups and iteration theory in several complex variables. In fact we recall that there is an important theorem due to J. E. Fornæss and N. Sibony (see [60]), stating that in the space of all rational maps of degree d in $\mathbb{P}^n_{\mathbb{C}}$, for $n \geq 2$, those whose Fatou set is Kobayashi hyperbolic form an open dense set with the Zariski topology. Similarly, in this chapter we see that under certain "generic" conditions, the region of equicontinuity of a complex Kleinian group in $\mathbb{P}^2_{\mathbb{C}}$ coincides with the Kulkarni region of discontinuity, and it is the largest open invariant set where the group acts properly discontinuously. And this region is Kobayashi hyperbolic.

Hence the results in this chapter, which are based on work by W. Barrera, A. Cano and J. P. Navarrete, contribute to getting a better understanding of the concept of "the limit set" in higher dimensions, and set down the first steps of a theory that points towards an analogy for Kleinian groups of the aforementioned theorem of Fornæss-Sibony.

6.1   Montel's theorem in higher dimensions

The first studies on iteration theory of holomorphic functions, due to Siegel, Cremer, Fatou, Bochner and others (see for instance[130]) were all of a local character, mostly focusing on the behaviour near the fixed points. A key step for getting global results was Montel's theorem, which provides a simple criterium granting that a given family of holomorphic maps on an open set in the Riemann sphere is a normal family.

Montel's theorem is a very powerful and useful tool for holomorphic dynamics in one complex variable, with significant applications for both, the theory of classical Kleinian groups (see for instance [146]), and for the iteration theory of rational maps, and also of more general endomorphisms of the Riemann sphere.

Theorem 6.1.1 (Montel). *Let $\Omega \subset \mathbb{P}^1_{\mathbb{C}}$ be a domain and $\mathcal{F}$ be a family of holomorphic functions defined on Ω with values in $\mathbb{P}^1_{\mathbb{C}}$. If $\bigcup_{f \in \mathcal{F}} f(\Omega)$ omits at least three points in $\mathbb{P}^1_{\mathbb{C}}$, then $\mathcal{F}$ is a normal family.*

There is a generalization of Montel's theorem to higher dimensions due to H. Cartan, see [135], [126], which has had significant applications to holomorphic dynamics in several complex variables (see for instance [60]). Notice that in Montel's theorem, points in the Riemann sphere can be regarded as being codimension 1 projective subspaces in $\mathbb{P}^1_{\mathbb{C}}$; this is the viewpoint that generalizes to higher dimensions.

Recall that $\mathbb{P}^n_{\mathbb{C}}$ can be identified with its dual $(\mathbb{P}^n_{\mathbb{C}})^* := Gr(n, n+1)$, which is the Grassmanian of n-planes in $\mathbb{C}^{n+1}$. In other words, a set of hyperplanes in $\mathbb{P}^n_{\mathbb{C}}$ can be also regarded as being points in $(\mathbb{P}^n_{\mathbb{C}})^*$. We may thus define:

Definition 6.1.2. Let H be a (possibly infinite) set of hyperplanes in $\mathbb{P}^n_{\mathbb{C}}$. We say that these hyperplanes are in general position if for each subset $\widetilde{H}$ in H with at most k elements, $k \leq n+1$, one has that the dimension of the projective subspace in $(\mathbb{P}^n_{\mathbb{C}})^*$ generated by $\widetilde{H}$ is $k-1$.

Now we can state Cartan's generalization of Montel's theorem:

Theorem 6.1.3 (H. Cartan). *Let $\Omega \subset \mathbb{P}^n_{\mathbb{C}}$ be a domain and $\mathcal{F}$ be a family of holomorphic functions defined in Ω with values in $\mathbb{P}^n_{\mathbb{C}}$. If $\bigcup_{f \in \mathcal{F}} f(\Omega)$ omits at least $2n+1$ hyperplanes in general position in $\mathbb{P}^n_{\mathbb{C}}$, then $\mathcal{F}$ is a normal family.*

It is worth saying that this theorem can be also stated in terms of algebraic curves in the projective space (see [60]).

Of course, if $n = 1$, we are back in Montel's theorem, granting that a family that omits at least three points is a normal family. In this chapter we focus on dimension 2 and in this case Cartan's theorem says that if a family of holomorphic functions omits five projective lines in general position, then the family is normal. This can be improved as follows:

Theorem 6.1.4 (Barrera-Cano-Navarrete). *Let $\mathcal{F} \subset \mathrm{PSL}(3, \mathbb{C})$ and $\Omega \subset \mathbb{P}^2_{\mathbb{C}}$ be a domain. If $\bigcup_{f \in \mathcal{F}} f(\Omega)$ omits at least three lines in general position in $\mathbb{P}^2_{\mathbb{C}}$, then $\mathcal{F}$ is a normal family in Ω.*

This result is essential for the classification theorems of two-dimensional complex Kleinian groups whose limit set has few lines, shown by W. Barrera, A. Cano and J. P. Navarrete (see the bibliography at the end). We refer to [14] for the proof of this theorem, which is based on the theory of pseudo-projective transformations introduced in [40] and explained in Chapter 7 of this monograph.

6.2 Lines and the limit set

We know from Chapter 1 that for classical Kleinian groups acting on the Riemann sphere, the equicontinuity region is also the maximal open subset in the Riemann sphere where the action is properly discontinuous. We also know that in this setting, if the limit set has finite cardinality, then it consists of at most two points. We now study the equivalent statements in the case of complex Kleinian groups in $\mathbb{P}^2_{\mathbb{C}}$.

Recall from Proposition 3.3.4, see page 85, that for complex Kleinian groups acting on higher-dimensional projective spaces, the Kulkarni limit set must contain at least one projective line. Recall too, that there are examples of Kleinian subgroups of $\mathrm{PSL}(3, \mathbb{C})$ such that their Kulkarni region of discontinuity is neither the maximal region where the action is properly discontinuous, nor coincides with the region of equicontinuity. Yet one has:

Theorem 6.2.1. *Let Γ be an infinite discrete subgroup of $\mathrm{PSL}(3, \mathbb{C})$ and let $\Omega \subset \mathbb{P}^2_{\mathbb{C}}$ be either its equicontinuity region, its Kulkarni region of discontinuity or a maximal region where the group acts properly discontinuously. Then:*

(i) *The number of lines contained in $\mathbb{P}^2_{\mathbb{C}} \setminus \Omega$ is either 1, 2, 3 or infinite.*

(ii) *The number of lines contained in $\mathbb{P}^2_{\mathbb{C}} \setminus \Omega$ lying in general position is either 1, 2, 3, 4 or infinite.*

There are examples showing that this result is best possible, in the sense that there are groups where the number of lines contained in $\mathbb{P}^2_{\mathbb{C}} \setminus \Omega$ is exactly 1, examples where this number is 2 and examples where this is 3. And similarly for the number of lines in general position. In fact it is also interesting to study and classify the complex Kleinian groups whose limit set has few lines. This is being done in a series of articles by Barrera, Cano and Navarrete (see references in the bibliography).

This theorem actually is a key step for showing that "generically", there is a well defined notion of the limit set, where "generically" means that, within the space of all complex Kleinian groups with d generators, we consider those whose Kulkarni limit set has at least three lines in general position. These turn out to form an open and dense subset with respect to a certain appropriate topology (compare with Remark 6.3.7).

6.3   The limit set is a union of complex projective lines

Let us define the following sets, which are used in the sequel:

Definition 6.3.1. Let Γ be a discrete subgroup of $\mathrm{PSL}(3, \mathbb{C})$.

(i) The set $C(\Gamma)$ is the union of the Kulkarni limit sets of all elements in Γ. That is, for each $\gamma \in \Gamma$, let $\Lambda_{\mathrm{Kul}}(\gamma)$ be the limit set of the corresponding cyclic group generated by γ, then $C(\Gamma) := \bigcup_{\gamma \in \Gamma} \Lambda_{\mathrm{Kul}}(\gamma)$.

(ii) The set $\mathcal{E}(\Gamma)$ consists of all the projective lines l for which there exists an element $\gamma \in \Gamma$ such that $l \subset \Lambda_{\mathrm{Kul}}(\gamma)$.

(iii) The set $E(\Gamma)$ is the subset of $\mathbb{P}^2_{\mathbb{C}}$ defined by $E(\Gamma) = \overline{\bigcup_{l \in \mathcal{E}(\Gamma)} l}$.

It is clear that $E(\Gamma) \subset C(\Gamma)$ and it is not hard to show that $E(\Gamma) = \bigcup_{l \in \overline{\mathcal{E}(\Gamma)}} l$.

The following lemma is used for showing that when the limit set has enough lines, the Kulkarni region of discontinuity coincides with the region of equicontinuity.

Lemma 6.3.2. *Assume that the Kulkarni limit set $\Lambda_{\mathrm{Kul}}(\Gamma)$ has at least three projective lines in general position. Then $\Lambda_{\mathrm{Kul}}(\gamma) \subset \Lambda_{\mathrm{Kul}}(\Gamma)$ for every $\gamma \in \Gamma$, and therefore $C(\Gamma) \subset \Lambda_{\mathrm{Kul}}(\Gamma)$.*

Proof. Let γ be an element in Γ, then there are three cases depending on whether γ is elliptic, parabolic or loxodromic. Assume first that the element is either parabolic or elliptic. If the Kulkarni limit set is empty, then there is nothing to prove, so we assume $\Lambda_{\mathrm{Kul}}(\Gamma) \neq \emptyset$. Then the Kulkarni limit set satisfies that:

(i) Either it is the whole space $\mathbb{P}^n_{\mathbb{C}}$; or

(ii) it consists of either a single line; or

(iii) it consists of a line and a point.

(iv) it consists of two lines

In the first case, one can show, see [39], that γ has a diagonalizable lift $\tilde{\gamma} \in SL(n+1, \mathbb{C})$ such that each of its proper values is a unitary complex number and at least one has infinite order, therefore $\Lambda(\Gamma) = \mathbb{P}^n_{\mathbb{C}}$. For the cases (ii) and (iii) we have that using the same arguments as in [161], we find that the limit set of each element in Γ necessarily shares a line with the limit set of Γ itself. In the last case, γ is loxodromic, we have several cases, but as we will see below, the only interesting one is when γ has a lift $\tilde{\gamma}$ whose normal Jordan form is

$$\begin{pmatrix} \lambda_1 & 0 & 0 \\ 0 & \lambda_2 & 0 \\ 0 & 0 & \lambda_3 \end{pmatrix}, \quad |\lambda_1| < |\lambda_2| < |\lambda_3|.$$

Let us assume that $\tilde{\gamma}$ is such a matrix. It follows that $\Lambda_{\mathrm{Kul}}(\gamma) = \overleftrightarrow{e_1, e_2} \cup \overleftrightarrow{e_2, e_3}$, and the Lambda-lemma 7.3.6 implies that

$$\overleftrightarrow{e_1, e_2} \subset \Lambda_{\mathrm{Kul}}(\Gamma) \text{ or } \overleftrightarrow{e_2, e_3} \subset \Lambda_{\mathrm{Kul}}(\Gamma).$$

Let us assume, without loss of generality, that $\overleftrightarrow{e_1, e_2} \subset \Lambda_{\mathrm{Kul}}(\Gamma)$, then there exists $l \subset \Lambda_{\mathrm{Kul}}(\Gamma)$ such that l does not pass through the point $[1 : 0 : 0]$, because $\Lambda_{\mathrm{Kul}}(\Gamma)$ contains at least three projective lines in general position. We note that the sequence of projective lines $\gamma^n(l)$ goes to the projective line $\overleftrightarrow{e_2, e_3}$ as $n \to \infty$, therefore $\overleftrightarrow{e_2, e_3} \subset \Lambda_{\mathrm{Kul}}(\Gamma)$. $\square$

The theorem below says that when the group has neither fixed points nor fixed lines, then there is a well-defined region of discontinuity.

Theorem 6.3.3. *Let $\Gamma \subset \mathrm{PSL}(3, \mathbb{C})$ be a group with four lines in general position in $\Lambda_{\mathrm{Kul}}(\Gamma)$. Then:*

(i) *The set $C(\Gamma)$ is the whole limit set: $C(\Gamma) = \Lambda_{\mathrm{Kul}}(\Gamma)$, and its complement $\Omega_{\mathrm{Kul}}(\Gamma)$ is the equicontinuity region $Eq(\Gamma)$.*

(ii) *The limit set $\Lambda_{\mathrm{Kul}}(\Gamma)$ is a union of projective lines.*

(iii) *If the group acts without fixed points or lines, then the equicontinuity set of Γ is the largest open set where the groups act properly discontinuously.*

Proof. To prove statement (i) notice that $\mathbb{P}^2_{\mathbb{C}} \setminus C(\Gamma)$ is contained in $Eq(\Gamma)$, by the theorem of Montel-Cartan. Similarly, Theorem 6.1.4 implies that we have $Eq(\Gamma) = \Omega_{\mathrm{Kul}}(\Gamma)$. The proof then follows from the quasi-minimality lemma for the Kulkarni limit set, Proposition 3.3.6.

For statement (ii) we notice that a point in $C(\Gamma)$ can be approximated by points belonging each to the limit set of some element in the group. We know, see [161], that such limit sets consist of either one line, two lines, or a line and a point, and in the latter case the corresponding isolated points are all attractors.

If the point in $C(\Gamma)$ can be approximated by points belonging each to a projective line in the limit set of some element in the group, then the limit of these lines is a union of lines and we have proved statement (ii). Otherwise the point in $C(\Gamma)$ can be approximated by points which are each an attractor for some element in the group. Then, by hypothesis we have enough lines in $C(\Gamma)$ in general position as close to the given point in $C(\Gamma)$. Thus we arrive at statement (ii).

The proof of (iii) is by inspection on the number of lines in the Kulkarni limit set. If $\Lambda_{\mathrm{Kul}}(\Gamma)$ only has one line, this line must be invariant, which is not possible by hypothesis. If $\Lambda_{\mathrm{Kul}}(\Gamma)$ has exactly two lines in general position, then the intersection point p of these lines is a fixed point of Γ, which, again, is not possible by hypothesis. Hence the group must have at least three lines in $\Lambda_{\mathrm{Kul}}(\Gamma)$. We will see in Chapter 8 that every such group necessarily is co-compact and its equicontinuity set $Eq(\Gamma)$ is the largest open set where the action is properly discontinuous. $\square$

Lemma 6.3.4. *If $\Gamma \subset \mathrm{PSL}(3,\mathbb{C})$ is a discrete group and $\mathcal{E}(\Gamma)$ contains at least four projective lines in general position, then $\overline{\mathcal{E}(\Gamma)} \subset (\mathbb{P}^2_{\mathbb{C}})^*$ is a perfect set. Hence, $C(\Gamma)$ is an uncountable union of complex projective lines.*

Proof. The proof of this lemma uses Lemma 6.3.5 below. Notice that it suffices to prove that each projective line in $\overline{\mathcal{E}(\Gamma)}$ is an accumulation line of projective lines lying in $\mathcal{E}(\Gamma)$. Furthermore, it is sufficient to prove that each projective line in $\mathcal{E}(\Gamma)$ is an accumulation line of projective lines lying in $\mathcal{E}(\Gamma)$. Let l_1 be a projective line in $\mathcal{E}(\Gamma)$, then there exists $\gamma \in \Gamma$ such that $l_1 \subset \Lambda_{\mathrm{Kul}}(\gamma)$. By Lemma 6.3.5 and given that $\mathcal{E}(\Gamma)$ contains at least four projective lines in general position, we have that there exists a projective line l in $\mathcal{E}(\Gamma)$ such that some of the sequences of distinct projective lines in $\mathcal{E}(\Gamma)$, $(\gamma^n(l))_{n\in\mathbb{N}}$ or $(\gamma^{-n}(l))_{n\in\mathbb{N}}$ go to l_1 as $n \to \infty$. $\square$

Lemma 6.3.5. *Let $\gamma \in \mathrm{PSL}(3,\mathbb{C})$ be any element. Let us assume that l_1 is a projective line contained in $\Lambda_{\mathrm{Kul}}(\gamma)$, then for every projective line l different from l_1 (except, maybe, for a family of projective lines in a pencil of projective lines), some of the sequences of distinct projective lines, $(\gamma^n(l))_{n\in\mathbb{N}}$ or $(\gamma^{-n}(l))_{n\in\mathbb{N}}$ goes to l_1, as $n \to \infty$.*

Now we arrive to the main theorem in this Chapter:

Theorem 6.3.6. *Let $\Gamma \subset \mathrm{PSL}(3,\mathbb{C})$ be an infinite discrete subgroup without fixed points nor invariant projective lines. Then one has:*

a) *$Eq(\Gamma) = \Omega_{\mathrm{Kul}}(\Gamma)$, is the maximal open set on which Γ acts properly and discontinuously. Moreover, if $\mathcal{E}(\Gamma)$ contains more than three projective lines, then every connected component of $\Omega_{\mathrm{Kul}}(\Gamma)$ is complete Kobayashi hyperbolic (compare with [16]).*

b) *The set*

$$\Lambda_{\mathrm{Kul}}(\Gamma) = \overline{\bigcup_{l\in\mathcal{E}(\Gamma)} l} = \bigcup_{l\in\overline{\mathcal{E}(\Gamma)}} l = \overline{\bigcup_{\gamma\in\Gamma} \Lambda_{\mathrm{Kul}}(\Gamma)}$$

is path-connected.

c) *If $\mathcal{E}(\Gamma)$ contains more than three projective lines, then $\overline{\mathcal{E}(\Gamma)} \subset (\mathbb{P}^2_{\mathbb{C}})^*$ is a perfect set, and it is the minimal closed Γ-invariant subset of $(\mathbb{P}^2_{\mathbb{C}})^*$.*

Proof. a) Clearly the set $\Lambda_{\mathrm{Kul}}(\Gamma) \cap C(\Gamma)$ contains at least three projective lines in general position, then $Eq(\Gamma) = \Omega_{\mathrm{Kul}}(\Gamma)$.

If $U \subset \mathbb{P}^2_{\mathbb{C}}$ is a Γ-invariant open set on which Γ acts properly and discontinuously, then there exists a projective line l contained in $\mathbb{P}^2_{\mathbb{C}} \setminus U$. Given that Γ does not have invariant projective lines nor fixed points, then $\mathbb{P}^2_{\mathbb{C}} \setminus U$ contains at least three projective lines in general position. Hence $U \subset Eq(\Gamma) = \Omega_{\mathrm{Kul}}(\Gamma)$. Therefore $\Omega_{\mathrm{Kul}}(\Gamma)$ is the maximal open set on which Γ acts properly and discontinuously. The proof that every connected component of $\Omega_{\mathrm{Kul}}(\Gamma)$ is complete Kobayashi hyperbolic is obtained by imitating the proof of Lemma 2.3 in [16], and of the main theorem in [16].

(b) The equality $\Lambda_{\mathrm{Kul}}(\Gamma) = C(\Gamma)$ follows easily. Convexity follows from the facts that two distinct projective lines always intersect and a projective line is path-connected.

(c) Lemma 6.3.4 implies that $\overline{\mathcal{E}(\Gamma)}$ is a perfect set. Now, if $\mathcal{D} \subset (\mathbb{P}^2_{\mathbb{C}})^*$ is a nonempty, closed Γ-invariant subset, then there is a projective line $l \in \mathcal{D}$ and the set $\{\gamma(l) \mid \gamma \in \Gamma\}$ contains at least three projective lines in general position. It cannot happen that $\{\gamma(l) \mid \gamma \in \Gamma\}$ contains only three projective lines because in such case $\Lambda_{\mathrm{Kul}}(\Gamma)$ would consist of only three projective lines, a contradiction. Therefore, there are more than three projective lines in the set $\{\gamma(l) \mid \gamma \in \Gamma\}$, so we obtain that $\{\gamma(l) \mid \gamma \in \Gamma\}$ contains four projective lines in general position. Therefore, $\mathcal{D}$ contains four projective lines in general position. Let γ be an element in Γ, applying Lemma 6.3.5 we deduce that every projective line contained in $\Lambda_{\mathrm{Kul}}(\gamma)$ is contained in $\mathcal{D}$, hence $\overline{\mathcal{E}(\Gamma)} \subset \mathcal{D}$. $\qquad\square$

Remark 6.3.7. The above discussion shows that for Kleinian groups in $\mathbb{P}^2_{\mathbb{C}}$ with "enough" lines in their Kulkarni limit set, the various concepts of a limit set discussed in this monograph coincide. It is natural to ask what can be said about groups with "few" lines in their limit set. The simplest cases of such groups are the cyclic groups, studied in Chapter 4, and we already know that for those groups the different notions of the limit set may not coincide. In [15] the authors study groups with exactly four lines in their limit set, and in [42] the authors give the complete classification of the groups with few lines in their limit set. It is proved that although the different notions of the limit set may not coincide, in all cases but an exceptional one, there is always a maximal region where the action is properly discontinuous, and there is a "simple" description of its complement. In that article the authors show too that except for some exceptional cases, complex Kleinian groups in dimension 2 always contain loxodromic elements. This has important consequences, as for instance the fact that the groups with "many lines" in their limit set, are Zariski dense. This relies on a certain "duality technique" introduced in [17], where the authors use Pappus' theorem in a similar way to [196], to show the existence of groups in $\mathrm{PSL}(3, \mathbb{C})$ with many lines in their limit set, which are not conjugate to groups in $\mathrm{PU}(2, 1)$.

There is yet another interesting notion of a limit set which has not been discussed in this monograph: The closure of the set of fixed lines of loxodromic elements. This is the analogue of considering, for hyperbolic groups, the closure of the fixed points of loxodromic elements. It looks likely that using the results of [42] one can study this set and show that, just as in dimension 1, its complement is an open set where the action is properly discontinuous. Moreover, in dimension 2 we should also have that generically, this set coincides with the Kulkarni region of discontinuity.

Chapter 7

On the Dynamics of Discrete Subgroups of $\mathrm{PU}(n,1)$

If G is a discrete subgroup of $\mathrm{PU}(n,1) \subset \mathrm{PSL}(n+1,\mathbb{C})$, then it acts on the complex projective space $\mathbb{P}^n_{\mathbb{C}}$ preserving the unit ball

$$\{[z_0 : z_2 : \ldots : z_n] \in P^n_{\mathbb{C}} \,|\, |z_1|^2 + \cdots + |z_n|^2 < |z_0|^2\},$$

which can be equipped with the Bergman metric and provides a model for the complex hyperbolic space $\mathbb{H}^n_{\mathbb{C}}$, with $\mathrm{PU}(n,1)$ as its group of holomorphic isometries.

In this chapter we look at the action of G globally, on all of $\mathbb{P}^n_{\mathbb{C}}$. We compare the limit set $\Lambda_{CG}(G)$ of G in the sense of Chen-Greenberg, which is a subset of $\partial \mathbb{H}^n_{\mathbb{C}}$, with Kulkarni's limit set of $\Lambda_{\mathrm{Kul}}(G)$ and with the complement of the region of equicontinuity $\mathrm{Eq}\,(G)$. This allows us to get information about the dynamics of G on all of $\mathbb{P}^n_{\mathbb{C}}$ from its behaviour on the ball $\mathbb{H}^n_{\mathbb{C}}$.

We consider first the case $n = 2$, following [161]. We prove that the equicontinuity region and the Kulkarni region of discontinuity coincide, and their complement, the Kulkarni limit set, is the union of all the complex projective lines in $\mathbb{P}^2_{\mathbb{C}}$ which are tangent to $\partial \mathbb{H}^2_{\mathbb{C}}$ at points in $\Lambda_{CG}(G)$. And if G is nonelementary (i.e., $\Lambda_{CG}(G)$ has more than two points), then $\mathrm{Eq}\,(G)$ is the largest open set on which G acts properly discontinuously. This proof uses geometric arguments from complex hyperbolic geometry.

We then look at the case $n \geq 2$ following [40]. This approach is analytic and it relies on the use of a class of maps called *pseudo-projective transformations*, introduced in [40], which provide a completion of the noncompact Lie group $\mathrm{PSL}(n+1,\mathbb{C})$. These are reminiscent of the *quasi-projective* transformations introduced by Furstenberg in [61] and studied in [1].

We prove in Section 7.5 that the complement of the region of equicontinuity $\mathrm{Eq}\,(G)$ is the union of all complex projective hyperplanes tangent to $\partial \mathbb{H}^n_{\mathbb{C}}$ at points in $\Lambda_{CG}(G)$, and G acts discontinuously on Eq (G). Furthermore, the set of accumulation points of the G-orbit of every compact set $K \subset$ Eq (G) is contained in $\Lambda_{CG}(G)$. However we have not yet been able to decide whether or not Eq (G) is the complement of $\Lambda_{\text{Kul}}(G)$ in higher dimensions.

We also have some geometric applications of our results, the main one being that if $G \subset$ PU$(n, 1)$ is discrete and such that $\Lambda_{CG}(G)$ does not lie in any k-chain, then each connected component of Eq (G) is a complete Kobayashi hyperbolic manifold, and therefore a holomorphy domain.

The methods we use in higher dimensions work also for $n = 2$, and in this case one can say more due to dimensional reasons. This leads to an alternative proof of the theorem in [161] that we mentioned above. This is done in Section 7.7.

7.1 Discrete subgroups of PU$(n, 1)$ revisited

As before, we denote by $[\ ]_n : \mathbb{C}^{n+1} \setminus \{0\} \to \mathbb{P}^n_{\mathbb{C}}$ the natural projection map. Recall that a nonempty set $H \subset \mathbb{P}^n_{\mathbb{C}}$ is said to be a projective subspace of dimension k (i.e., $\dim_{\mathbb{C}}(H) = k$) if there is a $\mathbb{C}$-linear subspace $\widetilde{H}$ of dimension $k + 1$ such that $[\widetilde{H}]_n = H$. Hyperplanes are the projective subspaces of dimension $n - 1$. Given distinct points $p, q \in \mathbb{P}^n_{\mathbb{C}}$, there is a unique projective line passing through p and q that we denote by $\overleftrightarrow{p, q}$. This line is the image under $[\ ]_n$ of a two-dimensional linear subspace of $\mathbb{C}^{n+1}$. Observe that if $n = 2$ and ℓ_1, ℓ_2 are different complex lines, then one has that $\ell_1 \cap \ell_2$ contains exactly one point.

We recall too that the group of projective automorphisms of $\mathbb{P}^n_{\mathbb{C}}$ is

$$\text{PSL}(n + 1, \mathbb{C}) := \text{GL}(n + 1, \mathbb{C})/(\mathbb{C}^*)^{n+1} \cong \text{SL}(n + 1, \mathbb{C})/\mathbb{Z}_{n+1},$$

where $(\mathbb{C}^*)^{n+1}$ is regarded as the group of diagonal matrices with a single nonzero eigenvalue, and $\mathbb{Z}_{n+1}$ is regarded as the roots of unity in $\mathbb{C}^*$. As before, we denote by $[[\ ]]_{n+1}$ the quotient map SL$(n + 1, \mathbb{C}) \to$ PSL$(n + 1, \mathbb{C})$.

As in Chapter 3, in what follows $\mathbb{C}^{n,1}$ is a copy of $\mathbb{C}^{n+1}$ equipped with a Hermitian form of signature $(n, 1)$ that we assume is given by

$$\langle u, v \rangle = \sum_{j=1}^{n} u_j \overline{v_j} - u_{n+1} \overline{v_{n+1}},$$

where $u = (u_0, u_1, \ldots, u_n)$ and $v = (v_0, v_1, \ldots, v_n)$. We let V_-, V_0 and V_+ be as before, the sets of negative, null and positive vectors, respectively. The image $[V_-]$ is a complex n-ball that serves as a model for complex hyperbolic space $\mathbb{H}^n_{\mathbb{C}}$. Its boundary $\partial \mathbb{H}^n_{\mathbb{C}}$ is the *sphere at infinity*.

We recall too, from Chapter 2, that an element in PU$(n, 1)$ is classified as:

- *elliptic*, if it has a fixed point in $\mathbb{H}^n_{\mathbb{C}}$;

- *parabolic*, if it has a unique fixed point in $\partial \mathbb{H}_\mathbb{C}^n$; and

- *loxodromic*, if it fixes a unique pair of points in $\partial \mathbb{H}_\mathbb{C}^n$.

As before, we set $\overline{\mathbb{H}}_\mathbb{C}^n = \mathbb{H}_\mathbb{C}^n \cup \partial \mathbb{H}_\mathbb{C}^n$, and recall that if G is a discrete subgroup of $\mathrm{PU}(n, 1)$, then its *region of discontinuity* in the hyperbolic space is the set $\Omega = \Omega(G)$ of all points in $\overline{\mathbb{H}}_\mathbb{C}^n$ which have a neighbourhood that intersects only finitely many copies of its G-orbit. The Chen-Greenberg limit set of G, denoted by $\Lambda_{CG}(G)$ or simply Λ_{CG}, is the set of accumulation points in $\overline{\mathbb{H}}_\mathbb{C}^n$ of orbits of points in $\mathbb{H}_\mathbb{C}^n$. We know from Chapter 2 that for nonelementary groups, this set is independent of the choice of orbit.

On the other hand, if we look at the action of G on all of $\mathbb{P}_\mathbb{C}^n$, then its *Kulkarni limit set* is $\Lambda_{\mathrm{Kul}}(G) = L_0(G) \cup L_1(G) \cup L_2(G)$, where: $L_0(G)$ is the closure of the set of points in $\mathbb{P}_\mathbb{C}^n$ with infinite isotropy; the set $L_1(G)$ is the closure of the set of cluster points of orbits of points in $\mathbb{P}_\mathbb{C}^n \setminus L_0(G)$; the set $L_2(G)$ is the closure of the set of cluster points of $\{G(K)\}_{G \in G}$, where K runs over all the compact subsets of $\mathbb{P}_\mathbb{C}^n \setminus \{L_0(G) \cup L_1(G)\}$. Its complement $\Omega_{\mathrm{Kul}}(G) := \mathbb{P}_\mathbb{C}^n \setminus \Lambda_{\mathrm{Kul}}(G)$ is the *Kulkarni region of discontinuity*.

We remark again that the domain of discontinuity as defined by Kulkarni is not necessarily the maximal domain of discontinuity. However, we know by Theorem 2.6.1 that $L_0(G) \cap \mathbb{H}_\mathbb{C}^n = \emptyset = L_1(G) \cap \mathbb{H}_\mathbb{C}^n$ whenever G is discrete. Moreover we have:

Proposition 7.1.1. *If $G \subset \mathrm{PU}(n, 1)$ is a discrete subgroup, then $L_2(G) \cap \mathbb{H}^2 = \emptyset$, and therefore $\mathbb{H}_\mathbb{C}^n \subset \Omega_{\mathrm{Kul}}(G)$.*

Proof. Let x be a point in $\mathbb{H}_\mathbb{C}^n$ and suppose that there exists a compact set $K \subset P_\mathbb{C}^n \setminus (L_0(G) \cup L_1(G))$, such that x is a cluster point of the family of compact sets $\{g(K)\}_{g \in G}$. Then, there exists a sequence g_n of distinct elements of G and a sequence k_n of elements in K such that $k_n \to k \in K$ and $g_n(k_n) \to x \in \mathbb{H}_\mathbb{C}^n$. Furthermore, we can suppose that $(k_n) \subset \mathbb{H}_\mathbb{C}^n$ and then $k \in \overline{\mathbb{H}}_\mathbb{C}^n$. If k is a point in $\mathbb{H}_\mathbb{C}^n$, then there exists a subsequence of $(g_n(k))$ which converges to $q \in \overline{\mathbb{H}}_\mathbb{C}^n$; this implies that $q \in \partial \mathbb{H}_\mathbb{C}^n$, because $L_1(G) \cap \mathbb{H}_\mathbb{C}^n = \emptyset$.

Now, using the Bergman metric, the length of the geodesic segment connecting k_n to k is equal to the length of the geodesic segment connecting $g_n(k_n)$ to $g_n(k)$. As $n \to \infty$ this length tends to the length of the geodesic segment connecting x to q, which is equal to infinity. This contradicts $k_n \to k$, so $k \in \partial \mathbb{H}_\mathbb{C}^n$.

By [45, Lemma 4.3.1] or Lemma 7.4.4 below, we can suppose, taking a subsequence if needed, that g_n^{-1} converges uniformly in compact subsets of $\mathbb{H}_\mathbb{C}^n$ to a constant function with value $p \in \partial \mathbb{H}_\mathbb{C}^n$. In particular, taking the compact subset of $\mathbb{H}_\mathbb{C}^n$ given by $\{g_n(k_n)\} \cup \{x\}$ we have that $k_n = g_n^{-1}(g_n(k_n)) \to p$, which means that $k = p \in L_1(G)$, a contradiction. Hence $L_2(G) \cap \mathbb{H}_\mathbb{C}^n = \emptyset$. $\qquad\square$

7.2   Some properties of the limit set

In this section we review some properties of the Chen-Greenberg limit set $\Lambda_{CG}(G)$ for subgroups of $\mathrm{PU}(2, 1)$ that we use in the sequel. Most of these statements hold also in higher dimensions and they were essentially discussed in the last section of Chapter 2. We include them here for completeness and to have these stated in exactly the way we use them.

Recall from Chapter 2 that if $z_1, z_2 \in \mathbb{H}^2_{\mathbb{C}}$ are two distinct points, then the *bisector* $\mathfrak{E}\{z_1, z_2\}$ is defined as the set $\{z \in \mathbb{H}^2_{\mathbb{C}} | \rho(z_1, z) = \rho(z_2, z)\}$, where ρ is the Bergman metric. The complex geodesic $\Sigma \subset \mathbb{H}^2_{\mathbb{C}}$ spanned by z_1 and z_2 is the *complex spine* of $\mathfrak{E}\{z_1, z_2\}$. By definition, the *spine* of $\mathfrak{E}\{z_1, z_2\}$ equals $\sigma = \mathfrak{E}\{z_1, z_2\} \cap \Sigma = \{z \in \Sigma | \rho(z_1, z) = \rho(z_2, z)\}$; this is the orthogonal bisector of the geodesic segment joining z_1 and z_2 in Σ (see [67]).

The slice decomposition of the bisector $\mathfrak{E}\{z_1, z_2\}$ says that

$$\mathfrak{E}\{z_1, z_2\} = \Pi_{\Sigma}^{-1}(\sigma),$$

where $\Pi_{\Sigma} : \mathbb{H}^2_{\mathbb{C}} \to \Sigma$ is the orthogonal projection onto Σ (see [63], [154]). This implies that the half-space $\{z \in \mathbb{H}^2_{\mathbb{C}} | \rho(z, z_1) \geq \rho(z, z_2)\}$ is equal to the set $\Pi_{\Sigma}^{-1}\{z \in \Sigma | \rho(z, z_1) \geq \rho(z, z_2)\}$. We use this equality to prove the following lemma, which is reminiscent of the *convergence property* for complex hyperbolic groups explained in Chapter 2 (see page 74).

Lemma 7.2.1. *Let (x_n) be a sequence of elements of $\mathbb{H}^2_{\mathbb{C}}$ such that $x_n \to q \in \partial\mathbb{H}^2_{\mathbb{C}}$. In order to simplify notation we identify $\mathbb{H}^2_{\mathbb{C}}$ with the unit ball of $\mathbb{C}^2$, and $\mathbf{0}$ with the origin of $\mathbb{C}^2$. Then:*

(i) *If S_n denotes the closed half-space $\{z \in \mathbb{H}^2_{\mathbb{C}} | \rho(z, \mathbf{0}) \geq \rho(z, x_n)\}$, and $\partial S_n \subset \partial\mathbb{H}^2_{\mathbb{C}}$ denotes its ideal boundary, then the Euclidean diameter of $S_n \cup \partial S_n$ goes to 0 as $n \to \infty$.*

(ii) *If (z_n) is a sequence such that $z_n \in S_n \cup \partial S_n$ for all $n \in \mathbb{N}$, then $z_n \to q$.*

Proof. We prove (i) first. We can assume that $x_n = (r_n, 0)$, $0 < r_n < 1$ for all n, because the elements of $U(2)$ are isometries for the Bergman metric. It follows that the complex spine, Σ_n, of $\mathfrak{E}\{\mathbf{0}, x_n\}$ is equal to the disc $\mathbb{H}^1_{\mathbb{C}} \times \{0\}$ for each n, and the orthogonal projection $\Pi_{\Sigma_n} : \mathbb{H}^2_{\mathbb{C}} \to \Sigma_n$ is given by $\Pi_{\Sigma_n}(z_1, z_2) = (z_1, 0)$. By the comments above concerning the slice decomposition we have that $S_n = \Pi_{\Sigma_n}^{-1}(\{z \in \mathbb{H}^1_{\mathbb{C}} \times \{0\} | \rho(z, \mathbf{0}) \geq \rho(z, x_n)\})$. We notice that the set $\{z \in \mathbb{H}^1_{\mathbb{C}} \times \{0\} | \rho(z, \mathbf{0}) \geq \rho(z, x_n)\}$ is contained in the set $\{(z_1, 0) \in \mathbb{H}^1_{\mathbb{C}} \times \{0\} | \mathrm{Re}(z_1) \geq m_n\}$ for each n, where m_n is the intersection point of the real axis in $\mathbb{H}^1_{\mathbb{C}} \times \{0\}$ and the spine, σ_n, of $\mathfrak{E}\{\mathbf{0}, x_n\}$. Then $S_n \subset \Pi_{\Sigma}^{-1}(\{(z_1, 0) \in \mathbb{H}^1_{\mathbb{C}} \times \{0\} | \mathrm{Re}(z_1) \geq m_n\}) = \{(z_1, z_2) \in \mathbb{H}^2_{\mathbb{C}} | \mathrm{Re}(z_1) \geq m_n\}$, therefore

$$S_n \cup \partial S_n \subset \{(z_1, z_2) \in \overline{\mathbb{H}^2_{\mathbb{C}}} | \mathrm{Re}(z_1) \geq m_n\},$$

and the Euclidean diameter of the set $\{(z_1, z_2) \in \mathbb{H}^2_{\mathbb{C}} \mid \mathrm{Re}(z_1) \geq m_n\}$ goes to zero when $n \to \infty$, because $m_n \to 1$ when $n \to \infty$. This proves (i). The proof of (ii) follows easily from (i). $\qquad\square$

Proposition 7.2.2. *If (g_n) is a sequence of distinct elements of a discrete subgroup G of $\mathrm{PU}(2, 1)$, then there exists a subsequence, still denoted by (g_n), and elements $x, y \in \Lambda_{CG}(G)$, such that $g_n(z) \to x$ uniformly on compact subsets of $\overline{\mathbb{H}}^2_{\mathbb{C}} \setminus \{y\}$.*

Proof. Discarding a finite number of terms if necessary, we assume that $g_n(\mathbf{0}) \neq \mathbf{0}$ for all n. There exists a subsequence, still denoted by g_n, such that $g_n(\mathbf{0}) \to x \in \Lambda_{CG}(G)$ and $g_n^{-1}(\mathbf{0}) \to y \in \Lambda_{CG}(G)$. Let

$$S_{g_n} = \{z \in \mathbb{H}^2_{\mathbb{C}} \mid \rho(z, \mathbf{0}) \geq \rho(z, g_n^{-1}(\mathbf{0}))\}$$

and

$$S_{g_n^{-1}} = \{z \in \mathbb{H}^2_{\mathbb{C}} \mid \rho(z, \mathbf{0}) \geq \rho(z, g_n(\mathbf{0}))\}.$$

The result now follows from statement (ii) in Proposition 7.2.1, together with the fact that the Euclidean diameters of the sets $S_{g_n} \cup \partial S_{g_n}$, $S_{g_n^{-1}} \cup \partial S_{g_n^{-1}}$ tend to zero as $n \to \infty$, and the maps g_n carry $\overline{\mathbb{H}}^2_{\mathbb{C}} \setminus (S_{g_n} \cup \partial S_{g_n})$ into $S_{g_n^{-1}} \cup \partial S_{g_n^{-1}}$. $\quad\square$

Our proof of Proposition 7.2.2 is inspired by the proof in [141] of the analogous result in real hyperbolic geometry. The sets $S_{g_n} \cup \partial S_{g_n}$ play a role similar to that of the isometric circle. In fact, in [67] the boundary of ∂S_{g_n} in $\partial \mathbb{H}^2_{\mathbb{C}}$ is the isometric sphere of g_n with respect to $\mathbf{0} \in \mathbb{H}^2_{\mathbb{C}}$. Some mild changes in these arguments yield the following result, which is essentially contained in [103].

Corollary 7.2.3. *Let G be as above. If (g_n) is a sequence of distinct elements of G, then there exists a subsequence (g_n) and points $x, y \in \Lambda_{CG}(G)$ such that $g_n(z) \to x$ uniformly on compact subsets of $\mathbb{H}^2_{\mathbb{C}} \setminus \{y\}$, and $g_n^{-1}(z) \to y$ uniformly on compact subsets of $\mathbb{H}^2_{\mathbb{C}} \setminus \{x\}$.*

Lemma 7.2.4. *If G is nonelementary, then there exist $x, y \in \Lambda_{CG}(G)$, $x \neq y$, and a sequence (g_n) of distinct elements of G such that $g_n(z) \to x$ uniformly on compact subsets of $\mathbb{H}^2_{\mathbb{C}} \setminus \{y\}$, and $g_n^{-1}(z) \to y$ uniformly on compact subsets of $\mathbb{H}^2_{\mathbb{C}} \setminus \{x\}$.*

Proof. Let x be a point of $\Lambda_{CG}(G)$ not fixed by every element of G, see [102]. By Corollary 7.2.3, there is a sequence g_n of distinct elements of G and a point $x' \in \Lambda_{CG}(G)$, such that $g_n(z) \to x$ uniformly on compact subsets of $\mathbb{H}^2_{\mathbb{C}} \setminus \{x'\}$ and $g_n^{-1}(z) \to x'$ uniformly on compact subsets of $\mathbb{H}^2_{\mathbb{C}} \setminus \{x\}$. If $x \neq x'$ we are done. If $x = x'$, then there exists $g \in G$ such that $y = g(x') \neq x$. The elements $x, y \in \Lambda_{CG}(G)$ and the sequence $(g_n g^{-1})$ satisfy the conditions of the lemma. $\quad\square$

Proposition 7.2.5. *Let G be a discrete subgroup of $\mathrm{PU}(2, 1)$ such that $\Lambda_{CG}(G)$ contains at least two distinct points, then there exists a loxodromic element in G.*

Proof. First assume $\Lambda_{CG}(G)$ has precisely two points, then we may consider G as a classical Fuchsian group, because the complex geodesic determined by the points in $\Lambda_{CG}(G)$ is G-invariant. Moreover, $\Lambda_{CG}(G)$ agrees with the classical limit set, so G has a loxodromic element.

If G is nonelementary, then we apply Lemma 7.2.4 and use its notation. Choose two disjoint open 3-balls $D_x, D_y \subset \partial\mathbb{H}^2_\mathbb{C}$, such that $x \in D_x$, $y \in D_y$. There exists n such that $g_n(\partial\mathbb{H}^2_\mathbb{C} \setminus D_y) \subset D_x$. The Brouwer fixed point theorem implies that g_n has a fixed point in D_x and a similar reasoning implies that g_n^{-1} has a fixed point in D_y. Now it is easy to see that g_n is a loxodromic element. $\qquad\square$

7.3   Comparing the limit sets $\Lambda_{\mathrm{Kul}}(G)$ and $\Lambda_{CG}(G)$

In this section we prove the following theorem from [161]:

Theorem 7.3.1. *Let G be a discrete subgroup of* $\mathrm{PU}(2,1)$, *then:*

a) *The limit set* Λ_{CG} *satisfies*

$$\Lambda_{CG} = L_0(G) \cap \partial\mathbb{H}^2_\mathbb{C} = L_1(G) \cap \partial\mathbb{H}^2_\mathbb{C} = L_2(G) \cap \partial\mathbb{H}^2_\mathbb{C} = \Lambda_{\mathrm{Kul}}(G) \cap \partial\mathbb{H}^2_\mathbb{C}.$$

b) *The limit set* $\Lambda_{\mathrm{Kul}}(G)$ *is the union of all complex projective lines* l_z *tangent to* $\partial\mathbb{H}^2_\mathbb{C}$ *at points in* Λ_{CG};

$$\Lambda_{\mathrm{Kul}}(G) = \bigcup_{z \in \Lambda_{CG}} l_z.$$

Furthermore, if G is nonelementary, then the orbit of each line l_z, $z \in \Lambda_{CG}$, *is dense in* $\Lambda_{\mathrm{Kul}}(G)$ *(though the G-action on* $\Lambda_{\mathrm{Kul}}(G)$ *is not minimal).*

The proof we now give uses methods from complex hyperbolic geometry. We remark that this proof also shows that the Kulkarni region of discontinuity of G in $\mathbb{P}^2_\mathbb{C}$ coincides with its region of equicontinuity, i.e., $\Omega_{\mathrm{Kul}}(G) = \mathrm{Eq}\,(G)$. In Section 7.7 we give an alternative proof of Theorem 7.3.1 using analytic methods, and we prove also the claim that $\Omega_{\mathrm{Kul}}(G) = \mathrm{Eq}\,(G)$.

We first need several lemmas.

Lemma 7.3.2. *Let* $(w_n) \subset \mathbb{P}^2_\mathbb{C} \setminus \overline{\mathbb{H}}^2_\mathbb{C}$ *be a sequence such that $w_n \to w$. Denote by* Σ_n *the complex geodesic which is polar to* w_n. *Assume* (v_n) *is a sequence of points such that each v_n is in* Σ_n *for all $n \in \mathbf{N}$, and $v_n \to v$.*

a) *If $w \in \partial\mathbb{H}^2_\mathbb{C}$, then $w = v$.*

b) *If $w \in \mathbb{P}^2_\mathbb{C} \setminus \overline{\mathbb{H}}^2_\mathbb{C}$, then $v \in \Sigma \cup \partial\Sigma$, where Σ denotes the complex geodesic which is polar to w. In particular, if $v \in \partial\mathbb{H}^2_\mathbb{C}$, then $w \in l_v$, where l_v is the only complex line which is tangent to $\partial\mathbb{H}^2_\mathbb{C}$ at v.*

Proof. Let $\langle \cdot, \cdot \rangle$ denote the Hermitian product in $\mathbb{C}^3$ given by

$$\langle (x_1, x_2, x_3), (y_1, y_2, y_3) \rangle = x_1 \bar{y}_1 + x_2 \bar{y}_2 - x_3 \bar{y}_3.$$

Let $\tilde{w}_n$ (respectively $\tilde{v}_n$) be a lift of w_n (respectively v_n) to $S^5 \subset \mathbb{C}^3$. Take subsequences such that $\tilde{v}_n \to \tilde{v}$ and $\tilde{w}_n \to \tilde{w}$. The elements $\tilde{v}$ and $\tilde{w}$ are lifts of v and w, respectively, to S^5; and the equations $\langle \tilde{v}_n, \tilde{w}_n \rangle = 0$, $n \in \mathbb{N}$, imply that $\langle \tilde{v}, \tilde{w} \rangle = 0$.

In order to prove a) we assume $w \in \partial \mathbb{H}^2_{\mathbb{C}}$, then the complex line in $\mathbb{P}^2_{\mathbb{C}}$, induced by the two-dimensional complex space $\{\tilde{w}\}^\perp$, is the only complex line which is tangent to $\partial \mathbb{H}^2_{\mathbb{C}}$ at w. Given that $\tilde{v} \in \{\tilde{w}\}^\perp$ and $v \in \overline{\mathbb{H}}^2_{\mathbb{C}}$, we have that $v = w$.

For b) we assume $w \in \mathbb{P}^2_{\mathbb{C}} \setminus \overline{\mathbb{H}}^2_{\mathbb{C}}$, then the equation $\langle \tilde{v}, \tilde{w} \rangle = 0$ implies that $v \in \Sigma \cup \partial \Sigma$, where Σ denotes the complex geodesic which is polar to w. In particular, when $v \in \partial \Sigma$, the equation $\langle \tilde{v}, \tilde{w} \rangle = 0$ implies $w \in l_v$. $\square$

Lemma 7.3.3. *If G is a discrete subgroup of* $\mathrm{PU}(2,1)$*, then* $\Lambda_{CG}(G) = L_0(G) \cap \partial \mathbb{H}^2_{\mathbb{C}}$*.*

Proof. If $L_0(G) \cap \partial \mathbb{H}^2_{\mathbb{C}} = \emptyset$, then Proposition 2.1.2 implies that G is finite, so $\Lambda_{CG}(G) = \emptyset$.

If $L_0(G) \cap \partial \mathbb{H}^2_{\mathbb{C}}$ consists of a single point, then this point is fixed by the whole group G, and Proposition 7.2.2 implies that it belongs to $\Lambda_{CG}(G)$. If $\Lambda_{CG}(G)$ has more than one point, then there exists a loxodromic element in G, whose fixed points are in $L_0(G)$, a contradiction to our assumption. Therefore $\Lambda_{CG}(G) = L_0(G) \cap \partial \mathbb{H}^2_{\mathbb{C}}$.

Finally, we assume $|L_0(G) \cap \partial \mathbb{H}^2_{\mathbb{C}}| \geq 2$. Since $L_0(G) \cap \partial \mathbb{H}^2_{\mathbb{C}}$ is a G-invariant closed set, Proposition 2.6.5 implies that $\Lambda_{CG}(G) \subset L_0(G) \cap \partial \mathbb{H}^2_{\mathbb{C}}$. The converse inclusion is easily obtained from Proposition 7.2.2. $\square$

Lemma 7.3.4. *If G is a discrete subgroup of* $\mathrm{PU}(2,1)$*, then* $\Lambda_{CG}(G) = L_1(G) \cap \partial \mathbb{H}^2_{\mathbb{C}}$*.*

Proof. If G is finite, then both sets in the equality are empty. Assume G is infinite and notice that the definition of $\Lambda_{CG}(G)$ and the fact that $L_0(G) \cap \mathbb{H}^2_{\mathbb{C}} = \emptyset$ show that $\Lambda_{CG}(G) \subset L_1(G) \cap \partial \mathbb{H}^2_{\mathbb{C}}$.

Let us show the converse, i.e., that $L_1(G) \cap \partial \mathbb{H}^2_{\mathbb{C}} \subset \Lambda_{CG}(G)$. Assume that $z \in \partial \mathbb{H}^2_{\mathbb{C}}$, suppose that there exist a sequence g_n of distinct elements of G and one point $\zeta \in P^2_{\mathbb{C}} \setminus L_0(G)$, such that $g_n(\zeta) \to z$. We consider the following three cases:

1. We suppose $\zeta \in \mathbb{H}^2_{\mathbb{C}} \setminus L_0(G)$, then by definition $z \in \Lambda_{CG}(G)$.

2. If $\zeta \in \partial \mathbb{H}^2_{\mathbb{C}} \setminus L_0(G)$, then Lemma 7.3.3 implies that $\zeta \in \partial \mathbb{H}^2_{\mathbb{C}} \setminus \Lambda_{CG}(G)$. By Proposition 7.2.2, we have that $z \in \Lambda_{CG}(G)$.

3. Finally we assume $\zeta \in P^2_{\mathbb{C}} \setminus (\overline{\mathbb{H}^2_{\mathbb{C}}} \cup L_0(G))$. Let Σ denote the complex geodesic which is polar to ζ, then the complex geodesic $g_n(\Sigma)$ is polar to $g_n(\zeta)$. We take a point $x \in \Sigma$, and a subsequence of g_n, still denoted by g_n, such that $g_n(x) \to q \in \Lambda_{CG}(G)$. Lemma 7.3.2 a) implies that $z = q \in \Lambda_{CG}(G)$. $\square$

Lemma 7.3.5. *If G is a discrete subgroup of* $\mathrm{PU}(2,1)$*, then* $\Lambda_{CG}(G) = L_2(G) \cap \partial \mathbb{H}^2_{\mathbb{C}}$*.*

Proof. We assume G is infinite, because when G is finite the equality is trivial. The definition of $\Lambda_{CG}(G)$ and the fact that $(L_0(G) \cup L_1(G)) \cap \mathbb{H}^2_\mathbb{C} = \emptyset$ show that $\Lambda_{CG}(G) \subset L_2(G) \cap \partial \mathbb{H}^2_\mathbb{C}$.

Let us show the converse, i.e., that $L_2(G) \cap \partial \mathbb{H}^2_\mathbb{C} \subset \Lambda_{CG}(G)$. Let $K \subset P^2_\mathbb{C} \setminus (L_0(G) \cup L_1(G))$ be a compact set. If $z \in \partial \mathbb{H}^2_\mathbb{C}$ is a cluster point of the orbit of K, then there exists a sequence $(k_n) \subset K$, such that $k_n \to k \in K$, and a sequence of distinct elements $(g_n) \subset G$ such that $g_n(k_n) \to z$. We consider four cases:

1. If $k \in \mathbb{H}^2_\mathbb{C} \setminus (L_0(G) \cup L_1(G))$, then we can assume that $k_n \in \mathbb{H}^2_\mathbb{C}$ for all n. By Proposition 7.2.2, there exists a subsequence, g_n, and elements $x, y \in \Lambda_{CG}(G)$ such that $g_n(\cdot) \to x$ uniformly on compact subsets of $\overline{\mathbb{H}^2_\mathbb{C}} \setminus \{y\}$. Then $z = x \in \Lambda_{CG}(G)$.

2. We assume $k \in \partial \mathbb{H}^2_\mathbb{C} \setminus (L_0(G) \cup L_1(G))$ and there exists a subsequence of k_n, still denoted by k_n, such that $k_n \in \overline{\mathbb{H}^2_\mathbb{C}}$ for each n. By Lemmas 7.3.3 and 7.3.4 we have that $k \in \partial \mathbb{H}^2_\mathbb{C} \setminus \Lambda_{CG}(G)$, and Proposition 7.2.2 implies that $z \in \Lambda_{CG}(G)$.

3. We assume $k \in \partial \mathbb{H}^2_\mathbb{C} \setminus (L_0(G) \cup L_1(G)) = \partial \mathbb{H}^2_\mathbb{C} \setminus \Lambda_{CG}(G)$ and there exists a subsequence of k_n, still denoted by k_n, such that $k_n \in P^2_\mathbb{C} \setminus \overline{\mathbb{H}^2_\mathbb{C}}$ for all n. We denote by Σ_n the complex geodesic which is polar to k_n, and let x_n be an element of Σ_n. We can assume, taking a subsequence, if needed, that $x_n \to x \in \overline{\mathbb{H}^2_\mathbb{C}}$. Lemma 7.3.2 a) implies that $x = k$. By Proposition 7.2.2 we can suppose, taking subsequences if needed, that $g_n(x_n) \to q \in \Lambda_{CG}(G)$.

   The complex geodesic $g_n(\Sigma_n)$ is polar to $g_n(k_n)$, and we know that $g_n(k_n) \to z$, thus Lemma 7.3.2 a) implies that $z = q \in \Lambda_{CG}(G)$.

4. Finally, if $k \in P^2_\mathbb{C} \setminus (\overline{\mathbb{H}^2_\mathbb{C}} \cup L_0(G) \cup L_1(G))$, then we can assume, discarding a finite number of terms, that $k_n \in P^2_\mathbb{C} \setminus \overline{\mathbb{H}^2_\mathbb{C}}$ for all n. Let Σ_n (respectively Σ) be the complex geodesic which is polar to k_n (respectively k), and x_n a point in Σ_n. We assume, taking a subsequence, if needed, that $x_n \to x \in \Sigma \cup \partial \Sigma$. Also, we can choose the sequence in such a way that $x \in \Sigma \subset \mathbb{H}^2_\mathbb{C}$. Once more, we apply Proposition 7.2.2 to find subsequences of g_n and x_n such that $g_n(x_n) \to q \in \Lambda_{CG}(G)$. Finally, the complex geodesic $g_n(\Sigma_n)$ is polar to $g_n(k_n)$, and we know that $g_n(k_n) \to z$, then Lemma 7.3.2 a) implies that $z = q \in \Lambda_{CG}(G)$. $\qquad \square$

It follows immediately from Lemmas 7.3.3, 7.3.4 and 7.3.5 that $\Lambda_{CG}(G) = \Lambda_{\mathrm{Kul}}(G) \cap \partial \mathbb{H}^2_\mathbb{C}$. We have thus proved the first statement of Theorem 7.3.1.

We need a few results in order to prove statement b). The first of these is reminiscent of the λ-lemma in [167]; we use it to prove that the limit set $\Lambda_{\mathrm{Kul}}(G)$ contains a line whenever G contains a loxodromic element. As before, we denote by $\overleftrightarrow{x,y}$ the complex projective line determined by the points $x, y \in \mathbb{P}^2_\mathbb{C}$.

Lemma 7.3.6 (λ-Lemma). *Let* $g \in \mathrm{PU}(2,1)$ *be a loxodromic element with fixed points* $z_r, z_s, z_a \in P^2_\mathbb{C}$, *where* z_r *is a repelling fixed point for* g, z_s *is a saddle, and*

z_a *is an attractor. Let* $S \subset \mathbb{P}^2_{\mathbb{C}}$ *be any 3-sphere that does not contain either* z_r *or* z_s *and meets transversally the complex projective line* $\overleftrightarrow{z_r, z_s}$ *in a circle. Then the set of cluster points of the family of compact sets* $\{g^n(S)\}_{n \in \mathbb{N}}$ *is equal to the whole complex projective line* $\overleftrightarrow{z_s, z_a}$.

Proof. We can assume that, up to conjugation in $\mathrm{SL}(3, \mathbb{C})$, g is induced by a matrix in $\mathrm{SL}(3, \mathbb{C})$ of the form

$$\begin{pmatrix} \lambda_1 & 0 & 0 \\ 0 & \lambda_2 & 0 \\ 0 & 0 & \lambda_3 \end{pmatrix},$$

where $0 < |\lambda_1| < |\lambda_2| = 1 < |\lambda_3|$, so we have $z_r = [1 : 0 : 0]$, $z_s = [0 : 1 : 0]$, and $z_a = [0 : 0 : 1]$. The set of cluster points of the family $\{g^n(S)\}_{n \in \mathbb{N}}$ is contained in $\overleftrightarrow{z_s, z_a}$ because the sequence of functions $\{d(g^n(\cdot), \overleftrightarrow{z_s, z_a})\}_{n \in \mathbb{N}}$ converges to zero uniformly on S, where $d(\cdot, \cdot)$ denotes the Fubini-Study metric in $\mathbb{P}^2_{\mathbb{C}}$.

Now we want to prove that every point in $\overleftrightarrow{z_s, z_a}$ is a cluster point of the family $\{g^n(S)\}_{n \in \mathbb{N}}$. For this, we take $z \in \overleftrightarrow{z_s, z_a} \setminus \{z_a\}$ and $\epsilon > 0$. Since the sequence of functions $\{d(g^n(\cdot), \overleftrightarrow{z_s, z_a})\}_{n \in \mathbb{N}}$ converges to zero uniformly on S, there exists $N_1 > 0$ such that $d(g^n(s), \overleftrightarrow{z_s, z_a}) < \epsilon$ for every $n > N_1$ and $s \in S$.

Let $\Pi : \mathbb{P}^2_{\mathbb{C}} \setminus \{z_r\} \to \overleftrightarrow{z_s, z_a}$ be the projection given by $[x_1 : x_2 : x_3] \mapsto [0 : x_2 : x_3]$. Given that $\Pi(S)$ is a neighbourhood of z_s in $\overleftrightarrow{z_s, z_a}$, we consider a disc D in $\overleftrightarrow{z_s, z_a}$ with centre z_s such that $D \subset \Pi(S)$. The transformation g acts on $\overleftrightarrow{z_s, z_a}$ as a classical loxodromic transformation with z_s as a repelling point, therefore there exists $N_2 \in \mathbb{N}$ such that $z \in g^{N_2}(D)$.

Now, let $n > \max(N_1, N_2)$, then $g^{N_2}(D) \subset g^n(D) \subset g^n(\Pi(S))$, thus $z = g^n(\Pi(s))$ for some $s \in S$, and

$$d(g^n(s), z) = d(g^n(s), g^n(\Pi(s))) = d(g^n(s), \Pi(g^n(s))) = d(g^n(s), \overleftrightarrow{z_s, z_a}) < \epsilon.$$

Then z is a cluster point of the family $\{g^n(S)\}_{n \in \mathbb{N}}$, and we have proved that the set $\overleftrightarrow{z_s, z_a} \setminus \{z_a\}$ is contained in the (closed) set of cluster points of $\{g^n(S)\}_{n \in \mathbb{N}}$, therefore the whole complex projective line $\overleftrightarrow{z_s, z_a}$ is contained in this set of cluster points. □

Lemma 7.3.7. *If G is a discrete subgroup of $\mathrm{PU}(2, 1)$ and $g \in G$ is a loxodromic element with fixed points $z_r, z_s, z_a \in \mathbb{P}^2_{\mathbb{C}}$, where z_r is a repelling fixed point, z_s is a saddle fixed point, and z_a is an attracting fixed point, then*

$$\overleftrightarrow{z_r, z_s} \subset \Lambda_{\mathrm{Kul}}(G) \quad or \quad \overleftrightarrow{z_s, z_a} \subset \Lambda_{\mathrm{Kul}}(G).$$

Proof. Clearly $\{z_r, z_s, z_a\} \subset L_0(G)$. If $\overleftrightarrow{z_r, z_s} \subset (L_0(G) \cup L_1(G))$, then we are done. If $\overleftrightarrow{z_r, z_s} \not\subset (L_0(G) \cup L_1(G))$, then $\overleftrightarrow{z_r, z_s} \setminus (L_0(G) \cup L_1(G)) \neq \emptyset$ is an open set in $\overleftrightarrow{z_r, z_s}$. Thus there exists a 3-sphere satisfying the hypothesis of Lemma 7.3.6, therefore $\overleftrightarrow{z_s, z_a} \subset \Lambda_{\mathrm{Kul}}(G)$. □

We remark, using the notation of Lemma 7.3.7, that $\overleftrightarrow{z_r, z_s}$ (respectively $\overleftrightarrow{z_s, z_a}$) is the only complex projective line tangent to $\partial \mathbb{H}^2_{\mathbb{C}}$ at the point z_r (respectively z_a).

Therefore, when $G \subset \mathrm{PU}(2,1)$ is a discrete group containing a loxodromic element, the limit set $\Lambda_{\mathrm{Kul}}(G)$ contains at least one complex projective line tangent to $\partial \mathbb{H}_{\mathbb{C}}^2$ at a point in $\Lambda_{CG}(G)$. In the sequel we denote by l_z the only complex projective line tangent to $\partial \mathbb{H}_{\mathbb{C}}^2$ at the point $z \in \partial \mathbb{H}_{\mathbb{C}}^2$.

Lemma 7.3.8. *If G is a discrete subgroup of $\mathrm{PU}(2,1)$, then $L_0(G) \subset \bigcup_{z \in \Lambda_{CG}(G)} l_z$.*

Proof. Let ζ be a point in $\mathbb{P}_{\mathbb{C}}^2$ whose isotropy subgroup has infinite order, and consider the two possible cases: either $\zeta \in \overline{\mathbb{H}}_{\mathbb{C}}^2$ or not. In the first case Proposition 7.1.1 and Theorem 7.3.1 a) imply that $\zeta \in L_0(G) \cap \partial \mathbb{H}_{\mathbb{C}}^2 = \Lambda_{CG}(G) \subset \bigcup_{z \in \Lambda_{CG}(G)} l_z$.

If $\zeta \in \mathbb{P}_{\mathbb{C}}^2 \setminus \overline{\mathbb{H}}_{\mathbb{C}}^2$, let Σ denote the complex geodesic polar to ζ. Observe that $\partial \Sigma$ is invariant under the isotropy subgroup of ζ, so Proposition 2.6.5 implies that the (nonempty) Chen-Greenberg's limit set of this isotropy subgroup is contained in $\partial \Sigma$. Hence there exists a point $z \in \Lambda_{CG}(G)$ such that $z \in \partial \Sigma$, then $\zeta \in l_z$ for some $z \in \Lambda_{CG}(G)$. $\qquad \square$

Lemma 7.3.9. *If G is a discrete subgroup of $\mathrm{PU}(2,1)$, then $L_1(G) \subset \bigcup_{z \in \Lambda_{CG}(G)} l_z$.*

Proof. Proposition 7.1.1 implies $L_0(G) \subset \mathbb{P}_{\mathbb{C}}^2 \setminus \mathbb{H}_{\mathbb{C}}^2$. If $L_0(G) = \mathbb{P}_{\mathbb{C}}^2 \setminus \mathbb{H}_{\mathbb{C}}^2$, then $L_1(G) = \Lambda_{CG}(G) \subset \bigcup_{z \in \Lambda_{CG}(G)} l_z$, and we have finished. Therefore, we assume that $L_0(G) \subsetneq \mathbb{P}_{\mathbb{C}}^2 \setminus \mathbb{H}_{\mathbb{C}}^2$. We take $x \in \mathbb{P}_{\mathbb{C}}^2 \setminus L_0(G)$ and there are two possible cases, according to whether $x \in \overline{\mathbb{H}}_{\mathbb{C}}^2 \setminus L_0(G) = \overline{\mathbb{H}}_{\mathbb{C}}^2 \setminus \Lambda_{CG}(G)$ or not. In the first case Proposition 7.2.2 implies that the cluster points of the orbit of x are contained in $\Lambda_{CG}(G) \subset \bigcup_{z \in \Lambda_{CG}(G)} l_z$ proving the lemma in this case.

If $x \in \mathbb{P}_{\mathbb{C}}^2 \setminus (\overline{\mathbb{H}}_{\mathbb{C}}^2 \cup L_0(G))$ and g_n is a sequence of different elements of G such that $g_n(x) \to \xi \in L_1(G)$, then one has two possibilities: either $\xi \in \overline{\mathbb{H}}_{\mathbb{C}}^2$ or not. In the first case Proposition 7.1.1 implies $\xi \in \partial \mathbb{H}_{\mathbb{C}}^2$, and Theorem 7.3.1 a) implies that $\xi \in \Lambda_{CG}(G) \subset \bigcup_{z \in \Lambda_{CG}(G)} l_z$ as claimed.

If $\xi \in \mathbb{P}_{\mathbb{C}}^2 \setminus \overline{\mathbb{H}}_{\mathbb{C}}^2$, let Σ be the complex geodesic which is polar to x. Then $g_n(\Sigma)$ is the complex geodesic polar to $g_n(x)$. Let Σ' be the complex geodesic polar to ξ. Using Proposition 7.2.2, take a subsequence of g_n, still denoted by g_n, such that $g_n(z) \to p \in \Lambda_{CG}(G)$ for all $z \in \mathbb{H}_{\mathbb{C}}^2$. In particular, if $\sigma \in \Sigma$, then $g_n(\sigma) \to p$, and $g_n(\sigma) \in g_n(\Sigma)$. Thus Lemma 7.3.2 b) implies $p \in \partial \Sigma'$ and therefore $\xi \in l_p$. $\qquad \square$

Lemma 7.3.10. *If G is a discrete subgroup of $\mathrm{PU}(2,1)$, then $L_2(G) \subset \bigcup_{z \in \Lambda_{CG}(G)} l_z$.*

Proof. By Proposition 7.1.1, we have $L_0(G) \cup L_1(G) \subset \mathbb{P}_{\mathbb{C}}^2 \setminus \mathbb{H}_{\mathbb{C}}^2$. If $L_0(G) \cup L_1(G)$ is equal to $\mathbb{P}_{\mathbb{C}}^2 \setminus \mathbb{H}_{\mathbb{C}}^2$, then $L_2(G) = \Lambda_{CG}(G) \subset \bigcup_{z \in \Lambda_{CG}(G)} l_z$ and we have finished.

Assume $L_0(G) \cup L_1(G) \subsetneq \mathbb{P}_{\mathbb{C}}^2 \setminus \mathbb{H}_{\mathbb{C}}^2$. Let z be a cluster point of the orbit of the compact set $K \subset \mathbb{P}_{\mathbb{C}}^2 \setminus (L_0(G) \cup L_1(G))$, then there exists a sequence $(k_n) \subset K$, such that $k_n \to k \in K$, and a sequence of distinct elements $(g_n) \subset G$ such that $g_n(k_n) \to z$.

If $z \in \overline{\mathbb{H}^2_{\mathbb{C}}}$, then Proposition 7.1.1 and Theorem 7.3.1 a) imply that $z \in L_2(G) \cap \partial\mathbb{H}^2_{\mathbb{C}} = \Lambda_{CG}(G) \subset \bigcup_{z \in \Lambda_{CG}(G)} l_z$. Otherwise, if $z \notin \overline{\mathbb{H}^2_{\mathbb{C}}}$, then $k \notin \mathbb{H}^2_{\mathbb{C}}$, and we consider two cases:

1. If $k \in \partial\mathbb{H}^2_{\mathbb{C}} \setminus (L_0(G) \cup L_1(G))$, then $k_n \in \mathbb{P}^2_{\mathbb{C}} \setminus \overline{\mathbb{H}^2_{\mathbb{C}}}$ for almost every $n \in \mathbb{N}$, for otherwise there would exist a subsequence, still denoted by k_n, such that $k_n \in \mathbb{H}^2_{\mathbb{C}}$ for all n, and given that $g_n(k_n) \to z$ we have $z \in \mathbb{H}^2_{\mathbb{C}}$, which is a contradiction.

   We denote by Σ_n the complex geodesic which is polar to k_n. Let x_n be a point in Σ_n; we can assume, taking a subsequence if needed, that $x_n \to x \in \overline{\mathbb{H}^2_{\mathbb{C}}}$. Lemma 7.3.2 a) implies that $x = k$. By Proposition 7.2.2, we can suppose, taking subsequences if needed, that $g_n(x_n) \to q \in \Lambda_{CG}(G)$. The complex geodesic $g_n(\Sigma_n)$ is polar to $g_n(k_n)$, and we know that $g_n(k_n) \to z$, then Lemma 7.3.2 b) implies that $z \in l_q$.

2. If $k \in P^2_{\mathbb{C}} \setminus (\overline{\mathbb{H}^2_{\mathbb{C}}} \cup L_0(G) \cup L_1(G))$, then we can assume, discarding a finite number of terms, that $k_n \in P^2_{\mathbb{C}} \setminus \overline{\mathbb{H}^2_{\mathbb{C}}}$ for all n. Let Σ_n (respectively Σ) denote the complex geodesic which is polar to k_n (respectively k). Let x_n be an element of Σ_n. We assume, taking a subsequence if necessary, that $x_n \to x \in \Sigma \cup \partial\Sigma$. Also, we can choose the sequence in such a way that $x \in \Sigma \subset \mathbb{H}^2_{\mathbb{C}}$. Once more, we apply Proposition 7.2.2 to find subsequences of g_n and x_n such that $g_n(x_n) \to q \in \Lambda_{CG}(G)$. Finally, the complex geodesic $g_n(\Sigma_n)$ is polar to $g_n(k_n)$, and we know that $g_n(k_n) \to z$, then Lemma 7.3.2 b) implies that $z \in l_q$. $\square$

We are now ready to finish the proof of Theorem 7.3.1 b). First notice that Lemmas 7.3.8, 7.3.9 and 7.3.10 imply that $\Lambda_{\mathrm{Kul}}(G) \subset \bigcup_{z \in \Lambda_{CG}(G)} l_z$, so it suffices to prove the converse inclusion. If $\Lambda_{CG}(G) = \emptyset$, then G is finite and there is nothing to do. So we split the proof in three cases according to whether the cardinality of $\Lambda_{CG}(G)$ is 1, 2 or more than 2. We begin with the last case which is the most interesting.

If $|\Lambda_{CG}(G)| > 2$, then G is nonelementary. Let $g_0 \in G$ be a loxodromic element, and $p \in \Lambda_{CG}(G)$ a fixed point of g_0 such that the line l_p, tangent to $\partial\mathbb{H}^2_{\mathbb{C}}$ at p, is contained in $\Lambda_{\mathrm{Kul}}(G)$ (Lemma 7.3.7). Observe that $g(l_p) = l_{g(p)}$ for each $g \in G \subset \mathrm{PU}(2,1)$. So $\bigcup_{g \in G} l_{g(p)}$ is contained in $\Lambda_{\mathrm{Kul}}(G)$, because $\Lambda_{\mathrm{Kul}}(G)$ is a G-invariant closed set. Thus,

$$\Lambda_{\mathrm{Kul}}(G) \supset \overline{\bigcup_{g \in G} l_{g(p)}} = \overline{\bigcup_{z \in \{g(p)\}_{g \in G}} l_z} = \bigcup_{z \in \overline{\{g(p)\}}_{g \in G}} l_z = \bigcup_{z \in \Lambda_{CG}(G)} l_z.$$

Now notice that, by Proposition 2.6.5, if $l_z \subset \Lambda_{\mathrm{Kul}}(G)$, then its orbit is dense on $\Lambda_{\mathrm{Kul}}(G)$. This proves Theorem 7.3.1 b) when G is nonelementary.

Assume now that $|\Lambda_{CG}(G)| = 2$ and denote by x and y the two points in $\Lambda_{CG}(G)$. Then G contains a loxodromic element and it is not difficult to see that $L_0(G) = \{x, y, z\} = L_1(G)$, where $z = l_x \cap l_y$, and $L_2(G) = l_x \cup l_y$.

Finally consider the case $|\Lambda_{CG}(G)| = 1$ and denote by x the only point in $\Lambda_{CG}(G)$. Proposition 2.1.2 implies that there exists an element $g_0 \in G$ of infinite order which is parabolic and fixes x, and one has $\Lambda_{\mathrm{Kul}}(G) = \Lambda_{\mathrm{Kul}}(\langle g_0 \rangle) = l_x$. $\square$

Corollary 7.3.11. *Let* G, G_1, G_2 *be discrete subgroups of* $\mathrm{PU}(2,1)$. *Then:*

a) *The limit set* $\Lambda_{\mathrm{Kul}}(G)$ *is path connected.*

b) *If* $\Lambda_{CG}(G)$ *is all of* $\partial \mathbb{H}^2_{\mathbb{C}}$, *then* $\Lambda_{\mathrm{Kul}}(G) = \mathbb{P}^2_{\mathbb{C}} \setminus \mathbb{H}^2_{\mathbb{C}}$.

c) *If* $G_1 \subset G_2$, *then* $\Lambda_{\mathrm{Kul}}(G_1) \subset \Lambda_{\mathrm{Kul}}(G_2)$.

d) *If* G *is nonelementary and* W *is a* G-*invariant open set such that* G *acts properly discontinuously on* W, *then* $W \subseteq \Omega_{\mathrm{Kul}}(G)$. *In other words,* $\Omega(G)$ *is the maximal open set on which* G *acts properly discontinuously.*

Proof.   a) Let x_1, x_2 be points in $\Lambda_{\mathrm{Kul}}(G)$. By Theorem 7.3.1 b) there exist $z_1, z_2 \in \Lambda_{CG}(G)$ such that $x_1 \in l_{z_1}$ and $x_2 \in l_{z_2}$. If $z_1 = z_2$, then we can join x_1 to x_2 by a path in $l_{z_1} \subset \Lambda_{\mathrm{Kul}}(G)$. Finally, if $z_1 \neq z_2$, then we can join x_1 to x_2 by a path in $l_{z_1} \cup l_{z_2} \subset \Lambda_{\mathrm{Kul}}(G)$, because any two complex lines in $\mathbb{P}^2_{\mathbb{C}}$ have nonempty intersection.

b) This claim follows immediately from Theorem 7.3.1 b).

c) This claim follows from Theorem 7.3.1 b) and the fact that for the Chen-Greenberg limit sets one obviously has $\Lambda_{CG}(G_1) \subset \Lambda_{CG}L(G_2)$.

d) If $W \not\subseteq \Omega_{\mathrm{Kul}}(G)$, then $W \cap l_z \neq \emptyset$ for every $z \in \Lambda_{CG}(G)$ (for otherwise there exists $z_0 \in \Lambda_{CG}(G)$ such that $l_{z_0} \subset (\mathbb{P}^2_{\mathbb{C}} \setminus W)$. Theorem 7.3.1 b) and the fact that $\mathbb{P}^2_{\mathbb{C}} \setminus W$ is a G-invariant closed set imply that $\Lambda_{\mathrm{Kul}}(G) = \overline{\bigcup_{g \in G} g(l_{z_0})} \subset \mathbb{P}^2_{\mathbb{C}} \setminus W$. Hence $W \subset \Omega_{\mathrm{Kul}}(G)$, which is a contradiction to our assumption). In particular, if $g_0 \in G$ is a loxodromic element having fixed points $z_1, z_2 \in \Lambda_{CG}(G)$, then $W \cap l_{z_1} \neq \emptyset \neq W \cap l_{z_2}$. But Lemma 7.3.6 says that G does not act properly discontinuously on W, so we get a contradiction and therefore $W \subseteq \Omega_{\mathrm{Kul}}(G)$. $\square$

7.4   Pseudo-projective maps and equicontinuity

In this section we develop some machinery that we use in the sequel in order to extend to higher dimensions the previous results. For this we look at a completion of the Lie group $\mathrm{PSL}(n, \mathbb{C})$ inspired by the space of quasi-projective maps introduced in [61] and studied in [1]. We call this completion of $\mathrm{PSL}(n+1, \mathbb{C})$ the space of pseudo-projective maps.

Let $\widetilde{M} : \mathbb{C}^{n+1} \to \mathbb{C}^{n+1}$ be a nonzero linear transformation which is not necessarily invertible. Let $\mathrm{Ker}(\widetilde{M})$ be its kernel and let $\mathrm{Ker}([[\widetilde{M}]]_{n+1})$ denote its projectivization. That is, $\mathrm{Ker}([[\widetilde{M}]]_{n+1}) := [\mathrm{Ker}(\widetilde{M}) \setminus \{0\}]_n$ where $[\ ]_n$ is the projection $\mathbb{C}^{n+1} \to \mathbb{P}^n_{\mathbb{C}}$.

Then $\widetilde{M}$ induces a map $[[\widetilde{M}]]_{n+1} : \mathbb{P}^n_{\mathbb{C}} \setminus \mathrm{Ker}(M) \to \mathbb{P}^n_{\mathbb{C}}$ given by

$$[[\widetilde{M}]]_{n+1}([v]_n) = [\widetilde{M}(v)]_n .$$

This is well defined because $v \notin \mathrm{Ker}(\widetilde{M})$. Moreover, the commutative diagram below implies that $[[\widetilde{M}]]_{n+1}$ is a holomorphic map:

$$
\begin{array}{ccc}
\mathbb{C}^{n+1} \setminus \mathrm{Ker}(\widetilde{M}) & \xrightarrow{\ \widetilde{M}\ } & \mathbb{C}^{n+1} \setminus \{0\} \\
{\scriptstyle [\ ]_n} \downarrow & & \downarrow {\scriptstyle [\ ]_n} \\
\mathbb{P}^n_{\mathbb{C}} \setminus \mathrm{Ker}(M) & \xrightarrow[{[[\widetilde{M}]]_{n+1}}]{} & \mathbb{P}^n_{\mathbb{C}}
\end{array} \quad .
$$

We call the map $M = [[\widetilde{M}]]_{n+1}$ a *pseudo-projective transformation*, and we denote by $\mathrm{QP}(n+1, \mathbb{C})$ the space of all *pseudo-projective transformations* of $\mathbb{P}^n_{\mathbb{C}}$. Thence $\mathrm{QP}(n+1, \mathbb{C})$ is the space

$$\{M = [[\widetilde{M}]]_{n+1} : \widetilde{M} \text{ is a nonzero linear transformation of } \mathbb{C}^{n+1}\}.$$

Clearly $\mathrm{PSL}(n+1, \mathbb{C}) \subset \mathrm{QP}(n+1, \mathbb{C})$.

A linear map $\widetilde{M} : \mathbb{C}^{n+1} \to \mathbb{C}^{n+1}$ is said to be *a lift of* the pseudo-projective transformation M if $[[\widetilde{M}]]_{n+1} = M$. Conversely, given a pseudo-projective transformation M and a lift $\widetilde{M}$ we define *the image of* M as

$$\Im(M) = [\widetilde{M}(\mathbb{C}^{n+1}) \setminus \{0\}]_n .$$

Notice that if M is in $\mathrm{QP}(n+1, \mathbb{C}) \setminus \mathrm{PSL}(n+1, \mathbb{C})$, then

$$\dim_{\mathbb{C}}(\mathrm{Ker}(M)) + \dim_{\mathbb{C}}(\Im(M)) = n - 1 ;$$

this will be used later.

Proposition 7.4.1. *Let $(\gamma_m)_{m \in \mathbb{N}} \subset \mathrm{PSL}(n+1, \mathbb{C})$ be a sequence of distinct elements, then there is a subsequence of $(\gamma_m)_{m \in \mathbb{N}}$, still denoted by $(\gamma_m)_{m \in \mathbb{N}}$, and $\gamma \in \mathrm{QP}(n+1, \mathbb{C})$ such that $\gamma_m \xrightarrow[m \to \infty]{} \gamma$ uniformly on compact sets of $\mathbb{P}^n_{\mathbb{C}} \setminus \mathrm{Ker}(\gamma)$.*

Proof. For each m, let $\tilde{\gamma}_m = (\gamma_{ij}^{(m)}) \in \mathrm{SL}(n+1, \mathbb{C})$ be a lift of γ_m. Define

$$|\gamma_m| = \max\{|\gamma_{ij}^{(m)}| : i, j = 1, \ldots, n+1\}.$$

Notice that $|\gamma_m|$ is independent of the choice of lift in $\mathrm{SL}(n+1, \mathbb{C})$ since the projection map $\mathrm{SL}(n+1, \mathbb{C}) \to \mathrm{PSL}(n+1, \mathbb{C})$ is determined by the action of $\mathbb{Z}_{n+1}$ regarded as the roots of unity. Notice also that $|\gamma_m|^{-1}\tilde{\gamma}_m$ is again a lift of γ_m. Since every bounded sequence in $M(n+1, \mathbb{C}) = \mathbb{C}^{n^2+2n+1}$, has a convergent subsequence, we deduce that there is a subsequence of $(|\gamma_m|^{-1}\tilde{\gamma}_m)$, still denoted by $(|\gamma_m|^{-1}\tilde{\gamma}_m)$, and a nonzero $((n+1)\times(n+1))$-matrix $\tilde{\gamma} = (\gamma_{ij}) \in M(n+1,\mathbb{C})$, such that for each entry γ_{ij} we have

$$|\gamma_m|^{-1}\gamma_{ij}^{(m)} \xrightarrow[m\to\infty]{} \gamma_{ij}.$$

This implies that, regarded as linear transformations, we have the following convergence in the compact-open topology

$$|\gamma_m|^{-1}\tilde{\gamma}_m \xrightarrow[m\to\infty]{} \tilde{\gamma}. \tag{7.4.1}$$

Now, let $K \subset \mathbb{P}^n_{\mathbb{C}} - \mathrm{Ker}(\gamma)$ be a compact set and let $\widehat{K} = \{k \in \mathbb{C}^{n+1} | [k]_n \in K\} \cap \mathbb{S}^n_{\mathbb{C}}$, where $\|\ \| : \mathbb{C}^n \to \mathbb{R}$ is the usual norm and $\mathbb{S}^n_{\mathbb{C}} = \{v \in \mathbb{C}^{n+1} : \|v\| = 1\}$. Clearly $\widehat{K}$ is a compact set which satisfies $[\widehat{K}]_n = K$. On the other hand, by the convergence described in equation (7.4.1), we get the following convergence in the compact-open topology,

$$|\gamma_m|^{-1}\tilde{\gamma}_m|_{\widehat{K}} \xrightarrow[m\to\infty]{} \tilde{\gamma}|_{\widehat{K}}. \tag{7.4.1}$$

Since $[[\,|\gamma_m|^{-1}\tilde{\gamma}_m\,]]_{n+1} = \gamma_m$ and $[\widehat{K}]_n = K$, it follows that equation (7.4.1) implies the following convergence in the compact-open topology,

$$\gamma_m|_K \xrightarrow[m\to\infty]{} [\tilde{\gamma}]_n|_K. \qquad\qquad \square$$

In what follows we will say that the sequence $(\gamma_m)_{m\in\mathbb{N}} \subset \mathrm{PSL}(n+1,\mathbb{C})$ converges to $\gamma \in \mathrm{QP}(n+1,\mathbb{C})$ in the sense of pseudo-projective transformations if $\gamma_m \xrightarrow[m\to\infty]{} \gamma$ uniformly on compact sets of $\mathbb{P}^n_{\mathbb{C}} \setminus \mathrm{Ker}(\gamma)$.

Proposition 7.4.2. *Let* $(\gamma_m)_{m\in\mathbb{N}} \subset \mathrm{PSL}(n+1,\mathbb{C})$ *be a sequence which converges to* $\gamma \in \mathrm{QP}(n+1,\mathbb{C})$ *such that* $\mathrm{Ker}(\gamma)$ *is a hyperplane,* $p \in \mathrm{Ker}(\gamma) \setminus \Im(\gamma)$, U *a neighborhood of* p *and* ℓ *a line such that* $\mathrm{Ker}(\gamma) \cap \ell = \{p\}$. *Then there is a subsequence of* $(\gamma_m)_{m\in\mathbb{N}}$, *still denoted by* $(\gamma_m)_{m\in\mathbb{N}}$, *and a line* $\mathcal{L}$, *such that for every open neighborhood* W *of* p *with compact closure in* U, *the set of cluster points of* $\{\gamma_m(\overline{W} \cap \ell)\}$ *is* $\mathcal{L}$.

Proof. Since $Gr_1(\mathbb{P}^n_{\mathbb{C}})$ is compact, there is a subsequence of $(\gamma_m)_{m\in\mathbb{N}}$, still denoted by $(\gamma_m)_{m\in\mathbb{N}}$, and a line $\mathcal{L}$ such that $\gamma_m(\ell) \xrightarrow[m\to\infty]{} \mathcal{L}$. On the other hand, since the sequence $(\gamma_m)_{m\in\mathbb{N}}$ converges to γ and $\ell \cap \mathrm{Ker}(\gamma) = \{p\}$ we conclude that $\mathcal{L} \cap \Im(\gamma) = \Im(\gamma)$ is a point, say q. Let $x \in \mathcal{L} \setminus \{q\}$, then there is a sequence $(y_m)_{m\in\mathbb{N}} \subset \ell$ such that $(y_m)_{m\in\mathbb{N}}$ is convergent and $\gamma_m(y_m) \xrightarrow[m\to\infty]{} x$. This implies that the limit point of $(y_m)_{m\in\mathbb{N}}$ lies in $\mathrm{Ker}(\gamma)$, thence such a point is p. In short $y_m \xrightarrow[m\to\infty]{} p$ and $\gamma_m(y_m) \xrightarrow[m\to\infty]{} x$. $\qquad\square$

We now recall that the *equicontinuity region* for a family G of endomorphisms of $\mathbb{P}^n_{\mathbb{C}}$, denoted by $\mathrm{Eq}\,(G)$, is defined to be the set of points $z \in \mathbb{P}^n_{\mathbb{C}}$ for which there is an open neighborhood U of z such that $G|_U$ is a normal family. (Where normal family means that every sequence of distinct elements has a subsequence which converges uniformly on compact sets.)

As a consequence of Proposition 7.4.2 one has:

Corollary 7.4.3. *Let $(\gamma_m)_{m\in\mathbb{N}} \subset \mathrm{PSL}(n+1,\mathbb{C})$ be a sequence which converges to $\gamma \in \mathrm{QP}(n+1,\mathbb{C})$. If $\mathrm{Ker}(\gamma)$ is a hyperplane, then the equicontinuity set is*

$$\mathrm{Eq}\left(\{\gamma_m : m \in \mathbb{N}\}\right) = \mathbb{P}^n_{\mathbb{C}} \setminus \mathrm{Ker}(\gamma).$$

Proof. Let us assume on the contrary, that there is

$$x \in \mathrm{Ker}(\gamma) \cap \mathrm{Eq}\left(\{\gamma_m : m \in \mathbb{N}\}\right).$$

Let $y_0 \in \mathbb{P}^n_{\mathbb{C}} \setminus \mathrm{Ker}(\gamma)$ and $\mathfrak{S}(\gamma) = \{w_0\}$. Define: $\ell = \langle\{x, y_0\}\rangle$ and $U = \mathrm{Eq}\left(\{\gamma_m : m \in \mathbb{N}\}\right)$. Then $\ell \cap \mathrm{Ker}(\gamma) = x$ and U is a neighborhood of x. By Proposition 7.4.2 there is a subsequence of $(\gamma_m)_{m\in\mathbb{N}}$, still denoted by $(\gamma_m)_{m\in\mathbb{N}}$, and a line $\mathcal{L}$ such that for every open neighborhood W of x with compact closure in U, the set of cluster points of $\{\gamma_m(\overline{W} \cap \ell)\}$ is $\mathcal{L}$. Let $z_0 \in \mathcal{L} \setminus \{w_0\}$; the previous argument yields that there is a sequence $(z_m) \subset U$ such that $z_m \xrightarrow[m\to\infty]{} x$ and $\gamma_m(z_m) \xrightarrow[m\to\infty]{} z_0$.

On the other hand, there is a subsequence of (γ_m), still denoted by (γ_m), and $\tau : U \to \mathbb{P}^n_{\mathbb{C}}$ such that $(\gamma_m\,|_U)$ converges to τ with respect to the compact open topology. Therefore τ is holomorphic.

Finally, by hypotesis it follows that $\tau\,|_{U\setminus\mathrm{Ker}(\gamma)} = w_0$, so τ is constant. In particular $\gamma_m(z_m) \xrightarrow[m\to\infty]{} w_0$. Which is a contradiction. $\qquad\square$

Now we have:

Lemma 7.4.4. *Let $G \subset \mathrm{PU}(n,1)$ be a discrete group, $(\gamma_m)_{m\in\mathbb{N}} \subset G$ a sequence of distinct elements and $\gamma \in \mathrm{QP}(n+1,\mathbb{C}) \setminus \mathrm{PSL}(n+1,\mathbb{C})$ such that $(\gamma_m)_{m\in\mathbb{N}}$ converges to γ in the sense of pseudo-projective transformations. Then:*

(i) *The image $\mathfrak{S}(\gamma)$ is a point in $\partial\mathbb{H}^n_{\mathbb{C}}$.*

(ii) *$\mathrm{Ker}(\gamma)^{\perp}$ is a point in $\partial\mathbb{H}^n_{\mathbb{C}}$.*

(iii) *One has $\mathrm{Ker}(\gamma) \cap \partial\mathbb{H}^n_{\mathbb{C}} \in \Lambda_{CG}(G)$.*

Proof. Let us prove (i) by contradiction. Since γ is holomorphic the set $\gamma(\mathbb{H}^n_{\mathbb{C}} \setminus \mathrm{Ker}(\gamma))$ is a relative open set in the projective subspace $\mathfrak{S}(\gamma)$. On the other hand, by Theorem 1.2.21 the set $\gamma(\mathbb{H}^n_{\mathbb{C}} \setminus \mathrm{Ker}(\gamma))$ is contained in $\mathfrak{S}(\gamma) \cap \partial\mathbb{H}^n_{\mathbb{C}}$, which as a subspace of $\mathfrak{S}(\gamma)$ has empty interior, which is a contradiction. In the rest of the proof the unique element in $\mathfrak{S}(\gamma)$ will be denoted by q.

Now let us prove (ii). From the previous part, it follows that $\mathrm{Ker}(\gamma)$ is a hyperplane, thus to get the proof it will be enough to show $\mathrm{Ker}(\gamma) \cap \mathbb{H}^n_{\mathbb{C}} = \emptyset$ and $\mathrm{Ker}(\gamma) \cap \overline{\mathbb{H}}^n_{\mathbb{C}} \neq \emptyset$.

Claim 1. $\mathrm{Ker}(\gamma) \cap \mathbb{H}^n_{\mathbb{C}} = \emptyset$. Let us assume, on the contrary, that $\mathrm{Ker}(\gamma) \cap \mathbb{H}^n_{\mathbb{C}} \neq \emptyset$. Thus there exists $x \in \left(\mathrm{Ker}(\gamma) \setminus \mathfrak{S}(\gamma)\right) \cap \mathbb{H}^n_{\mathbb{C}}$. Applying Proposition 7.4.2 to

$(\gamma_m)_{m\in\mathbb{N}}$, γ, x and $\mathbb{H}^n_{\mathbb{C}}$, it follows that there is a line ℓ such that each point contained in ℓ is a cluster point of $\{\gamma_m(\mathbb{H}^n_{\mathbb{C}}) : m \in \mathbb{N}\}$. Since $(\gamma_m)_{m\in\mathbb{N}} \subset \Gamma$ and $\Gamma \subset \mathrm{PU}(1,n)$, it follows that $\ell \subset \overline{\mathbb{H}}^n_{\mathbb{C}}$, which is a contradiction. Therefore $\mathrm{Ker}(\gamma) \cap \mathbb{H}^n_{\mathbb{C}} = \emptyset$.

Claim 2. $\mathrm{Ker}(\gamma) \cap \overline{\mathbb{H}}^n_{\mathbb{C}} \neq \emptyset$. Assume, on the contrary, that $\mathrm{Ker}(\gamma) \cap \overline{\mathbb{H}}^n_{\mathbb{C}} = \emptyset$. From this and (i) of the present lemma we conclude that γ_m converges uniformly to the constant function q on $\overline{\mathbb{H}}^n_{\mathbb{C}}$. Let $x \in \mathbb{H}^n_{\mathbb{C}}$ and U be a neighborhood of q in $\overline{\mathbb{H}}^n$ such that $U \cap \mathbb{H}^n_{\mathbb{C}} \subset \mathbb{H}^n_{\mathbb{C}} \setminus \{x\}$. The uniform convergence implies that there is a natural number n_0 such that $\gamma_m(\overline{\mathbb{H}}^n_{\mathbb{C}}) \subset U \cap \mathbb{H}^n_{\mathbb{C}} \subset \mathbb{H}^n_{\mathbb{C}} \setminus \{x\}$ for each $m > n_0$. This is a contradiction since each γ_m is a homeomorphism.

Let us prove (iii). By Proposition 7.4.1 we can assume that there is $\tau \in \mathrm{QP}(n+1,\mathbb{C})$ such that $(\gamma_m^{-1})_{m\in\mathbb{N}}$ converges to τ in the sense of pseudo-projective transformations. Thus by (i) of the present lemma we have that $\Im(\tau)$ is a point p in $\Lambda_{CG}(G)$. We claim that $\{p\} = \Im(\tau) = \mathrm{Ker}(\gamma) \cap \partial\mathbb{H}^n_{\mathbb{C}}$. Assume this does not happen; let $x \in \mathbb{H}^n_{\mathbb{C}}$, then $\gamma_m^{-1}(x) \xrightarrow[m\to\infty]{} p$, thus $\{\gamma_m^{-1}(x) : m \in \mathbb{N}\} \cup \{p\}$ is a compact set which lies in $\mathbb{P}^n_{\mathbb{C}} \setminus \mathrm{Ker}(\gamma)$, thus $x = \gamma_m(\gamma_m^1(x)) \xrightarrow[m\to\infty]{} q$. Which is a contradiction. $\square$

From the proof of the previous result one gets:

Corollary 7.4.5. *Let* $G \subset \mathrm{PU}(1,n)$ *be a discrete group,* $(\gamma_m)_{m\in\mathbb{N}} \subset G$ *a sequence of distinct elements and* $\gamma \in \mathrm{QP}(n+1,\mathbb{C})$ *such that* $(\gamma_m)_{m\in\mathbb{N}}$ *converges to* γ *in the sense of pseudo-projective transformations. Then there are a subsequence of* $(\gamma_m)_{m\in\mathbb{N}}$*, still denoted by* $(\gamma_m)_{m\in\mathbb{N}}$*, and an element* $\tau \in \mathrm{QP}(n+1,\mathbb{C})$ *such that:*

(i) *the sequence* (γ_m^{-1}) *converges to* τ *in the sense of pseudo-projective transformations;*

(ii) *the image* $\Im(\tau)$ *and* $\mathrm{Ker}(\tau)^\perp$ *are points in* $\partial\mathbb{H}^n_{\mathbb{C}}$*;*

(iii) *one has* $\Im(\tau) = \mathrm{Ker}(\gamma) \cap \partial\mathbb{H}^n_{\mathbb{C}}$*;*

(iv) *also* $\Im(\gamma) = \mathrm{Ker}(\tau) \cap \partial\mathbb{H}^n_{\mathbb{C}}$*.*

7.5 On the equicontinuity region

Let $V \subset \mathbb{P}^n_{\mathbb{C}}$ be a proper and nonempty projective subspace, set $\widetilde{V} = \{v \in \mathbb{C}^{n+1} \setminus \{0\} | [v]_n \in V\}$, and define (its orthogonal complement):

$$V^\perp = [\{w \in \mathbb{C}^{n+1} \setminus \{0\}| < v, w >= 0 \text{ for all } w \in \widetilde{V}\}]_n.$$

When V consists of a single point $v \in \partial\mathbb{H}^n_{\mathbb{C}}$, one has that $V^\perp$ is the hyperplane tangent to $\partial\mathbb{H}^n_{\mathbb{C}}$ at v.

Now, given a discrete subgroup $G \subset \mathrm{PU}(n,1)$ let us define

$$\mathcal{C}(G) := \bigcup_{p \in \Lambda_{CG}(G)} \{p\}^{\perp}.$$

In other words, for each point $p \in \partial \mathbb{H}^n_{\mathbb{C}}$, the space $\{p\}^{\perp}$ is the projective hyperplane in $\mathbb{P}^n_{\mathbb{C}}$ tangent to partial $\mathbb{H}^n_{\mathbb{C}}$ at p. Clearly $\mathcal{C}(G)$ is a closed G-invariant subset of $\mathbb{P}^n_{\mathbb{C}}$. The following lemma proves part of Theorem 7.5.3, which is the main result in this section.

Lemma 7.5.1. *The equicontinuity region of G is*

$$\mathrm{Eq}\,(G) = \mathbb{P}^n_{\mathbb{C}} \setminus \mathcal{C}(G).$$

Proof. Since G is infinite and discrete, it contains at least a parabolic or a loxodromic element γ, by (2.1.2). Let x_0 be a fixed point of γ. By Corollary 7.4.5 we can ensure that $\{x_0\}^{\perp}$ is contained in $\mathbb{P}^n_{\mathbb{C}} \setminus \mathrm{Eq}\,(G)$. On the other hand by Proposition 2.6.5, the closure of the orbit of x_0 is $\Lambda_{CG}(G)$. Thus $\mathrm{Eq}\,(G) \subset \mathbb{P}^n_{\mathbb{C}} \setminus \mathcal{C}(G)$. Let us now prove $\mathbb{P}^n_{\mathbb{C}} \setminus \mathcal{C}(G) \subset \mathrm{Eq}\,(G)$. Let $p \in \mathbb{P}^n_{\mathbb{C}} \setminus \mathcal{C}(G)$ and $(\gamma_m)_{m \in \mathbb{N}}$ a sequence of distinct elements. By Lemma 7.4.4 there are points $p, q \in \Lambda_{CG}(G)$ and a subsequence of $(\gamma_m)_{m \in \mathbb{N}}$, still denoted by $(\gamma_m)_{m \in \mathbb{N}}$, such that $\gamma_m \xrightarrow[m \to \infty]{} q$ uniformly on compact sets of $\mathbb{P}^n_{\mathbb{C}} \setminus \{p\}^{\perp}$. This completes the proof. $\qquad \square$

Corollary 7.5.2. *Let $G \subset \mathrm{PU}(n,1)$ be a discrete subgroup. Then G acts discontinuously on $\mathrm{Eq}\,(G)$ and, moreover, for every compact set $K \subset \mathrm{Eq}\,(G)$ the cluster points of the orbit GK lie in $\Lambda_{CG}(G)$.*

Proof. Assume on the contrary that G does not act discontinuously on $\mathrm{Eq}\,(\Gamma)$. Then there is a compact set K and a sequence of distinct elements $(\gamma_m)_{m \in \mathbb{N}} \subset G$, such that $\gamma_m(K) \cap K \neq \emptyset$. By Proposition 7.4.1, there is a subsequence of $(\gamma_m)_{m \in \mathbb{N}}$, still denoted by $(\gamma_m)_{m \in \mathbb{N}}$, and $\gamma \in \mathrm{QP}(n+1, \mathbb{C})$, such that $(\gamma_m)_{m \in \mathbb{N}}$ converges to γ in the sense of pseudo-projective transformations. Moreover, by Lemma 7.4.4, $\Im(\gamma)$ is a point p in $\partial \mathbb{H}^n_{\mathbb{C}}$ and $\mathrm{Ker}(\gamma)^{\perp}$ is a point in $\partial \mathbb{H}^n_{\mathbb{C}}$. Therefore there is a neighborhood U of p disjoint from K and a natural number n_0 such that $\gamma_m(K) \subset U$ for all $m > n_0$. This implies $\gamma_m(K) \cap K = \emptyset$, which is a contradiction. Therefore Γ acts discontinuously on $\mathrm{Eq}\,(G)$.

From the previous argument we deduce also that for every compact set $K \subset \mathrm{Eq}\,(G)$ the cluster points of GK lie in $\Lambda_{CG}(G)$. $\qquad \square$

Summarising the above results, we arrive at the following theorem of [40]:

Theorem 7.5.3. *Let $G \subset \mathrm{PU}(n,1)$ be a discrete subgroup and let $\mathrm{Eq}\,(G)$ be its equicontinuity region in $\mathbb{P}^n_{\mathbb{C}}$. Then $\mathbb{P}^n_{\mathbb{C}} \setminus \mathrm{Eq}\,(G)$ is the union of all complex projective hyperplanes tangent to $\partial \mathbb{H}^n_{\mathbb{C}}$ at points in $\Lambda_{CG}(G)$, and G acts properly discontinuously on $\mathrm{Eq}\,(G)$.*

Remark 7.5.4. By the previous theorem and Proposition 3.3.6 we have $\mathrm{Eq}\,(G) \subset \Lambda_{\mathrm{Kul}}(G)$ for every discrete group $G \subset \mathrm{PU}(n,1)$.

7.6   Geometric Applications

7.6.1   The Kobayashi Metric

Let M be a (nonnecessarily connected) complex manifold and d the distance on $\mathbb{H}^1_{\mathbb{C}}$ defined by the Poincaré metric. Given two points p and q of M, choose the following objects:

(i) Points $p = p_0, p_1, \dots, p_{k-1}, p_k = q$ of M.

(ii) Points $a_1, \dots, a_k, b_1, \dots, b_k \in \mathbb{H}^1_{\mathbb{C}}$.

(iii) Holomorphic maps $f_1, \dots, f_k$ of $\mathbb{H}^1_{\mathbb{C}}$ into M such that

$$f_j(a_j) = p_{j-1}, \; f_j(b_j) = p_j \text{ for each } j \in \{1, \dots, k\}.$$

For each choice of points and mappings satisfying (i) and (ii), consider the number $\sum_{j=1}^{k} d(a_j, b_j)$, where d is the hyperbolic distance in $\mathbb{H}^1_{\mathbb{C}}$. Let $\rho_M(p, q)$ be the infimum of the numbers obtained in this manner for all possible choices. Then ρ_M is a pseudometric on M. When ρ_M is a metric, M is called Kobayashi hyperbolic. The Kobayashi distance can be considered as a generalisation of the Poincaré distance because both agree on the unit disc.

The following important result from [121] is used in the sequel.

Theorem 7.6.1. *Let $D \subset \mathbb{P}^n_{\mathbb{C}}$ be a nonempty open set, then:*

(i) *Every holomorphic map of M is norm decreasing with respect ρ_M.*

(ii) *If D is the complement of $2n + 1$ or more hyperplanes in general position in $\mathbb{P}^n_{\mathbb{C}}$, then the Kobayashi metric is complete.*

(iii) *Let $D_i \subset \mathbb{P}^n_{\mathbb{C}}$, $i \in I$, be connected open sets of Y, i.e. domains, such that $D = \bigcap_{i \in I} X_i$. If every D_i is complete Kobayashi hyperbolic, so is D.*

(iv) *If D is a domain that omits a hyperplane and is complete Kobayashi hyperbolic, then D is a holomorphy domain, see [129].*

7.6.2   Complex Hyperbolic groups and k-chains

Recall from Chapter 2 that given $0 < k < n$ we say that $V \subset \partial \mathbb{H}^n_{\mathbb{C}}$ is a k-chain if $V = W \cap [N_-]_n$ where W is a projective subspace of dimension k, and V contains more than two points. Clearly every k-chain is diffeomorphic to $\mathbb{S}^{2k-1}$.

Proposition 7.6.2. *Let $G \subset \mathrm{PU}(n,1)$ be a discrete group. Then the limit set $\Lambda_{CG}(G)$ lies in a k-chain if and only if there is a proper G-invariant projective subspace which is nonempty.*

Proof. If $\Lambda_{CG}(G)$ lies in a k-chain, then there is a projective subspace $V \nsubseteq \mathbb{P}^n_{\mathbb{C}}$ such that $\Lambda_{CG}(G) \subset V$, thus $\langle \Lambda_{CG}(G) \rangle \subset V$, and clearly $\langle \Lambda_{CG}(G) \rangle$ is a G-invariant projective subspace. On the other hand, if there is a G-invariant projective proper subspace V which is nontrivial, then either $V \cap \mathbb{H}_{\mathbb{C}}^n \neq \emptyset$ or $V^{\perp} \cap \mathbb{H}_{\mathbb{C}}^n \neq \emptyset$. Notice that $V^{\perp}$ is also G-invariant, due to G-invariance of the Hermitian form $\langle \ \rangle$. In any case there is a G-invariant space such that $W \cap \mathbb{H}_{\mathbb{C}}^n \neq \emptyset$. Thus given $z \in W \cap \mathbb{H}_{\mathbb{C}}^n$ the set of accumulation points of Gz lies in W, which concludes the proof. $\qquad\square$

Corollary 7.6.3. *If $G \subset \mathrm{PU}(n,1)$ is a discrete subgroup which does not leave invariant any k-chain, then $\Lambda_{CG}(G)$ is the unique minimal closed set for the action of G in $\mathbb{P}_{\mathbb{C}}^n$.*

Proof. Let $\mathcal{C}$ be a closed G-invariant set and $c \in \mathcal{C}$.

Claim 1. There is $\gamma \in G$ and $x \in \mathrm{Fix}(\gamma|_{\partial \mathbb{H}^n})$ such that $c \in \{x\}^{\perp}$. Otherwise, let us set

$$\mathrm{FLP}(G) = \{p \in \partial \mathbb{H}^n | p \in \mathrm{Fix}(G) \text{ for some loxodromic element} \gamma \in G\},$$

then $c \in \bigcap_{y \in \mathrm{FLP}(G)} \{y\}^{\perp}$. That is $c \in \langle \mathrm{FLP}(G) \rangle^{\perp}$ and therefore $\langle \mathrm{FLP}(G) \rangle \neq \mathbb{P}_{\mathbb{C}}^n$. Thus $\Lambda_{CG}(G)$ is contained in a k-chain.

Now let $\gamma \in G$ be a loxodromic element and $x \in \partial \mathbb{H}^n$ be a fixed point by γ such that $c \notin \{x\}^{\perp}$. Then one has that there is a sequence (n_m) such that $\gamma^{n_m}(c) \xrightarrow[m \to \infty]{} q$, where q is the other fixed point by γ on $\partial \mathbb{H}^n$. Hence $q \in \mathcal{C}$, which concludes the proof. $\qquad\square$

Lemma 7.6.4. *Let $G \subset \mathrm{PU}(n,1)$ be a discrete group such that $\Lambda_{CG}(G)$ does not lie in a k-chain and $x \in \Lambda_{CG}(G)$. Then for each $m \in \mathbb{N}$, $m \geq n+1$, there is a set $A_m = \{\gamma_1, \ldots, \gamma_m\} \subset G$, such that the points in $A_m x$ are in general position.*

Proof. By induction on m. Let us show the case $m = n+1$. Let $\{\gamma_m | m \in \mathbb{N}\}$ be an enumeration for G and for each $m \in \mathbb{N}$, set $C_m = \langle \gamma_1(x), \ldots, \gamma_m(x) \rangle$ and $k_m = \dim_{\mathbb{C}}(C_m)$. Then $C_m \subset C_{m+1}$ and $k_m \leq K_{m+1} \leq n$. Define $K_0 = \max\{k_m | m \in \mathbb{N}\}$ and $m_0 = \min\{m \in \mathbb{N} | k_m = k_0\}$. Thus $C_m = C_{m_0}$ for all $m \geq m_0$. Hence $\langle Gx \rangle = C_{m_0}$, so that C_{m_0} is G-invariant. Therefore $C_{m_0} = \mathbb{P}_{\mathbb{C}}^n$. Then there is a set $A = \{\tau_1, \ldots, \tau_{n+1}\}$ such that $\langle Ax \rangle = \mathbb{P}_{\mathbb{C}}^n$, which concludes this part of the proof.

Now assume that there is $m \geq n+1$ and $A_m = \{\gamma_1, \ldots, \gamma_m\} \subset G$ such that the points in $A_m x$ are in general position. Set $B = \{B \subset A_m | \mathrm{card}(B) = n\}$, $\kappa : Gx \to 2^B$ given by $\kappa(y) = \{\beta \in B | y \in \langle \beta \rangle\}$ and $\phi : Gx \to \mathbb{N} \cup \{0\}$ given by $\phi(y) = \mathrm{card}(\kappa(y))$. Clearly the result will follow by proving $l = \min\{\phi(y) | y \in Gx\} = 0$.

Claim 1. $l = 0$. Assume on the contrary that $l > 0$. Let $y_0 \in Gx$ be such that $\phi(y_0) = l$. Set $r_0 = \min\{d(y_0, \langle \beta \rangle) | \beta \in B \setminus \kappa(y_0)\}$. Thus there is a loxodromic element $\gamma_0 \in G$ such that $\mathrm{Fix}(\gamma_0|_{\overline{\mathbb{H}_{\mathbb{C}}^n}}) \subset W = B_{r_0}(y_0) \cap \overline{\mathbb{H}_{\mathbb{C}}^n}$. Thus for $l \in \mathbb{N}$ large $\gamma_0^l(A_m x) \subset W$. Then for each $z \in \gamma_0^l(A_m x)$ it follows that $\kappa(z) \subset \kappa(y_0)$. Since $l = \min\{\phi(y) | y \in Gx\} = \mathrm{card}(\kappa(y_0))$ we have that $\kappa(z) = \kappa(y_0)$ for each

$z \in \gamma_0^l(A_m x)$. Thus $\gamma_0^l(A_m x) \subset \bigcap_{\beta \in \kappa(y)} \langle \beta \rangle$, which is a contradiction since $\gamma_0^l(A_m x)$ is a set with m points, $m \geq n+1$, and $\dim_{\mathbb{C}}(\bigcap_{\beta \in B(y)} \langle \beta \rangle) < n$. Thus $l = 0$. $\qquad \square$

Now we get an application of the previous results to the geometry of Eq (G):

Theorem 7.6.5. *Let $G \subset \mathrm{PU}(n,1)$ be a discrete group such that $\Lambda_{CG}(G)$ does not lie in any k-chain. Then each connected component of* Eq (G) *is complete Kobayashi hyperbolic, and therefore a holomorphy domain.*

Recall that a holomorphy domain means an open set which is maximal in the sense that there exists a holomorphic function on this set which cannot be extended to a bigger one.

Proof. For each point $p \in \Lambda_{CG}(G)$, let $\gamma_{p,1}, \ldots, \gamma_{p,2n} \in G$ be such that the points in the set $\{\gamma_{p,j}(p) | j = 0, \ldots, 2n\}$, where $\gamma_{p,0} = \mathrm{Id}$, are in general position. Define $X_p = \mathbb{P}_{\mathbb{C}}^n \setminus (\bigcup_{j=0}^{2n} \{\gamma_{p,j}(p)\}^{\perp})$. Thus by Theorem 7.6.1 we have that X_p is complete Kobayashi hyperbolic. Moreover, since Eq $(G) = \bigcap X_p$, it follows that each connected component of Eq (G) is complete Kobayashi hyperbolic. $\qquad \square$

Remark 7.6.6. In the 2-dimensional case the previous theorem asserts that if $G \subset \mathrm{PU}(2,1)$ is a discrete group without fixed points, then each connected component of $\Omega_{\mathrm{Kul}}(G)$ is a Kobayashi hyperbolic space and a holomorphy domain (compare with [162]).

7.7   The two-dimensional case revisited

We now use the results of the previous sections to give an alternative proof, using pseudo-projective transformations, of the theorem of Navarrete discussed in Section 7.3 and stated below:

Theorem 7.7.1. *Let G be a discrete subgroup of* $\mathrm{PU}(2,1)$, *then:*

(i) *If G is nonelementary and W is a G-invariant open set such that G acts properly discontinuously on W, then $W \subseteq$ Eq (G). In other words, Eq (G) is the largest open set on which G acts properly discontinuously.*

(ii) Eq $(G) = \Omega_{\mathrm{Kul}}(G)$.

(iii) *The limit set $\Lambda_{\mathrm{Kul}}(G)$ is the union of all complex projective lines l_z tangent to $\partial \mathbb{H}_{\mathbb{C}}^2$ at points in $\Lambda_{CG}(G)$, that is,*

$$\Lambda_{\mathrm{Kul}}(G) = \bigcup_{z \in \Lambda_{CG}(G)} l_z.$$

Of course statement (iii) follows from (ii) together with Theorem 7.5.3. We also notice that by Proposition 3.1.1 the first statement in this theorem is false for elementary groups.

We have:

Lemma 7.7.2. *Let $c \in \mathbb{C}^*$ be a nonunitary complex number, $\theta \in \mathbb{R}$ and $\gamma_c \in$ PSL$(3, \mathbb{C})$ induced by the matrix*

$$\begin{pmatrix} c & 0 & 0 \\ 0 & ce^{2\pi i\theta} & 0 \\ 0 & 0 & c^{-2} \end{pmatrix}.$$

Then γ cannot be conjugate to a transformation in PU$(n, 1)$ by a homeomorphism of $\mathbb{P}^2_{\mathbb{C}}$.

Proof. Taking the inverse of γ if necessary, we can assume that $|c| > 1$. Define

$$\tilde{\gamma}_m = \begin{pmatrix} 1 & 0 & 0 \\ 0 & e^{2\pi i\theta m} & 0 \\ 0 & 0 & c^{-3m} \end{pmatrix}; \ \tilde{\gamma}_\vartheta = \begin{pmatrix} 1 & 0 & 0 \\ 0 & e^{2\pi i\vartheta} & 0 \\ 0 & 0 & 0 \end{pmatrix}, \ \text{where } \vartheta \in \mathbb{R}.$$

Thus $\tilde{\gamma}_m$ is a lift of γ^m and for any subsequence (τ_m) of (γ^m) which converges to $\tau \in \text{QP}(n, \mathbb{C})$, one has that $\tau = \gamma_\vartheta$ for some $\vartheta \in \mathbb{R}$. On the other hand, if γ is topologically conjugate to an element of PU$(2, 1)$, by Lemma 7.4.4 we deduce that for any sequence $(\vartheta_m) \subset \langle\gamma\rangle$ which converges to $\vartheta \in \text{QP}(2, 1)$, we must have that $\Im(\vartheta)$ is a single point, which is a contradiction. □

Now we have:

Proof of Theorem 7.7.1. Let us prove (i) by contradiction. Then by Theorem 7.5.3 there is $x \in \Lambda_{CG}(G)$ such that the line ℓ_x tangent to $\partial\mathbb{H}^2_{\mathbb{C}}$ at x, is not contained in $\mathbb{P}^2_{\mathbb{C}} \setminus W$. Now let: $p \in W \cap \ell_x$, ℓ be a line different from ℓ_x and contained in $\mathbb{P}^2 \setminus \mathbb{H}^2_{\mathbb{C}}$. Let $(\gamma_m) \subset G$ be a sequence of distinct elements, $\gamma \in \text{QP}(n, \mathbb{C})$ such that $\text{Ker}(\gamma) = \ell_x$ and (γ_m) converges to γ. Applying Proposition 7.4.2, to $\ell, p, W, (\gamma_m)$ and γ, we get a line ℓ_p such that each point in ℓ_p is a cluster point of $(\gamma_m(K))$, where K is a compact set which varies on $W \cap \ell$. Therefore $\ell_p \subset (\mathbb{P}^2_{\mathbb{C}} \setminus W) \cap (\mathbb{P}^2 \setminus \mathbb{H}^2_{\mathbb{C}})$. By Corollary 2.6.8 we conclude that $\mathbb{P}^2 \setminus \text{Eq}(G) \subset \overline{G\ell_p} \subset \mathbb{P}^2 \setminus W$, which is a contradiction.

Let us prove (ii). By Remark 7.5.4 one has $\Omega_{\text{Kul}}(G) \subset \text{Eq}(G)$. So it is enough to prove $\text{Eq}(G) \subset \Omega_{\text{Kul}}(G)$ in the elementary case. We have two cases.

Case 1. Assume $\Lambda_{\text{Kul}}(G)$ is a point q. Let ℓ be the unique line in $\mathbb{P}^2_{\mathbb{C}} \setminus \text{Eq}(G)$. We claim that $L_0(G) \cup L_1(G) \subset \ell$. If $L_0(G) \cup L_1(G) \neq \ell$, let $p \in \ell \setminus (L_0(G) \cup L_1(G))$, $(\gamma_m) \subset G$ be an infinite sequence and $\gamma \in \text{QP}(n, \mathbb{C})$ such that (γ_m) converges to γ, $\Im(\gamma) = \{q\}$ and $\text{Ker}(\gamma) = \ell$. Applying Proposition 7.4.2 to $\ell, p, \mathbb{P}^2_{\mathbb{C}} \setminus (L_0(G) \cup L_1(G))$, (γ_m) and γ, we get a line ℓ_p such that each point in ℓ_p is a cluster point of $(\gamma_m(K))$, where K is a compact set which varies on $W \cap \ell$. Hence $\ell_p \subset \Lambda_{\text{Kul}}(G)$, concluding the proof.

Case 2. The set $\Lambda_{\text{Kul}}(G)$ consists of exactly two points q, p. Let ℓ_p, ℓ_q be the tangent lines to $\mathbb{H}^2_{\mathbb{C}}$ at p and q respectively. In this case we claim that $L_0(G) \cup L_1(G) =$

$\{p,q\} \cup (l_p \cap l_q)$. Since $\{p,q\} \cup (\ell_p \cap \ell_q)$ is trivially contained in the set of fixed points of γ, for any $\gamma \in G$ with infinite order, and by Lemma 7.7.2 we conclude that $L_0(G) = \{p,q\} \cup (\ell_p \cap \ell_q)$. Let $\tau \in \mathrm{PSL}(3,\mathbb{C})$ such that $\tau(p) = e_1$, $\tau(q) = e_3$ and $\tau(\ell_p \cap \ell_q) = e_2$, then for any $\gamma \in G$ one has that $\tau\gamma\tau^{-1}$ has one of the following lifts:

$$\begin{pmatrix} 0 & 0 & f \\ 0 & e & 0 \\ d & 0 & 0 \end{pmatrix} \quad \text{where } -def = 1 \text{ and } e,d,f \text{ are roots of unity,}$$

$$\begin{pmatrix} a & 0 & 0 \\ 0 & b & 0 \\ 0 & 0 & c \end{pmatrix} \quad \text{where } abc = 1 \text{ and } |a| < |b| < |c| \,, \tag{7.7.2}$$

depending on whether γ interchanges p and q or not. Now let $z \in L_1(G) \setminus L_0(G)$. Then there are $x \in \mathbb{P}^2_{\mathbb{C}} \setminus L_0(G)$ and $(\gamma_m) \subset G$ a sequence of distinct elements in G such that $(\gamma_m(x))$ converges to z. Thus there is an infinite sequence of (γ_m), still denoted by (γ_m), such that γ_m leaves invariant p, for every m, for otherwise by equation (7.7.2) one has that (γ_m) is eventually constant, which is a contradiction. Now, by Proposition 7.4.1, Lemma 7.4.4 and equation (7.7.2) there is a subsequence of (γ_m), still denoted by (γ_m), and sequences (a_m), (b_m), $(c_m) \in \mathbb{C}$, such that either

$$\tau\gamma_m\tau^{-1} = \begin{bmatrix} a_m & 0 & 0 \\ 0 & b_m & 0 \\ 0 & 0 & c_m \end{bmatrix} \xrightarrow{m\to\infty} \begin{bmatrix} 0 & 0 & 0 \\ 0 & 0 & 0 \\ 0 & 0 & 1 \end{bmatrix};$$

$$\tau\gamma_m\tau^{-1} = \begin{bmatrix} a_m^{-1} & 0 & 0 \\ 0 & b_m^{-1} & 0 \\ 0 & 0 & c_m^{-1} \end{bmatrix} \xrightarrow{m\to\infty} \begin{bmatrix} 1 & 0 & 0 \\ 0 & 0 & 0 \\ 0 & 0 & 0 \end{bmatrix} \quad \text{or}$$

$$\tau\gamma_m\tau^{-1} = \begin{bmatrix} a_m & 0 & 0 \\ 0 & b_m & 0 \\ 0 & 0 & c_m \end{bmatrix} \xrightarrow{m\to\infty} \begin{bmatrix} 1 & 0 & 0 \\ 0 & 0 & 0 \\ 0 & 0 & 0 \end{bmatrix};$$

$$\tau\gamma_m\tau^{-1} = \begin{bmatrix} a_m^{-1} & 0 & 0 \\ 0 & b_m^{-1} & 0 \\ 0 & 0 & c_m^{-1} \end{bmatrix} \xrightarrow{m\to\infty} \begin{bmatrix} 0 & 0 & 0 \\ 0 & 0 & 0 \\ 0 & 0 & 1 \end{bmatrix}.$$

In any case, it follows that $z \in \{p,q\} \cup \ell_p \cap \ell_q$, which proves our claim.

Finally, let $\gamma \in G$ be any element with infinite order, thus by equation (7.7.2) we deduce that γ is a strongly loxodromic element. Thus applying the argument used in the calculation of the limit set (see page 111) in the Kulkarni sense for strongly loxodromic elements to γ, we conclude that $\ell_p \cup \ell_q \subset \Lambda_{\mathrm{Kul}}(G)$. This completes the proof. $\qquad\square$

Chapter 8

Projective Orbifolds and Dynamics in Dimension 2

Köbe's retrosection theorem says that every compact Riemann surface is isomorphic to an orbit space Ω/Γ, where Ω is an open set in the Riemann sphere $\mathbb{S}^2 \cong \mathbb{P}^1_{\mathbb{C}}$ and Γ is a discrete subgroup of $PSL(2,\mathbb{C})$ that leaves Ω invariant; in fact Γ is a Schottky group. It is thus natural to go one dimension higher and ask which compact orbit spaces one gets as quotients of an open set Ω in $\mathbb{P}^2_{\mathbb{C}}$ divided by some discrete subgroup $\Gamma \subset PSL(3,\mathbb{C})$. That is the topic we explore in this chapter.

We study discrete subgroups Γ of $PSL(3,\mathbb{C})$ whose action on $\mathbb{P}^2_{\mathbb{C}}$ leaves invariant a nonempty open invariant set Ω with compact quotient $M = \Omega/\Gamma$; we call such groups *quasi-cocompact*. The surface M is then an orbifold naturally equipped with a projective structure, and the open set Ω is a divisible set in Benoist's sense.

We determine the subgroups one has in $PSL(3,\mathbb{C})$ with this property. In each case we describe its region of discontinuity, the corresponding limit set and the kind of surfaces one gets as quotients. Of course among these are the complex hyperbolic groups $\Gamma \subset PU(2,1)$ with compact quotient $\mathbb{H}^2_{\mathbb{C}}/\Gamma$. This is based on [41] together with previous work by Kobayashi, Ochiai, Inoue, Klingler and Navarrete.

Although our main focus is on the complex case, there is a lot of interesting related literature on the topic of real projective structures on manifolds, and particularly on real surfaces. Thence before going into the complex case we have a glance at the real setting. This should serve as both an introduction to the topic we envisage in this chapter, and also as a source of inspiration for further research.

We begin the chapter with a fast review of geometric structures on manifolds and we briefly describe some of the work done by Goldman, Benoist and others, about projective structures on real manifolds, mainly on surfaces. We then look at compact complex surfaces with a projective structure, following the work on the topic by Kobayashi, Ochiai, Inoue and Klingler. Section 4 begins with a summary of basic results and ideas of Thurston about orbifolds, and then gives an extension of the work of Kobayashi et al for complex surfaces, to the case of compact, complex 2-dimensional orbifolds with a projective structure. Sections 5 to 7 are an outline of the results in [41], improving the aforementioned results of Kobayashi et al, and carrying them to the level of group actions. We include a classification of the quasi-cocompact discrete subgroups of $PSL(3, \mathbb{C})$. Finally we describe in each case, the region of discontinuity, the limit set and the quotient orbifold.

8.1 Geometric structures and the developing map

In his celebrated Erlangen talk in 1872, Felix Klein defined geometry as the study of those properties of figures that remain invariant under a particular group of transformations. Thus for instance, Euclidean geometry is the study of properties such as length, volume and angle which are all invariants of the group of Euclidean motions. By considering other groups of transformations, one obtains other geometries, such as hyperbolic geometry, conformal geometry, projective geometry, etc. This line of thought naturally leads to studying homogeneous spaces G/H of a Lie group G: These are manifolds which are "locally modeled" by the Lie group G. More generally, as described by W. Thurston, one may consider a simply connected manifold X equipped with some Riemannian metric, together with a group G of isometries of X acting transitively on this space; one then looks at manifolds of the form X/Γ, where Γ is a discrete subgroup of G that acts freely on X. Manifolds like this are said to have an (G, X)-structure.

In fact this way for describing geometric structures goes back to the classical Riemann-Koebe uniformisation theorem of Riemann surfaces, saying that the universal cover or every compact Riemann surface is conformally equivalent to either the Riemann sphere, the complex line, or the open unit 1-disc. In real dimension 3, the analogous question is given by Thurston's geometrisation conjecture, now proved by Perelman.

Here we are interested in the "uniformisation problem for complex manifolds", studied by many authors. It is well-known that the set of complex structures on a nice contractible higher-dimensional complex manifold is huge and cannot be classified in a simple manner. Therefore, in order to find a simple analogue of the aforementioned uniformisation theorem, one must impose a certain type of restrictions on the manifold.

In his "Lectures on Riemann surfaces", R. Gunning discusses holomorphic affine and projective structures and connections on Riemann surfaces. He shows there that a compact Riemann surface of any genus always admits projective structures while it admits affine structures only when the genus is 1. He then suggested looking at complex manifolds equipped with a projective structure, aiming to have for this class of manifolds a uniformisation theorem in the vein of the classical theorem for Riemann surfaces. This has been successfully done by Kobayashi, Ochiai and others in the case of compact complex surfaces, i.e., complex manifolds of complex dimension 2, and that is the subject we envisage in this chapter.

In complex dimension 3 there are several interesting articles by M. Kato on the subject (see for instance [107], [109], [108],[110]) where the author looks at complex 3-manifolds equipped with a projective structure, and also the results from [203], that we explain in Chapter 9, where the authors look at manifolds of the form Ω/Γ where Ω is an open subset of $\mathbb{P}^3_{\mathbb{C}}$ which is invariant under the free action of Γ, a complex Schottky subgroup of $\mathrm{PSL}(4,\mathbb{C})$.

Of course there is the real counter-part, of looking at real projective spaces on smooth manifolds. This is a rich area of current research, with remarkable recent work by W. Goldman, Y. Benoist and others. We briefly discuss this below.

In the sequel we say more about the previous discussion.

8.1.1 Projective structures on manifolds

Consider a real analytic manifold X and a Lie group G acting transitively on X.

Definition 8.1.1. A smooth manifold M has a (G, X)-*structure* if it has a maximal atlas by coordinate charts with values in X and all the local changes of coordinates are restriction of elements in G. In this case we also say that M is a (G, X)-manifold.

In other words, a (G, X)-*structure* on M means that M is locally modeled on X and the gluing maps are all in G.

Examples 8.1.2.   (i) A *real projective structure* on a real manifold M of dimension n means a $(\mathbb{P}^n_{\mathbb{R}}, \mathrm{PGL}(n+1, \mathbb{R})$-structure on M.

 (ii) If $\mathrm{Aff}\,(\mathbb{C}^n) \cong \mathrm{GL}\,(n, \mathbb{C}) \ltimes \mathbb{C}^n$ is the complex affine group, then an *affine structure* on a complex n-manifold, denoted as an $(\mathrm{Aff}\,(\mathbb{C}^n), \mathbb{C}^n)$-structure, means a maximal atlas by coordinate charts where the transition functions are local affine maps in $\mathbb{C}^n$, i.e., restriction of elements in $(\mathrm{Aff}\,(\mathbb{C}^n)$. Since every affine map is a projective map, one has that affine structures are a special class of projective structures. Also notice that every Euclidean isometry in $\mathbb{C}^n$ is an affine map, thence Euclidean structures are a special class of affine structures.

(iii) A *complex hyperbolic structure* on a complex manifold M means a maximal atlas modelled on the complex hyperbolic space $\mathbb{H}^n_{\mathbb{C}}$; the local changes of coordinates are now restriction of elements in $\mathrm{PU}(n, 1)$. Examples of this type are manifolds of the form $\mathbb{H}^n_{\mathbb{C}}/\Gamma$ where $\mathbb{H}^n_{\mathbb{C}}$ is the complex hyperbolic n-space (diffeomorphic to a $2n$-ball) and $\Gamma \subset \mathrm{PU}(n, 1)$ is a discrete subgroup of isometries of $\mathbb{H}^n_{\mathbb{C}}$.

(iv) A *complex projective structure* on a complex manifold M of dimension n is a maximal atlas for M modelled on open subsets of the projective space $\mathbb{P}^n_{\mathbb{C}}$, so that for any two overlapping charts, the corresponding change of coordinates is restriction of a projective transformation of $\mathbb{P}^n_{\mathbb{C}}$. Since the groups in the previous examples, $\mathrm{PU}(n, 1)$ and $\mathrm{Aff}\,(\mathbb{C}^n)$, are both subgroups (canonically) of $\mathrm{PSL}(n+1, \mathbb{C})$, it follows that all complex affine and complex hyperbolic structures are automatically projective. A similar statement holds in the real setting.

On a manifold with a complex projective structure one has a local projective geometry which coincides with the local geometry of $\mathbb{P}^n_{\mathbb{C}}$. Projective structures arise in many areas of mathematics, including differential geometry, mathematical physics, topology, and analysis. See [68], [66] for clear accounts on projective structures on real manifolds, and we refer to [83] for more on the topic for complex manifolds.

Examples 8.1.3. (i) Let M be a compact Riemann surface of genus $g > 1$. Then the *Fuchsian uniformisation* is the representation of M as the quotient of the hyperbolic disc $\Delta \cong \mathbb{H}^1$ by a Fuchsian group Γ. In this case one has (see Remark 8.4.10 below) that Δ is the universal covering space of M and Γ is isomorphic to the fundamental group $\pi_1(M)$ acting on Δ by deck transformations.

(ii) Now consider an arbitrary compact Riemann surface (any genus). By Koebe's theorem (see also [83]) M has a uniformisation as the quotient of a domain in $\mathbb{P}^1_{\mathbb{C}}$ by a Schottky group (Retrosection Theorem). This provides a projective structure on M whose developing map (see below) is not injective, although the holonomy group is discrete. For surfaces of genus > 1 this projective structure is in general different from the previous one.

(iii) Let $\Gamma \subset \mathrm{PSL}(n+1, \mathbb{C})$ be a complex Kleinian group acting freely and discontinuously on an invariant open subset $\Omega \subset \mathbb{P}^n_{\mathbb{C}}$. Then the quotient space Ω/Γ is obviously a projective manifold. Examples of this type are given for instance by the complex Schottky groups of Chapter 9.

(iv) If Λ is a lattice in $\mathbb{C}^2$, then $M = \mathbb{C}^2/\Lambda$ is an affine complex manifold.

(v) If $g \in \mathrm{GL}(2, \mathbb{C})$ is a contracting automorphism, then the Hopf surface $(\mathbb{C} \setminus \{0\})/\langle g \rangle$ is an affine manifold.

Remark 8.1.4. Observe that since a complex projective structure is locally modelled on the complex projective space, with the coordinate changes being restrictions of projective transformations, every projective structure determines an underlying complex structure.

8.1.2   The developing map and holonomy

Coordinate atlases may be a bit unwieldy, so it may be convenient to work instead with the developing map, that we now introduce. As noticed by W. Thurston, if one has a geometric structure on a manifold, then one has a developing map $\mathcal{D} : \widetilde{M} \to X$, from the universal covering $\widetilde{M}$ of M into the model space X, which globalises the coordinate charts. This is a concept that formalises the idea of "unrolling" a manifold, just as the 2-torus that can be "unrolled" on the plane.

To define the developing map $\mathcal{D}$ we start with some geometry (G, X) on a smooth manifold M and consider the universal cover $\pi : \widetilde{M} \to M$. We choose a base point $x_0 \in M$ and a lifting $\tilde{x}_0$ of it to $\widetilde{M}$. We consider also an atlas $\{(U_\alpha, \phi_\alpha)\}$ for M with transition functions

$$g_{ij} = \phi_i \circ \phi_j^{-1} : \phi_j(U_i \cap U_j) \to \phi_i(U_i \cap U_j).$$

By definition of a geometric structure, each g_{ij} is the restriction of an element in G and so defines a locally constant map from $U_i \cap U_j$ into G.

We now choose a coordinate chart (U_0, ϕ_0) around x_0, and a corresponding chart $(\widetilde{U}_0, \tilde{\phi}_0)$ around $\tilde{x}_0$. This gives a local diffeomorphism between $\widetilde{U}_0$ and an open set in X, which can be extended along any path in $\widetilde{M}$ by analytic continuation. This process is well defined because $\widetilde{M}$ is simply connected. We thus get a map

$$\mathcal{D} : \widetilde{M} \to X,$$

which is *the developing map* of the (G, X) manifold M. This is a local diffeomorphism.

The manifold $\widetilde{M}$ has a unique (G, X)-structure making the projection $\pi : \widetilde{M} \to M$ a (G, X)-map. The choice of a base point x_0 for M determines an action of the fundamental group $\Gamma := \pi_1(M, x_0)$ on $\widetilde{M}$, which is by deck transformations and preserves the (G, X)-structure. One gets in this way a natural group morphism

$$\mathcal{H} : \Gamma \to G$$

with respect to which the map $\mathcal{D}$ is equivariant. The image $\mathcal{H}(\Gamma)$ is called *the holonomy* of M.

We remark that up to multiplication by an element in G, the (G, X)-structure on M is fully determined by its developing map. In particular one has the following important result due to Thurston:

Theorem 8.1.5. *If M has a (G, X)-structure which is complete, then the developing map $\mathcal{D}$ is a covering projection and if X is simply connected, then Γ can be identified with the holonomy subgroup $\mathcal{H}(\Gamma)$ and M is the quotient space X/Γ as a (G, X)-manifold.*

8.2 Real Projective Structures and Discrete Groups

Notice that every Euclidean structure is an affine structure, and the embedding

$$(\mathbb{R}^n, \mathrm{Aff}(R^n)) \to (\mathbb{P}^n_{\mathbb{R}}, \mathrm{PGL}(n+1, \mathbb{R}))$$

gives an embedding of affine geometry into projective geometry. Similarly, the inclusion of the projective orthogonal group $\mathrm{PO}(n+1)$ in $\mathrm{PGL}(n+1; \mathbb{R})$ gives an embedding of elliptic geometry into projective geometry. Also, the Klein model of hyperbolic geometry $(\mathbb{H}_{\mathbb{R}}^n, \mathrm{PO}(n,1))$ gives an embedding of hyperbolic geometry into projective geometry. In the particular case $n = 2$ we know, by the uniformisation theorem, that every closed oriented surface admits either an elliptic, a Euclidean or a hyperbolic structure. Therefore we have that every such manifold admits a projective structure. With a little more work we can show that every one of Thurston's eight geometries for 3-manifolds also leads to a $\mathbb{P}_{\mathbb{R}}^3$-structure.

There is a large literature with interesting results concerning real projective structures and discrete group actions. We now say a few words about two such lines of research.

8.2.1   Projective structures on real surfaces

The basic general question concerning geometric structures on manifolds is: Given a topological manifold M and a geometry (G, X), does there exist a (G, X)-structure on M, and if so, can we classify all (G, X)-structures on M up to (some type of) equivalence? In other words we would like to have for the geometry in question something like the Teichmüller space of a Riemann surface. That is, a topological space whose points correspond to equivalence classes of (G, X)-structures on the manifold M. We refer to the interesting recent article [69] for a survey on this subject.

The first basic results on $\mathbb{P}_{\mathbb{R}}^2$-structures on closed surfaces are due to Kuiper, Benzécri, Koszul, Kobayashi and Vey, among others. More recently, W. Goldman has been exploring this topic thoroughly, obtaining deep results, see for instance [68], [66], [47], [48]. We refer to these articles, and the bibliography in them, for clear accounts on the subject. Here we only sketch some of the main points.

The most important projective structures are the convex structures. A $\mathbb{P}_{\mathbb{R}}^n$-manifold M^n *is convex* if its universal covering is equivalent to a convex domain Ω in an affine patch of $\mathbb{P}_{\mathbb{R}}^n$. In that case the fundamental group of M is represented as a discrete group of projective transformations acting properly and freely on Ω.

If a closed surface S has a $\mathbb{P}_{\mathbb{R}}^2$-structure, then one defines the space $\mathbb{RP}^2(S)$ of equivalence classes of real projective structures on a surface S, in a way similar to the way the Teichmüller space of the surface S is defined in terms of equivalence classes of hyperbolic structures.

The following theorem is proved in [66]:

Theorem 8.2.1. *Let S be a closed surface with genus $g \geq 2$. Then the deformation space $\mathbb{RP}^2(S)$ is a real analytic manifold of dimension $16g - 16$. Moreover, the deformation space $\mathcal{C}(S)$ of convex $\mathbb{P}_{\mathbb{R}}^2$-structures on S is an open set in $\mathbb{RP}^2(S)$, diffeomorphic to an open cell of dimension $16g - 16$.*

In [66] there is also an extension of this theorem for compact surfaces with boundary.

As mentioned above, a convex $\mathbb{P}_{\mathbb{R}}^2$-structure on a closed surface S is a representation of S as a quotient Ω/Γ where $\Omega \subset \mathbb{P}_{\mathbb{R}}^2$ is a convex domain and

$\Gamma \subset \mathrm{PSL}(3, \mathbb{R})$ is a discrete group of automorphisms of $\mathbb{P}^2_{\mathbb{R}}$ acting properly discontinuously and freely on Ω. The space $\mathcal{C}(S)$ of projective equivalence classes of convex $\mathbb{P}^2_{\mathbb{R}}$-structures on S embeds, via the holonomy map, as an open subset in the space of equivalence classes of representations of its fundamental group $\pi_1(S)$ in $\mathrm{SL}(3, \mathbb{R})$.

Consider the space

$$X(S) = \mathrm{Hom}(\pi_1(S), \mathrm{PGL}(3, \mathbb{R})) / \mathrm{PGL}(3, \mathbb{R}),$$

and recall that every such equivalence class of representations of $\pi_1(S)$ corresponds to an isomorphism class of a flat $\mathbb{R}^3$-bundle over S. In [96] the author gives precise topological information concerning the deformation space $X(S)$. In particular he shows that $X(S)$ has exactly three connected components:

1. C_0, the component containing the class of the trivial representation;

2. C_1, the component consisting of classes of representations which do not lift to the double covering of $\mathrm{PGL}(3, \mathbb{R})$; or in other words, those for which the corresponding flat $\mathbb{R}^3$-bundle over S has nonvanishing 2nd Stiefel-Whitney class;

3. C_2, the component containing discrete faithful representations into $\mathrm{SO}(2, 1)$. These correspond to the convex real projective structures determined by a hyperbolic structure on S, so C_2 contains the classical Teichmüller space $\mathcal{T}(S)$ of hyperbolic structures on S, which is a cell of dimension $6(g-1)$. The component C_2 is known as the *Hitchin-Teichmüller* space of S.

Hitchin showed that C_2 is homeomorphic to $\mathbb{R}^{16g-16}$ and he conjectured that C_2 is $\mathcal{C}(S)$, the space of projective equivalence classes of convex $\mathbb{P}^2_{\mathbb{R}}$-structures on S. Hitchin's conjecture is proved in [47]:

Theorem 8.2.2 (Choi-Goldman). *The holonomy map defines a diffeomorphism from the space $\mathcal{C}(S)$ onto the (Hitchin-Teichmüller) space C_2 of $X(S)$.*

8.2.2 On divisible convex sets in real projective space

Y. Benoist has written a series of interesting articles about "divisible convex subsets". In this subsection we give a glimpse of some of Benoist's work.

A subset Ω of the real projective space $\mathbb{P}^n_{\mathbb{R}}$ is *convex* if its intersection with each projective line in $\mathbb{P}^n_{\mathbb{R}}$ is convex. A convex set Ω is called *strictly convex* if the intersection of each projective line in $\mathbb{P}^n_{\mathbb{R}}$ with its boundary $\partial\Omega$ contains at most two points, i.e., $\partial\Omega$ contains no nontrivial segment. It is *properly convex* if there exists a projective hyperplane in $\mathbb{P}^n_{\mathbb{R}}$ that does not meet its closure $\Omega \cup \partial\Omega$.

An open convex set is *divisible* if there exists a discrete subgroup $\Gamma \subset \mathrm{PSL}(n+1, \mathbb{R})$ which leaves Ω invariant and acts on it freely and properly discontinuously, so that the quotient $M := \Omega / \Gamma$ is compact. Of course in this case M is a smooth manifold equipped with a projective structure and Γ is its fundamental group.

A typical example of a strictly convex divisible set is the hyperbolic n-space, regarded via the Klein model.

The following remarkable theorem is proved in [21]:

Theorem 8.2.3. *Let Ω be a properly convex open subset of the real projective space $\mathbb{P}_{\mathbb{R}}^n$ which is divisible by a discrete subgroup $\Gamma \subset \mathrm{PGL}(n+1,\mathbb{R})$. Then the following conditions are equivalent:*

(i) *The set Ω is strictly convex.*

(ii) *The boundary $\partial\Omega$ is of class C^1.*

(iii) *The group Γ is hyperbolic (in the sense of Gromov, see for instance [78]).*

(iv) *The geodesic flow on $\Gamma \backslash \Omega$ is Anosov (and if this happens, then the flow is topologically mixing).*

In [20] the author studies the Zariski closure of a group Γ in $\mathrm{SL}(m,\mathbb{R})$ that divides a properly convex open cone C in $\mathbb{R}^m$. Recall that a subset $C \subset \mathbb{R}^m$ is a *convex cone* if it is closed under linear combinations with positive coefficients. The cone is called:

- *Properly convex* (or salient) if it contains no affine line, i.e., if there is no nonzero vector x in C such that its opposite $-x$ is also in C.

- *Symmetric* if for every $x \in C$ there is an automorphism s of C such that $s^2 = 1$ and the set of fixed points of s is the line containing x.

- A *product* if there is a direct sum decomposition $\mathbb{R}^m = \mathbb{R}^p \oplus \mathbb{R}^q$, $p, q \geq 1$, such that $C = C_1 \times C_2$ with $C_1 \subset \mathbb{R}^p$ and $C_2 \subset \mathbb{R}^q$ being cones.

- *Divisible* if there is a discrete subgroup Γ of $\mathrm{GL}(m,\mathbb{R})$ preserving C such that $\Gamma \backslash C$ is compact.

The main theorem in [20] says that if a discrete subgroup $\Gamma \subset \mathrm{GL}(m,\mathbb{R})$ divides a properly convex open cone C which is not a product nor symmetric, then Γ is Zarisky dense in $\mathrm{GL}(m,\mathbb{R})$.

In [22] the author proves a closedness property of the space of deformations of divisible convex sets in $\mathbb{P}_{\mathbb{R}}^n$, and in [23] the author describes the structure of the boundary $\partial\Omega$ of such divisible sets, and of the quotient $M = \Omega/\Gamma$, in the case when Ω is not strictly convex and it is indecomposable, i.e., its corresponding affine cone is not a product.

We remark that the Schottky groups we envisage in Chapter 9 give a natural way of constructing convex sets in complex projective spaces, which are divisible by the corresponding Schottky group. This construction is extended in [92] to real projective spaces, producing interesting examples of real manifolds with a projective structure.

8.3 Projective structures on complex surfaces

As mentioned before, every compact Riemann surface admits projective structures. Now we look at compact complex surfaces which admit a projective structure, following the work of Kobayashi, Ochiai and Inoue. We already know that all affine and all complex hyperbolic structures are also projective. It turns out that for compact nonsingular surfaces, the converse statement also holds. This is the theorem of Kobayashi-Ochiai proved in [120]:

Theorem 8.3.1. *Let M be a compact complex 2-dimensional manifold M that admits a projective structure. Then $M \cong \mathbb{P}_{\mathbb{C}}^2$ or else it admits either an affine or a complex hyperbolic structure.*

We refer to [120] (see also [116]) for the proof of this theorem. This depends on Kodaira's classification of surfaces, the work in [83] giving topological obstructions for a surface to admit a projective structure, and the article [100] where the authors build on Vitter's work in [230] and classify the compact complex surfaces that admit a holomorphic affine connection.

As we said before, an affine structure on a complex surface M is a maximal atlas of holomorphic charts into $\mathbb{C}^2$ such that the coordinate changes are complex-affine automorphisms of $\mathbb{C}^2$. In [100] it is proved that there are three types of such structures, up to finite coverings. First, there are complex solvmanifolds, which are homogeneous spaces of the form $\Gamma \backslash G$, where G is a 4-dimensional (real) Lie group with a left-invariant complex structure and $\Gamma \subset G$ is a lattice. Secondly, there are Hopf manifolds $(\mathbb{C}^2 - \{0\})/\Gamma$, where Γ is a cyclic group of linear expansions. Finally, to every projective structure on a Riemann surface S, there are associated fibrations by elliptic curves over S which admit complex affine structures.

Theorem 8.3.1 states that in the case of projective structures, aside from $\mathbb{P}_{\mathbb{C}}^2$ and the complex affine structures, the only other nonsingular, compact, complex projective surfaces are complex hyperbolic surfaces.

In [116] the author studies not only which compact surfaces admit a projective structure, but he actually looks at the more refined question of studying the deformations of such a structure. Since, by Theorem 8.1.5, geometric structures are determined by the corresponding developing map $\mathcal{D}$ and its holonomy $\mathcal{H}(\pi_1(M))$, it is essential for Klingler's work to determine $\mathcal{D}$ and the holonomy.

In order to say more about the classification in [120, 100, 116] of surfaces with a projective structure, their developing maps and the corresponding holonomy, we now introduce some notation. We identify the complex space $\mathbb{C}^2$ with the open chart $\{[z : w : 1] \in \mathbb{P}_{\mathbb{C}}^2 \mid z, w \in \mathbb{C}\}$, and we identify the complex affine group with a subgroup of the projective group by

$$\text{Aff}\,(\mathbb{C}^2) = \{g \in \text{PSL}(3, \mathbb{C}) : g\mathbb{C}^2 = \mathbb{C}^2\}.$$

We now set:

$$\mathrm{Sol}_0^4 = \left\{ \begin{pmatrix} \lambda & 0 & a \\ 0 & |\lambda|^{-2} & b \\ 0 & 0 & 1 \end{pmatrix} : (\lambda, a, b) \in \mathbb{C}^* \times \mathbb{C} \times \mathbb{R} \right\};$$

$$\mathrm{Sol}_1^4 = \left\{ \begin{pmatrix} \epsilon & a & b \\ 0 & \alpha & c \\ 0 & 0 & 1 \end{pmatrix} : \alpha, a, b, c \in \mathbb{R}, \alpha > 0, \epsilon = \pm 1 \right\};$$

$$\mathrm{Sol'}_1^4 = \left\{ \begin{pmatrix} 1 & a & b + i\log\alpha \\ 0 & \alpha & c \\ 0 & 0 & 1 \end{pmatrix} : \alpha, a, b, c \in \mathbb{R}, \alpha > 0 \right\};$$

$$A_1 = \left\{ \begin{pmatrix} 1 & 0 & b \\ 0 & a & 0 \\ 0 & 0 & 1 \end{pmatrix} : (a, b) \in \mathbb{C}^* \times \mathbb{C} \right\};$$

$$A_2 = \left\{ \begin{pmatrix} a & b & 0 \\ 0 & a & 0 \\ 0 & 0 & 1 \end{pmatrix} : (a, b) \in \mathbb{C}^* \times \mathbb{C}; \right\}.$$

With this notation, we can state the following stronger version of Theorem 8.3.1, given in [116, 117]):

Theorem 8.3.2. *Let M be a compact complex projective 2-manifold, let $\mathcal{D}$ be its developing map and $\mathcal{H}$ its holonomy. Then one of the following facts occurs:*

(i) *$\widetilde{M} = \mathbb{P}_{\mathbb{C}}^2$, $\mathcal{D} = \mathrm{Id}$ and the holonomy is finite.*

(ii) *$\widetilde{M} = \mathbb{H}_{\mathbb{C}}^2$. In this case $\mathcal{D} = \mathrm{Id}$ and $\mathcal{H}(\pi_1(M)) \subset \mathrm{PU}(2,1)$.*

(iii) *$\widetilde{M} = \mathbb{C}^2 - \{0\}$. In this case $\mathcal{D} = \mathrm{Id}$, $\mathcal{H}(\pi_1(M)) \subset \mathrm{Aff}(\mathbb{C}^2)$ contains a cyclic group of finite index and generated by a contraction, and M has a finite covering (possibly ramified) which is a primary Hopf surface.*

(iv) *$\mathcal{D}(\widetilde{M}) = \mathbb{C}^* \times \mathbb{C}^*$. In this case $\mathcal{H}(\pi_1(M)) \subset \mathrm{Aff}(\mathbb{C}^2)$ and M has a finite covering (possibly ramified) which is biholomorphic to a complex torus $\mathbb{T}^1 \times \mathbb{T}^1 \times \mathbb{T}^1 \times \mathbb{T}^1$.*

(v) *$\mathcal{D}(\widetilde{M}) = \mathbb{C} \times \mathbb{C}^*$. In this case $\mathcal{H}(\pi_1(M)) \subset \mathrm{Aff}(\mathbb{C}^2)$ has a subgroup of finite index contained in A_1 or A_2, and M has a finite covering (possibly ramified) which is a complex surface biholomorphic to a torus $\mathbb{T}^1 \times \mathbb{T}^1 \times \mathbb{T}^1 \times \mathbb{T}^1$.*

(vi) *$\mathcal{D}(\widetilde{M}) = \mathbb{C}^2$. In this case $\mathcal{H}(\pi_1(M)) \subset \mathrm{Aff}(\mathbb{C}^2)$ contains a unipotent subgroup of finite index and M has a finite covering (possibly ramified) which is a surface biholomorphic to a complex torus or a primary Kodaira surface.*

(vii) *$\widetilde{M} = \mathbb{C} \times \mathbb{H} = \mathcal{D}(\widetilde{M})$. In this case M is an Inoue Surface and $\mathcal{H}(\pi_1(M)) \subset \mathrm{Aff}(\mathbb{C}^2)$ is a torsion free group contained in one of the groups Sol_0^4, Sol_1^4 or $\mathrm{Sol'}_1^4$.*

(viii) $\widetilde{M}$ *is biholomorphic to* $\mathbb{C} \times \mathbb{H}$. *In this case* $\mathcal{D}(\widetilde{M}) = \mathbb{C}^* \times \Omega$ *where* Ω *is a hyperbolic domain contained in* $\mathbb{P}^1_{\mathbb{C}}$ *and* M *has a finite covering (possibly ramified) which is an elliptic affine surface.*

Notice that in cases (iii) to (viii) the corresponding structures are all affine.

8.4 Orbifolds

In this section we follow [41] and extend the previous results for complex surfaces to the case of orbifolds with a projective structure. We begin with a brief discussion about well-known basic facts on orbifolds, mostly due to W. Thurston.

Orbifolds have arisen several times in various contexts: In Satake's study of automorphic forms in the 1950s; he called them *V-manifolds* (see [190]); in Thurston's work on the geometry and topology of 3-manifolds in the 1970s, when he coined the name *orbifold* (for "orbit-manifold"); and in Haefliger's work in the 1980s in the context of Gromov's programme on $CAT(k)$-spaces, under the name *orbihedron*: a slightly more general notion of orbifold, convenient for applications in geometric group theory (see [85]). More recently, orbifolds appear also in algebraic topology and mathematical physics, mainly in the context of string theory (see [2]).

An orbifold is a generalisation of a manifold. It consists of a topological space (called the underlying space) with an orbifold structure (see below). Like a manifold, an orbifold is specified by local conditions; however, instead of being locally modelled on open subsets of $\mathbb{R}^n$, an orbifold is locally modelled on quotients of open subsets of $\mathbb{R}^n$ by finite group actions. The structure of an orbifold encodes not only that of the underlying quotient space, which need not be a manifold, but also that of the isotropy subgroups. For instance, every manifold with boundary carries a natural orbifold structure, since it is the quotient of its double by an action of $\mathbb{Z}/2\mathbb{Z}$.

Orbifolds arose first as surfaces with "marked" (or "singular") points long before they were formally defined. One of the first classical examples comes from the theory of modular forms with the action of the modular group $\mathrm{SL}(2, Z)$ on the upper half-plane. Orbifolds also play a key role in the theory of Seifert fibre spaces, initiated by Seifert in the 1930s.

Orbifolds appear naturally in the context of this monograph, as quotient spaces of a manifold under the action of a discrete group of diffeomorphisms with finite isotropy subgroups.

8.4.1 Basic notions on orbifolds

In what follows, G is a Lie group acting effectively, transitively and locally faithfully on a smooth manifold X. We restrict this discussion to orbifolds with a geometric structure in Thurston's sense. These are the orbifolds that appear in the sequel.

Definition 8.4.1. A (G, X)-*orbifold* $\mathcal{O}$ is a topological Hausdorff space $X_{\mathcal{O}}$, called the underlying space, with a countable basis and equipped with a collection $\{\widetilde{U}_i, \Gamma_i, \phi_i, U_i\}_{i \in I}$, where the $\{U_i\}$ are an open cover of $X_{\mathcal{O}}$, and the $\widetilde{U}_i$, called *folding charts*, are open subsets of X. For each $\widetilde{U}_i$ there is a finite group $\Gamma_i \subset G$ acting on $\widetilde{U}_i$, and a homeomorphism $\phi_i : U_i \to \widetilde{U}_i/\Gamma_i$, called a *folding map*. These charts must satisfy a certain compatibility condition (as they do for manifolds). Namely, whenever $U_i \subset U_j$, there is a group morphism $f_{ij} : \Gamma_i \hookrightarrow \Gamma_j$ and $\widetilde{\phi}_{ij} \in G$ with $\widetilde{\phi}_{ij}(U_i) \subset U_j$ and $\widetilde{\phi}_{ij}(\gamma x) = f_{ij}(\gamma)\widetilde{\phi}_{ij}(x)$ for every $\gamma \in \Gamma_i$, such that the diagram below commutes:

$$
\begin{array}{ccccc}
\widetilde{U}_i & \longrightarrow & \widetilde{U}_i/\Gamma_i & \xrightarrow{\ \phi_i^{-1}\ } & U_i \\
\Big\downarrow{\scriptstyle \widetilde{\phi}_{ij}} & & \Big\downarrow & & \Big\downarrow \\
\widetilde{U}_j & \longrightarrow & \widetilde{U}_j/f_{ij}(\Gamma_i) & \longrightarrow \ \widetilde{U}_j/\Gamma_j \xrightarrow{\ \phi_j^{-1}\ } & U_j.
\end{array}
\tag{8.4.1}
$$

Remark 8.4.2. If in the previous definition we require only that the action of the groups Γ_i be locally faithful and that $\widetilde{\phi}_{ij}$ be an equivariant embedding with respect to f_{ij}, dropping the conditions about being elements in G, then the space $\mathcal{O}$ is also called an orbifold, but not a geometric one.

Example 8.4.3.　(i) If $\Gamma \subset G$ is a finite group of automorphisms of a manifold X, then $M = X/\Gamma$ is a (G, X)-orbifold.

(ii) Consider a subgroup $\Gamma \subset G$ acting properly discontinuously on an open subset $\Omega \subset X$, such that the isotropy subgroups are finite. The points with nontrivial isotropy subgroup form a discrete subset of Ω, and the space of orbits $M = \Omega/\Gamma$ is a (G, X)-orbifold for which the groups Γ_i appearing in the above definition are the corresponding isotropy subgroups.

(iii) If (M, F) is a foliated manifold with compact leaves and finite holonomy, then the space of leaves M/F has the structure of an orbifold. This is a consequence of the local Reeb Stability Theorem. ·

A point x in a (G, X)-orbifold $X_{\mathcal{O}}$ is called *singular* if there is a folding map at x, say $\phi : U_i \to \widetilde{U}_i/\Gamma_i$ with $x \in U_i$, and a point $y \in \widetilde{U}_i$ such that $y = \phi(x)$ and the isotropy $\mathrm{Isot}(y, \Gamma_i)$ is nontrivial. The set $\Sigma_{\mathcal{O}} = \{y \in X_{\mathcal{O}} : \text{y is a singular point}\}$ is called the *singular locus* of $X_{\mathcal{O}}$. We say that $\mathcal{O}$ is a (G, X)-*manifold* if $\Sigma_{\mathcal{O}} = \emptyset$.

Remark 8.4.4.　(i) Observe that if M is as in parts (i) and (ii) of the above Example 8.4.3, then the singular set corresponds to the set $\{x \in X : \mathrm{Isot}(x, \Gamma)$ is nontrivial$\}$.

(ii) We notice that, topologically, in the particular cases of orbifolds of real dimension $n = 2, 3$ one has that orbifolds are locally modelled by an n-disc $\mathbb{D}^n$ modulo a finite subgroup Γ of $SO(3)$, and it happens that in these dimensions every such quotient $\mathbb{D}^n/\Gamma$ is again homeomorphic to an n-disc. Therefore every orbifold of dimension $2, 3$ is a topological manifold, and thence homeomorphic to a smooth manifold, but just as manifolds, not preserving the orbifold structure. For instance, the Riemann sphere with at least three marked points is a hyperbolic orbifold. And of course if we forget the marked points, then the Riemann sphere is not a hyperbolic manifold!

Now, given M, N two (G, X)-orbifolds, a continuous function $f : M \to N$ is called a (G, X)-*map* if for each point $x \in N$, a folding map for x, $\phi_i : U_i \to \widetilde{U}_i/\Gamma_i$, and $y \in f^{-1}(x)$, there is a folding map of y, $\phi_j : U_j \to \widetilde{U}_j/\Gamma_j$, and $\vartheta \in G$ with $\vartheta(\widetilde{U}_j) \subset U_i$, inducing f equivariant with respect to a morphism $\psi : \Gamma_j \to \Gamma_i$.

The map f is called a (G, X)-*equivalence* if f is bijective and f, f^{-1} are (G, X)-maps. The map f is called a (G, X)-*covering orbifold map* if f is a surjective (G, X)-map such that each point $x \in N$ has a folding map $\phi_j : U_j \to \widetilde{U}_j/\Gamma_j$ so that each component V_i of $f^{-1}U_j$ has a homeomorphism $\widetilde{\phi}_i : V_i \to \widetilde{U}_j/\Gamma_i$ (in the orbifold structure) where $\Gamma_i \subset \Gamma_j$. We require that the quotient map $\widetilde{U}_j \to V_i$ induced by $\widetilde{\phi}_i$ composed with f should be the quotient map $\widetilde{U}_j \to U_j$ induced by ϕ.

Also, if $p : M \to N$ is a covering orbifold map with M a manifold, we say that p is a *ramified* covering if $\Sigma_N \neq \emptyset$. Otherwise we say that the covering is *unramified*. The covering $p : M \to N$ is finite if $p^{-1}(y)$ is finite for every $y \in N$.

Example 8.4.5. If M is as part (ii) of Example 8.4.3, then the quotient map $q : \Omega \to M$ is a (G, X)-covering orbifold map.

Every covering map between manifolds is a covering orbifold map.

Definition 8.4.6. An orbifold M is a *good orbifold* (respectively a *very good orbifold*) if there exists a covering orbifold map $p : \widetilde{M} \to M$ such that $\widetilde{M}$ is a manifold (respectively a compact manifold).

Example 8.4.7. (i) An interesting example is the teardrop, depicted in Figure 8.1. Its underlying topological space is the sphere $\mathbb{S}^2$. Its singular set $\Sigma_{\mathcal{O}}$ consists of a single point, whose neighbourhood is modelled on $\mathbb{R}^2/\mathbb{Z}_n$, where $\mathbb{Z}_n$ acts by rotations. One can show that the teardrop is not a good orbifold. If we had one more singular point of the same type in the "drop", then this would be a very good orbifold, corresponding to the quotient $\mathbb{S}^2/\mathbb{Z}_n$.

(ii) If Γ is a discrete subgroup of $\mathrm{PSL}(2, \mathbb{C})$, then $M = \mathbb{H}^3/\Gamma$ is a good orbifold, because by Selberg's theorem, every such group contains a subgroup $\widetilde{\Gamma}$ of finite index that acts freely on $\mathbb{H}^3$. Thus the quotient $\widetilde{M} = \mathbb{H}^3/\widetilde{\Gamma}$ is a manifold and the natural projection $\widetilde{M} = \mathbb{H}^3/\widetilde{\Gamma} \to \mathbb{H}^3/\Gamma = M$ is a covering orbifold map. If M is compact, then $\widetilde{M}$ is also compact and therefore M is a very good orbifold.

One has:

Proposition 8.4.8. *Let Γ be a discrete subgroup of $\mathrm{PSL}(n + 1, \mathbb{C})$ acting properly discontinuously on an invariant open subset Ω of $\mathbb{P}^n_{\mathbb{C}}$. Then $M = \Omega/\Gamma$ is a good orbifold.*

Figure 8.1: The teardrop

This proposition follows from the fact that all isotropy groups must be finite, since the action is proper and the group is G discrete.

If M is a simply connected (G, X)-manifold, then one has the developing map $\mathcal{D} : M \to X$ which is a (G, X)-map. One can show that if $D : M \to X$ is any other (G, X)-map, then there is a unique $g \in G$ such that $\mathcal{D} = g \circ D$. There is also a unique group morphism $\mathcal{H}_{\mathcal{D}} : \mathrm{Aut}_{(G,X)}(M) \to G$, where $\mathrm{Aut}_{(G,X)}(M)$ denotes the group of (G, X)-equivalences of M, that verifies $\mathcal{D} \circ g = \mathcal{H}_{\mathcal{D}}(g) \circ \mathcal{D}$. Trivially, if D is another (G, X)-map, then there is a unique $g \in G$ such that $\mathcal{D} = g \circ D$ and $\mathcal{H}_{\mathcal{D}} = g \circ \mathcal{H}_D \circ g^{-1}$. Every (G, X)-map $\mathcal{D} : M \to X$ will be called a *developing map* and the corresponding $\mathcal{H}_{\mathcal{D}}$ is the *holonomy morphism* associated to $\mathcal{D}$.

Theorem 8.4.9 (Thurston, see [46]). *If M is an orbifold which admits a geometric structure, then its universal cover is a simply connected manifold. More precisely, if M is a (G, X)-orbifold, then M is a good orbifold. Moreover, there is a simply connected manifold $\widetilde{M}$ and an a (G, X)-covering orbifold map, with the property that if $q : \mathcal{O} \to M$ is another (G, X)-covering orbifold map, $* \in X_M$, $*' \in X_{\widetilde{M}}$ and $z \in X_{\mathcal{O}}$ satisfy $q(z) = p(*') = *$, then there is a (G, X)-covering orbifold map $p' : \widetilde{M} \to \mathcal{O}$ such that $p = q \circ p'$ and $p'(*') = z$.*

It is natural to call $\widetilde{M}$ the *universal covering orbifold* of M, the map p is the universal covering orbifold map and $\pi_1^{\mathrm{Orb}}(M) = \{g \in \mathrm{Aut}_{(G,X)}(M) : p \circ g = p\}$ the *orbifold fundamental group* of M. Also, if $\mathcal{D} : \widetilde{M} \to X$ is a developing map and $\mathcal{H}$ the holonomy morphism associated to $\mathcal{D}$, then $(\mathcal{D}, \mathcal{H}|_{\pi_1^{\mathrm{Orb}}(M)})$ will be called the *developing pair* associated to M.

Remark 8.4.10. Consider for instance a Fuchsian group $\Gamma \subset \mathrm{PSL}(2, \mathbb{R})$ of the second kind. Then the quotient of the upper half-plane $\mathbb{H}^+ = \{\Im z > 0\}$ by Γ is a manifold with a hyperbolic structure. It is easy to see that in this case the image of the corresponding developing map is all of $\mathcal{H}$ and the holonomy actually coincides with Γ. More generally, one can show (see [41]) that given a subgroup $\Gamma \subset G$ acting discontinuously over a nonempty Γ-invariant domain $\Omega \subset X$, there is a developing pair $(\mathcal{D}, \mathcal{H})$ for $M = \Omega/\Gamma$ such that $\mathcal{D}(\widetilde{M}) = \Omega$ and $\mathcal{H}(\pi_1(M)) = \Gamma$.

Corollary 8.4.11. *Let Γ, Ω, M, $\mathcal{D}$ and $\mathcal{H}$ be as in Remark 8.4.10 and q the natural quotient map. Then we have the following exact sequence of groups:*

$$0 \longrightarrow \pi_1(\Omega) \stackrel{i}{\longrightarrow} \pi_1^{\mathrm{Orb}}(M) \stackrel{\mathcal{H}}{\longrightarrow} \Gamma \longrightarrow 0$$

where i is the inclusion given by:

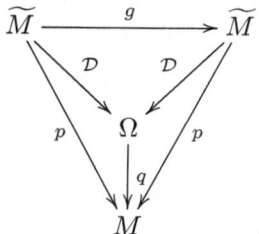

8.4.2 Description of the compact $(\mathbb{P}_{\mathbb{C}}^2, \mathrm{PSL}_3(\mathbb{C}))$-orbifolds

From now on, the material in this chapter is an outline of the material in [41], and we refer to that article for details. Using the fact that, by the aforementioned theorem of Thurston, every compact $(\mathbb{P}_{\mathbb{C}}^n, \mathrm{PSL}_{n+1}(\mathbb{C}))$-orbifold is a very good orbifold, the classification of $(\mathbb{P}_{\mathbb{C}}^2, \mathrm{PSL}_3(\mathbb{C}))$-manifolds given by Kobayashi, Ochiai, Inoue and Klingler (Theorem 8.3.2) can be easily extended to compact $(\mathbb{P}_{\mathbb{C}}^2, \mathrm{PSL}_3(\mathbb{C}))$-orbifolds as follows. For this we set

$$\mathrm{Aut}(\mathbb{C} \times \mathbb{C}^*) = \{g \in \mathrm{PSL}(3, \mathbb{C}) : g(\mathbb{C} \times \mathbb{C}^*) = \mathbb{C} \times \mathbb{C}^*\};$$

and

$$\mathrm{Aut}(\mathbb{C}^* \times \mathbb{C}^*) = \{g \in \mathrm{PSL}(3, \mathbb{C}) : g(\mathbb{C}^* \times \mathbb{C}^*) = \mathbb{C}^* \times \mathbb{C}^*\}$$

The following classification theorem strengthens Theorem 8.3.2.

Theorem 8.4.12. *Let M be a compact $(\mathbb{P}^2_{\mathbb{C}}, \mathrm{PSL}_3(\mathbb{C}))$-orbifold. Then M is either affine, complex hyperbolic, the quotient of $\mathbb{P}^2_{\mathbb{C}}$ by a finite group of automorphisms, or it has a (possibly ramified) finite covering which is a surface biholomorphic to a torus. More precisely, one has the following eight possibilities, the last four of them being affine.*

  (i) $\widetilde{M} = \mathbb{P}^2_{\mathbb{C}}$. *In this case $\mathcal{D} = \mathrm{Id}$ and $\mathcal{H}(\pi^{\mathrm{Orb}}_1(M))$ is finite. Hence M is of the form $\mathbb{P}^2_{\mathbb{C}}/\Gamma$ where Γ is a finite subgroup of $\mathrm{PSL}(3, \mathbb{C})$.*

 (ii) $\widetilde{M} = \mathbb{H}^2_{\mathbb{C}}$. *In this case $\mathcal{D} = \mathrm{Id}$ and $\mathcal{H}(\pi^{\mathrm{Orb}}_1(M)) \subset \mathrm{PU}(2, 1)$. Thence $M = \mathbb{H}^2_{\mathbb{C}}/\mathcal{H}(\pi^{\mathrm{Orb}}_1(M))$ and M has a complex hyperbolic structure.*

(iii) $\mathcal{D}(\widetilde{M}) = \mathbb{C}^* \times \mathbb{C}^*$. *In this case $\mathcal{H}(\pi^{\mathrm{Orb}}_1(M)) \subset \mathrm{Aut}(\mathbb{C}^* \times \mathbb{C}^*)(\mathbb{C}^2)$ and M has a finite covering (possibly ramified) which is a surface biholomorphic to a complex torus. Moreover, if $\Sigma_M = \emptyset$, then $\mathcal{H}(\pi_1(M)) \subset \mathrm{Aff}(\mathbb{C}^2)$.*

(iv) $\mathcal{D}(\widetilde{M}) = \mathbb{C} \times \mathbb{C}^*$. *In this case $\mathcal{H}(\pi^{\mathrm{Orb}}_1(M)) \subset \mathrm{Aut}(\mathbb{C} \times \mathbb{C}^*)(\mathbb{C}^2)$ has a subgroup of finite index contained on A_1 or A_2 and M has a finite covering (possibly ramified) which is a complex surface biholomorphic to a torus. Moreover, if $\Sigma_M = \emptyset$, then $\mathcal{H}(\pi_1(M)) \subset \mathrm{Aff}(\mathbb{C}^2)$.*

 (v) $\widetilde{M} = \mathbb{C}^2 - \{0\}$. *In this case $\mathcal{D} = \mathrm{Id}$, $\mathcal{H}(\pi^{\mathrm{Orb}}_1(M)) \subset \mathrm{Aff}(\mathbb{C}^2)$ contains a cyclic group of finite index, generated by a contraction, and M has a finite covering (possibly ramified) which is a Hopf surface.*

(vi) $\mathcal{D}(\widetilde{M}) = \mathbb{C}^2$. *In this case $\mathcal{H}(\pi^{\mathrm{Orb}}_1(M)) \subset \mathrm{Aff}(\mathbb{C}^2)$ contains an unipotent subgroup of finite index and M has a finite covering (possibly ramified) which is a surface biholomorphic to a complex torus or a primary Kodaira surface.*

(vii) $\widetilde{M} = \mathbb{C} \times \mathbb{H} = \mathcal{D}(\widetilde{M})$. *In this case M is an Inoue Surface and $\mathcal{H}(\pi_1(M)) \subset \mathrm{Aff}(\mathbb{C}^2)$ is a torsion free group contained on Sol^4_0 or Sol^4_1 or $\mathrm{Sol'}^4_1$. Thus in these case there there are not new quotients.*

(viii) $\widetilde{M}$ *is biholomorphic to $\mathbb{C} \times \mathbb{H}$. In this case $\mathcal{D}(\widetilde{M}) = \mathbb{C}^* \times \Omega$ where Ω is a hyperbolic domain contained in $\mathbb{P}^1_{\mathbb{C}}$ and M has a finite covering (possibly ramified) which is an elliptic affine surface.*

Remark 8.4.13. We notice that in the cases *i, ii, v* and *vii* above, the corresponding orbifold is uniformisable; this follows from Klingler's work in the first three cases, and by [98] and [231] in case vii. Conversely, Thurston proved that the orbifolds in vi are not uniformisable, and this same statement is proved in [116] for the orbifolds in case *viii*.

8.5   Discrete groups and divisible sets in dimension 2

In the previous sections we studied several types of discrete subgroups of $\mathrm{PSL}(3, \mathbb{C})$ acting on $\mathbb{P}^2_{\mathbb{C}}$ with a largest region of discontinuity Ω with compact quotient Ω/Γ.

One may call these actions *cocompact*. We now introduce a larger class of group actions that we call quasi-cocompact, and their study takes the rest of this chapter.

Definition 8.5.1. Let Γ be a discrete subgroup of $\mathrm{PSL}(3, \mathbb{C})$. The group Γ is *quasi-cocompact* if there is an invariant, nonempty, open set $\Omega \subset \mathbb{P}^2_{\mathbb{C}}$ where Γ acts properly discontinuously and the quotient Ω/Γ contains at least one connected component which is compact.

Example 8.5.2.   (i) Let $\Gamma \subset \mathrm{PSL}(2, \mathbb{C})$ be discrete. We say that Γ is quasi-cocompact if its region of discontinuity $\Omega(\Gamma)$ contains an open invariant set with compact quotient. Now suppose Γ is quasi-cocompact and $\Omega(\Gamma)/\Gamma$ has at least one connected component which is compact. Then the suspension of Γ with respect to $G = \{2^n : n \in \mathbb{Z}\}$ is a quasi-cocompact group with a connected component which is compact.

  (ii) Let $\Gamma \subset \mathrm{PSL}(2, \mathbb{C})$ be a Kleinian group, then by Theorem 8.3.2 its full suspension is not quasi-cocompact.

We now have the following result which extends Theorem 8.3.1 to orbifolds and to groups acting on open invariant sets of $\mathbb{P}^2_{\mathbb{C}}$. Furthermore, this theorem says that in dimension 2 there is not a simultaneous uniformisation theorem.

Theorem 8.5.3. *Let $\Gamma \subset \mathrm{PSL}(3, \mathbb{C})$ be a quasi-cocompact group. Then Γ is elementary, complex hyperbolic or affine.*

We recall (see Definition 3.3.9) that elementary means that its limit set in the Kulkarni sense has finitely many connected components.

We refer to [41] for the proof of this theorem; here we only sketch the main points in it. We remark that the proof of this theorem relies heavily on the work developed in Chapter 5 about controllable groups and (possibly nondiscrete) subgroups of $\mathrm{PSL}(2, \mathbb{C})$.

Outline of the proof. Let $p : \Omega \longrightarrow \Omega/\Gamma$ be the quotient map and let $M \subset \Omega/\Gamma$ be a compact connected component. Then $\widehat{\Omega} = p^{-1}(M)$ has the form $\Gamma(\Omega_0)$, where Ω_0 is a connected component of Ω. Set $\Gamma_0 = \mathrm{Isot}(\Omega_0, \Gamma)$, then by Remark 8.4.10 there is a developing pair $(\mathcal{D}, \mathcal{H})$ such that $\mathcal{D}(\widetilde{M}) = \Omega_0$ and $\mathcal{H}(\pi_1^{\mathrm{Orb}}(M)) = \Gamma_0$. Moreover by Theorem 8.4.12 one of the following cases must occur:

Case 1. One has $\Omega_0 = \mathbb{H}^2_{\mathbb{C}}$. From Theorem 7.3.1 we deduce that Γ is complex hyperbolic and $\mathbb{H}^2_{\mathbb{C}} = \Omega = \Omega_{\mathrm{Kul}}(\Gamma)$ is the largest open set where Γ acts discontinuously.

Case 2. The set Ω_0 is $\mathbb{C}^2 - \{0\}$, $\mathbb{C}^* \times \mathbb{C}^*$, $\mathbb{C}^* \times \mathbb{C}$, or $\mathbb{C}^2$. In all these cases the group Γ is obviously elementary.

Case 3. Let $\mathbb{H}^+$ be the upper half-plane in $\mathbb{C}$. One has $\Omega_0 = \bigcup_{z \in \mathbb{H}^+} \overleftrightarrow{e_1, z} - \{e_1\}$. Let $\ell_1, \ell_2 \subset \partial \Omega_0 \subset \mathbb{P}^2_{\mathbb{C}} - \Omega_\Gamma$ be distinct complex lines and $g \in \Gamma$. Then $\{e_1\} = \ell_1 \cap \ell_2$ and $g(\ell_1), g(\ell_2) \subset \mathbb{P}^2_{\mathbb{C}} - \Omega_\Gamma$ are different complex lines. Since $g(\ell_i) \cap \bigcup_{z \in \mathbb{H}^+} \overleftrightarrow{e_1, z} \neq \emptyset$

we deduce $\{e_1\} \subset g(\ell_i)$. Hence $g(e_1) = e_1$ and therefore we have a control group as in Chapter 5, $\Pi = \Pi_{p,\ell}$ where $p = e_1$ and $\ell = \overleftrightarrow{e_2, e_3}$.

Now let $p_1, p_2, p_3 \in \mathbb{R} \subset \ell$ be distinct elements and K a fundamental region for the action of Γ_0 on Ω_0. Since Ω/Γ is compact we deduce that p_i, $1 \leq i \leq 3$, is a cluster point of $\Gamma_0 \overline{K}$. Hence each p_i, $1 \leq i \leq 3$, is a cluster point of $\Pi(\Gamma_0)(\pi(\overline{K}))$. Thus from Theorem 5.1.3 and Corollaries 5.4.5, 5.4.9 we deduce that $\Pi(\Gamma_0)$ is nondiscrete and $\Lambda_{CG}(\Pi(\Gamma_0)) = \widehat{\mathbb{R}}$.

Now, since $\Pi(\Gamma(\Omega))$ is an open set which omits $\widehat{\mathbb{R}}$ and contains $\mathbb{H}^+$ we conclude that $\Lambda_{CG}(\Pi(\Gamma)) \neq \ell$. Finally, since $\Lambda_{CG}(\Pi(\Gamma_0)) \subset \Lambda_{CG}(\Pi(\Gamma))$ with $\Lambda_{CG}(\Pi(\Gamma_0))$ a circle in ℓ, by Theorem 5.1.3 we conclude that $\Lambda_{CG}(\Pi(\Gamma)) = \widehat{\mathbb{R}}$. Moreover from (vii) of Theorem 8.4.12 and Corollary 5.4.5 we deduce that $\Pi(\Gamma)\infty = \infty$, which ends this part, proving that the line $\overleftrightarrow{e_1, e_2}$ is invariant. Hence the group is affine.

Case 4. By hypothesis Ω_0 is of the form $\Omega \times \mathbb{C}^*$. So we can write $\Omega_0 = \bigcup_{z \in \Omega} \overleftrightarrow{z, e_1} - (\Omega \cup \{e_1\})$. Let $g \in \Gamma$. Since Ω is hyperbolic, in the sense of hyperbolic sets of $\mathbb{P}^1_{\mathbb{C}}$, we can assume that there exist $\ell_1, \ell_2 \subset \partial \Omega_0 \subset \mathbb{P}^2_{\mathbb{C}} - \Omega_\Gamma$ distinct complex lines such that $e_1 \in \ell_1 \cap \ell_2$ and $g(\ell_1) \neq \overleftrightarrow{e_3, e_2} \neq g(\ell_2)$. Let $z_1, z_2 \in \Omega$, then $\xi_{ij} = \overleftrightarrow{z_i, e_1} \cap g(\ell_j) \subset \mathbb{P}^2_{\mathbb{C}} - \Omega$. From this we get $\xi_{i,j} \in \{e_1, z_j\}$. Thus $e_1 \in g(\ell_i)$ ($i = 1, 2$) and therefore $g(e_1) = e_1$, so the group is weakly semi-controllable. Moreover, we have shown that any line in $\mathbb{P}^2_{\mathbb{C}} - \Omega_\Gamma$ which does not contain e_1 is equal to $\overleftrightarrow{e_3, e_2}$. Now we can consider $\Pi = \Pi_{p,\ell}$, where $p = e_1$ and $\ell = \overleftrightarrow{e_3, e_2}$. Thus we must consider the following cases:

(i) The group $\Pi(\Gamma_0)$ is nondiscrete. In this case, we must have $\overleftrightarrow{e_3, e_2} \subset \mathbb{P}^2_{\mathbb{C}} - \Omega$ and $e_1 \notin g(\overleftrightarrow{e_3, e_2}) \subset \mathbb{P}^2_{\mathbb{C}} - \Omega_\Gamma$ for every $g \in \Gamma$. That is $\Gamma(\overleftrightarrow{e_3, e_2}) = \overleftrightarrow{e_3, e_2}$. Hence in this case we have an invariant line and therefore the group is affine.

(ii) The group $\Pi(\Gamma_0)$ is discrete. In this case by (i) of Theorem 5.8.2 and Proposition 2.1.2 we deduce that $\mathrm{Ker}(\Pi)$ is infinite. By Proposition 2.1.2 there is $\gamma \in \mathrm{Ker}(\Pi)$ with infinite order. Since $\gamma(\ell \cup \{e_1\}) = \ell \cup \{e_1\}$ we conclude that

$$\gamma = \begin{pmatrix} a^{-2} & 0 & 0 \\ 0 & a & 0 \\ 0 & 0 & a \end{pmatrix} \quad \text{where } |a| \neq 1.$$

This implies $e_1 \notin \overleftrightarrow{e_3, e_2} \subset L_0(\Gamma) \subset \mathbb{P}^2_{\mathbb{C}} - \Omega_\Gamma$ and therefore $\Gamma(\overleftrightarrow{e_3, e_2}) = \overleftrightarrow{e_3, e_2}$. Hence we also have an invariant line in this case, so the group is affine. $\square$

8.6   Elementary quasi-cocompact groups of $\mathrm{PSL}(3, \mathbb{C})$

In this section we classify the elementary quasi-cocompact groups of $\mathrm{PSL}(3, \mathbb{C})$. We also describe their maximal region of discontinuity, the corresponding divisible sets in $\mathbb{P}^2_{\mathbb{C}}$, and the Kulkarni limit sets. First we state without proof the following two technical lemmas.

Lemma 8.6.1 ([41]). *Let* $\Gamma \subset (\mathrm{PSL}(3,\mathbb{C})$ *be a group,* $\ell \subset \mathbb{P}^2_\mathbb{C}$ *a complex line invariant under* Γ *and* $p \in \mathbb{P}^2_\mathbb{C} \setminus \ell$ *a fixed point by* Γ. *If there is a domain* $\Omega_0 \subset \ell$ *such that* Γ *acts discontinuously on* $\Omega \times \mathbb{C}^*$ *with compact quotient, then* $\mathrm{Ker}(\Pi \mid_{\Gamma_0})$ *is infinite.*

Lemma 8.6.2 ([221]). *Let* F *be a free abelian group acting on* $\mathbb{C}^2$ *freely and discontinuously. If the rank of* F *is less than or equal to 3, then the quotient space of* $\mathbb{C}^2/F$ *cannot be compact.*

Recall that in Section 8.3 we introduced the affine groups

$$A_1 = \left\{ \begin{pmatrix} 1 & 0 & b \\ 0 & a & 0 \\ 0 & 0 & 1 \end{pmatrix} : (a,b) \in \mathbb{C}^* \times \mathbb{C} \right\};$$

$$A_2 = \left\{ \begin{pmatrix} a & b & 0 \\ 0 & a & 0 \\ 0 & 0 & 1 \end{pmatrix} : (a,b) \in \mathbb{C}^* \times \mathbb{C}; \right\}.$$

We have

Theorem 8.6.3. *Let* $\Gamma \subset \mathrm{PSL}(3,\mathbb{C})$ *be a quasi-cocompact elementary group. If* Ω/Γ *is not a Hopf surface, then* $\Omega_{\mathrm{Kul}}(\Gamma)$ *is the largest open set on which* Γ *acts properly discontinuously, and up to projective equivalence we are in one of the following three cases:*

(i) $\Omega_{\mathrm{Kul}}(\Gamma) = \mathbb{C}^2$. *Then* Γ *is affine and it is a finite extension of a unipotent group. The quotient* $\Omega_{\mathrm{Kul}}(\Gamma)/\Gamma$ *has a finite covering (possibly ramified) which is a surface biholomorphic to a complex torus or a primary Kodaira surface. The Kulkarni limit set is a line.*

(ii) $\Omega_{\mathrm{Kul}}(\Gamma) = \mathbb{C} \times \mathbb{C}^*$. *Then the group* Γ *is a finite extension of a group isomorphic to* $\mathbb{Z}^3 = \mathbb{Z} \oplus \mathbb{Z} \oplus \mathbb{Z}$ *and it is contained in* A_1 *or* A_2. *Moreover,* $\Omega_{\mathrm{Kul}}(\Gamma)/\Gamma$ *has a finite covering (possibly ramified) which is a surface biholomorphic to a complex torus. The Kulkarni limit set consists of two lines.*

(iii) $\Omega_{\mathrm{Kul}}(\Gamma) = \mathbb{C}^* \times \mathbb{C}^*$. *Then* Γ *is a finite extension of a group isomorphic to* $\mathbb{Z}^2$ *and where each element is a diagonal matrix. Moreover,* $\Omega_{\mathrm{Kul}}(\Gamma)/\Gamma$ *has a finite covering (possibly ramified) which is a surface biholomorphic to a complex torus. The Kulkarni limit set now consists of three lines.*

Remark 8.6.4. If the quotient Ω/Γ is a Hopf surface, then the statement in Theorem 8.6.3 is false whenever there is a finite index cyclic subgroup G of Γ generated by a strongly loxodromic element. When this happens, we have two maximal regions of discontinuity and none of them coincides with Kulkarni's region of discontinuity, which equals the equicontinuity region.

Proof. Let us sketch the proof of this theorem. We refer to [41] for details. By the proof of Theorem 8.5.3 we must consider the following cases:

1) $\Omega = \mathbb{C}^* \times \mathbb{C}^*$;

2) $\Omega_0 = \mathbb{C} \times \mathbb{C}^*$;

3) $\Omega_0 = \mathbb{C}^2$.

These are the three cases envisaged in the statement of the theorem. Let us look at each of these separately:

Case 1) One has $\Omega = \mathbb{C}^* \times \mathbb{C}^*$. In this case Γ and Ω have the following properties.

Property 1. *The set Ω is the largest open set on which Γ acts properly discontinuously.*

The idea for this is as follows. By Selberg's lemma, we can assume (by simplicity) that Γ acts freely, so the quotient is a compact manifold. We can assume also that the complement of Ω consists of the three lines $\overleftrightarrow{e_1, e_2}$, $\overleftrightarrow{e_3, e_1}$ and $\overleftrightarrow{e_3, e_2}$. Now, for each of these lines, let $\Gamma_{\overleftrightarrow{e_i, e_j}}$ be the corresponding isotropy group, i.e., the set of elements in Γ that leave the line invariant.

Now we set: $\Gamma_1 = \Gamma_{\overleftrightarrow{e_1, e_2}} \cap \Gamma_{\overleftrightarrow{e_3, e_1}} \cap \Gamma_{\overleftrightarrow{e_3, e_2}}$. Since $\mathbb{C}^* \times \mathbb{C}^*$ is Γ-invariant, we deduce that Γ_1 is a normal subgroup of Γ and Γ/Γ_1 is a subgroup of S_3, the group of permutations in three elements. By Selberg's lemma there is a normal torsion free subgroup $\Gamma_0 \subset \Gamma_1$ with finite index. Thus Ω_0/Γ_0 is a compact manifold and $\Gamma_0 \, e_j = e_j$ $(1 \le j \le 3)$.

Now let $\Pi_i = \Pi_{e_i, \overleftrightarrow{e_j, e_k}}$ where $j, k \in \{1, 2, 3\} - \{i\}$ be the stabiliser of the point and the line in question. Then by Theorem 5.8.2 we deduce that $\mathrm{Ker}(\Pi_i)$ is either finite or nondiscrete. In both cases we have $\overleftrightarrow{e_1, e_2} \cup \overleftrightarrow{e_3, e_1} \cup \overleftrightarrow{e_3, e_2} \subset L_0(\Gamma) \cup L_1(\Gamma)$. Therefore $\overleftrightarrow{e_1, e_2} \cup \overleftrightarrow{e_3, e_1} \cup \overleftrightarrow{e_3, e_2}$ is contained in the complement of every open set Ω where Γ acts discontinuously. Thus Ω is the largest open set on which Γ acts discontinuously.

Property 2. *The group Γ contains a subgroup Γ_0 with finite index, isomorphic to $\mathbb{Z}^2 := \mathbb{Z} \oplus \mathbb{Z}$ and every element in it is a diagonal matrix.*

For this let $\Gamma_0 \subset \Gamma$ be a torsion free subgroup with finite index and such that the lines $\overleftrightarrow{e_1, e_2}, \overleftrightarrow{e_1, e_3}$ and $\overleftrightarrow{e_3, e_2}$ are Γ_0-invariant. Let $(\mathcal{D}, \mathcal{H})$ be a developing pair for Ω_0/Γ_0. By Remark 8.4.10 we can assume that $\Im(\mathcal{D}) = \Omega_0$ and $\Gamma_0 = \Im(\mathcal{H})$. By (iii) of Theorem 8.4.12 we can assume that $M = \Omega_0/\Gamma_0$ is a complex torus. Since $\pi_1(M) = \mathbb{Z}^4$, $\mathcal{H} : \pi_1(\Omega_0) \longrightarrow \Gamma_0$ is an epimorphism we conclude that Γ contains a free abelian subgroup $\check{\Gamma}$ of finite index and rank $k = 0, 1, 2, 3$ or 4. From here we deduce that $\mathcal{H}^{-1}(\check{\Gamma})$ is a free abelian group of rank 4, see [185]. Let $(\check{\mathcal{D}}, \check{\mathcal{H}})$ be the developing pair for $\check{M} = \Omega_0/\check{\Gamma}$ given by Remark 8.4.10. Then by Corollary 8.4.11 we have the following exact sequence of groups:

$$0 \longrightarrow \mathbb{Z}^2 = \pi_1(\Omega_0) \xrightarrow{\check{q}_*} \mathbb{Z}^4 = \pi_1(\check{M}) \xrightarrow{\check{\mathcal{H}}} \mathbb{Z}^k = \check{\Gamma} \longrightarrow 0 \qquad (8.6.4)$$

where $\check{q}_*$ is the group morphism induced by the quotient map $\check{q} : \Omega_0 \longrightarrow \check{M}$. Since (8.6.4) is a sequence of free abelian groups, we deduce that $\mathbb{Z}^4 = \mathbb{Z}^2 \oplus \mathbb{Z}^k$ (see [187]) and therefore $k = 2$, which ends the proof.

Case 2) If $\Omega_0 = \mathbb{C} \times \mathbb{C}^*$, then, as in the previous case, we can show that there is a subgroup $\Gamma_0 \subset \Gamma$ with finite index, isomorphic to $\mathbb{Z}^3$ and $\Gamma_0 \subset A_1$ or $\Gamma_0 \subset A_2$. Let us consider the case $\Gamma_0 \subset A_2$, the other case being similar. Then $\overleftrightarrow{e_1, e_2}$ and $\overleftrightarrow{e_1, e_3}$ are Γ_0-invariant.

Now define group morphisms $D_i : \Gamma_0 \longrightarrow \text{Bihol}(\overleftrightarrow{e_1, e_i})$, $i = 2, 3$, in the obvious way: $D_i(\gamma)$ equals the restriction of γ to the corresponding line, which is Γ_0-invariant. That is: $D_i(\gamma) = \gamma \mid_{\overleftrightarrow{e_1, e_i}}$. For simplicity we assume that $\text{Ker}(D_2)$ and $\text{Ker}(D_3)$ are trivial. Then $D_2(\Gamma_0)$ and $D_3(\Gamma_0)$ are isomorphic to $\mathbb{Z}^3$. On the other hand we observe that:

$$D_2(g_{ij})_{i,j=1}^3 (z) = z + g_{22}^{-1} g_{12},$$
$$D_3(g_{ij})_{i,j=1}^3 (z) = g_{11} z.$$

This implies that $D_2(\Gamma_0)$ is an additive subgroup of $\mathbb{C}$ and $D_3(\Gamma_0)$ is a multiplicative subgroup of $\mathbb{C}^*$. Thus $D_2(\Gamma_0)$ and $D_3(\Gamma_0)$ are nondiscrete groups, see [185]. Therefore $\overleftrightarrow{e_1, e_2} \cup \subset L_0(\Gamma) \cup L_1(\Gamma)$, which concludes the proof of this case.

Case 3) Now consider the case $\Omega_0 = \mathbb{C}^2$. By Selberg's lemma there is a normal torsion free subgroup $\Gamma_1 \subset \Gamma$ with finite index. Since $\overleftrightarrow{e_1, e_2}$ is Γ_1-invariant, we deduce that $D : \Gamma \longrightarrow \text{Bihol}(\overleftrightarrow{e_1, e_2})$ given by $D(\gamma) = \gamma \mid_{\overleftrightarrow{e_1, e_2}}$ is a group morphism.

Observe that if $\text{Ker}(D)$ is nontrivial or $D(\Gamma)$ is nondiscrete, then we have that $\overleftrightarrow{e_1, e_2} \subset L_0(\Gamma) \cup l_1(\Gamma)$. Therefore we can assume that $\text{Ker}(D)$ is trivial and $D(\Gamma)$ is discrete. From this we have that D is an isomorphism and every element in Γ_1 is unipotent (see (vi) of Theorem 8.4.12). Thus we conclude that every element in $\Gamma_1 - \{\text{Id}\}$ has a lift with the normal Jordan form

$$\begin{pmatrix} 1 & 1 & 0 \\ 0 & 1 & 1 \\ 0 & 0 & 1 \end{pmatrix}.$$

By Corollary 4.3.4 we deduce that $D(\Gamma_1)$ contains only parabolic elements. Since $D(\Gamma_1)$ is discrete we conclude that $D(\Gamma_1)$ is isomorphic to $\mathbb{Z}$ or $\mathbb{Z} \oplus \mathbb{Z}$, and since D is an isomorphism, by Lemma 8.6.2 we conclude that Ω_0/Γ_1 is noncompact, which is a contradiction. Therefore Ω_0 is the largest open set on which Γ acts discontinuously. $\qquad \square$

8.7 Nonelementary affine groups

We now look at the nonelementary quasi-cocompact affine groups. Recall that a subgroup $G \subset \text{PSL}(2, \mathbb{C})$ is quasi-cocompact if its discontinuity region contains an invariant open set with compact quotient.

Theorem 8.7.1. *Let $\Gamma \subset \text{PSL}(3, \mathbb{C})$ be quasi-cocompact, affine and nonelementary. Then $\Omega_{\text{Kul}}(\Gamma)$ is the largest open set on which Γ acts properly discontinuously, and up to projective equivalence, one of the following assertions applies:*

(i) $\Omega_{\mathrm{Kul}}(\Gamma) = \mathbb{C} \times (\mathbb{H}^- \cup \mathbb{H}^+)$ *and* $\Gamma_{\mathbb{C}\times\mathbb{H}}$ *is a torsion free group of* Sol_0^4 *or* Sol_1^4 *or* $\mathrm{Sol}_1'^4$. *In addition,* $\Omega_{\mathrm{Kul}}(\Gamma)/\Gamma$ *is equal to* M *or* $M \sqcup M$ *where* $M = (\mathbb{C} \times \mathbb{H})/\Gamma_{\mathbb{C}\times\mathbb{H}}$ *is an Inoue Surface and* $\sqcup$ *denotes disjoint union.*

(ii) $\Omega_{\mathrm{Kul}}(\Gamma) = \Omega \times \mathbb{C}^*$ *where* $\Omega \subset \mathbb{P}_{\mathbb{C}}^1$ *is the discontinuity region of a quasi-cocompact group* $G_\Gamma \subset \mathrm{M\ddot{o}b}(\widehat{\mathbb{C}})$, Γ *is a controllable group with quasi-cocompact control group and infinite kernel. In addition,* $\Omega_{\mathrm{Kul}}(\Gamma)/\Gamma = \bigsqcup_{i\in\mathcal{I}} N_i$ *where* $\mathcal{I}$ *is at most countable, the* N_i *are orbifolds whose universal covering orbifold is biholomorphic to* $\mathbb{H} \times \mathbb{C}$ *and every compact connected component is a finite covering (possibly ramified) of an elliptic affine surface.*

Proof. Take Ω_0, Γ_0 as in Theorem 8.5.3. Then we must consider the following two cases: $\Omega_0 = \mathbb{C} \times \mathbb{H}$ and $\Omega_0 = \Omega \times \mathbb{C}^*$.

Case 1. One has $\Omega_0 = \mathbb{C} \times \mathbb{H}$. Take Π, π and ℓ as in Theorem 8.5.3. Let $[z; r; 1] \in \partial\Omega_0 - \{e_1\}$, then $r \in \mathbb{R}$ and $z \in \mathbb{C}$. Since Ω/Γ is compact we deduce that there is a sequence $(g_n = [(\phi_{i,j}^{(n)})_{i,j=1}^3]_2)_{n\in\mathbb{N}} \subset \Gamma_0$ of distinct elements and $(k_n = [z_n; w_n; 1])_{n\in\mathbb{N}} \subset \Omega_0$ such that $z_n \xrightarrow[n\to\infty]{} z_0 \in \mathbb{C}$, $w_n \xrightarrow[n\to\infty]{} w_0 \in \mathbb{H}$ and $g_n(k_n) \xrightarrow[n\to\infty]{} [z; r; 1]$.

Since $(\Pi(g_n))_{n\in\mathbb{N}} \subset \mathrm{PSL}(2,\mathbb{R})$ and $\Pi(g_n)(\pi(k_n)) \xrightarrow[n\to\infty]{} r$, we can assume that the sequence $\Pi(g_n)$ converges uniformly on compact sets of $\pi(\Omega_0)$, i.e., $\Pi(g_n) \xrightarrow[n\to\infty]{} r$. From this we deduce $\phi_{22}^{(n)} \xrightarrow[n\to\infty]{} 0$ and $\phi_{23}^{(n)} \xrightarrow[n\to\infty]{} r$. Now consider the following three possible subcases:

1. $\Gamma_0 \subset \mathrm{Sol}_0^4$. In this case by Proposition 9.1 of [231] there is $M \in \mathrm{SL}(3,\mathbb{Z})$ with eigenvalues $\beta, \overline{\beta}, |\beta|^{-2}$, $\beta \neq \overline{\beta}$, a real eigenvector (a_1, a_2, a_3) belonging to $|\beta|^{-2}$ and an eigenvector (b_1, b_2, b_3) belonging to β such that $\Gamma_0 = \langle g_0, g_1, g_2, g_3 \rangle$, where g_i has a lift $\widetilde{g}_i$ given by

$$\widetilde{g}_0 = \begin{pmatrix} \beta & 0 & 0 \\ 0 & |\beta|^{-2} & 0 \\ 0 & 0 & 1 \end{pmatrix} ; \quad \widetilde{g}_i = \begin{pmatrix} 1 & 0 & b_i \\ 0 & 1 & a_i \\ 0 & 0 & 1 \end{pmatrix} \quad (1 \leq i \leq 3).$$

Let $i \in \{1, 2, 3\}$ be such that $b_i \neq 0$. Set $h_{n,m} = (\widetilde{g}_0{}^n \widetilde{g}_i \widetilde{g}_0{}^{-n})^m$. Then

$$h_{k,m} \circ h_{l,n} = \begin{pmatrix} 1 & 0 & b_i(n\beta^l + m\beta^k) \\ 0 & 1 & a_i(n|\beta|^{-2l} + m|\beta|^{-2k}) \\ 0 & 0 & 1 \end{pmatrix}, \quad k, l \in \mathbb{N}, \ n, m \in \mathbb{Z}.$$

Since β^n, β^{n+1} are linearly independent we have $b_i^{-1}(z - z_0) = r_n\beta^n + s_n\beta^{n+1}$ for some $r_n, s_n \in \mathbb{R}$. Then $b_i([r_n]\beta^n + [s_n]\beta^{n+1}) \xrightarrow[n\to\infty]{} z - w$.

If W is an open neighbourhood of the point $[z; r; 1]$, then there is $\epsilon \in \mathbb{R}^+$ such that:

$$W_0 = \{[w_1 : w_2 : 1] \mid |w_1 - z| < \epsilon \text{ and } |w_2 - r| < \epsilon\} \subset W.$$

Let $n_0 \in \mathbb{N}$ be such that

$$|b_i([r_{n_0}]\beta^{n_0} + [s_{n_0}]\beta^{n_0+1}) - z + z_0| < \epsilon,$$

where $[x]$ denotes the integer part of x. Define

$$\tau_n = [(\tau_{ij}^{(n)})]2)_{i,j=1}^3 = g_n \circ h_{n_0,[r_{n_0}]}^{-1} \circ h_{n_0+1,[s_{n_0}]}^{-1},$$

and $\tilde{k}_n = h_{n_0+1,[s_{n_0}]}(h_{n_0,[r_{n_0}]}(k_n)) = [\tilde{z}_n; \tilde{w}_n; 1]$. Then

$$\tilde{k}_n = [z_n + b_i([r_{n_0}]\beta^{n_0} + [s_{n_0}]\beta^{n_0+1}); w_n + a_i([r_{n_0}]|\beta|^{-2n_0} + [s_{n_0}]|\beta|^{-2(n_0+1)})].$$

Thus $\tau_n \tilde{k}_n \xrightarrow[n \to \infty]{} [z; r; 1]$ and

$$\tilde{z}_n \xrightarrow[n \to \infty]{} z_0 + b_i([r_{n_0}]\beta^{n_0} + [s_{n_0}]\beta^{n_0+1}),$$

$$\tilde{w}_n \xrightarrow[n \to \infty]{} w_0 + a_i([r_{n_0}]|\beta|^{-2n_0} + [s_{n_0}]|\beta|^{-2(n_0+1)}).$$

As in the case of the sequence $(g_n)_{n \in \mathbb{N}}$ we can deduce that $|\tau_{11}^{(n)}|^{-2} \xrightarrow[n \to \infty]{} 0$ and $\tau_{23}^{(n)} \xrightarrow[n \to \infty]{} r$. From here we deduce that

$$p_n = [\tau_{13}^{(n)}(1 - \tau_{11}^n)^{-1}; \tau_{23}^{(n)}(1 - \tau_{22}^n)^{-1}; 1]\} \in \mathrm{Fix}(g_n),$$

and

$$p_n \xrightarrow[n \to \infty]{} [z_0 + b_i([r_{n_0}]\beta^{n_0} + [s_{n_0}]\beta^{n_0+1}); r; 1] \in W_0 \cap L_0(\Gamma),$$

which completes the proof in this case.

2. $\Gamma_0 \subset \mathrm{Sol}_1^4$. In this case $\phi_{11}^{(n)} = \pm 1$ and

$$g_n(k_n) = [\pm z_n + \phi_{12}^{(n)} w_n + \phi_{13}^{(n)}; \phi_{22}^{(n)} w_n + \phi_{23}^{(n)}; 1].$$

Thus $\phi_{12}^{(n)} w_n + \phi_{13}^{(n)} \xrightarrow[n \to \infty]{} z \neq z_0$. Now we claim that $(\phi_{12}^{(n)})_{n \in \mathbb{N}}$ and $(\phi_{13}^{(n)})_{n \in \mathbb{N}}$ are bounded. Otherwise we have the following possibilities:

Possibility 1. $(\phi_{12}^{(n)})_{n \in \mathbb{N}}$ and $(\phi_{13}^{(n)})_{n \in \mathbb{N}}$ are unbounded. In this case we may assume that $\phi_{12}^{(n)}, \phi_{13}^{(n)} \xrightarrow[n \to \infty]{} \infty$. Then $(\phi_{12}^{(n)}(\phi_{13}^{(n)})^{-1})_{n \in \mathbb{N}}$ or $((\phi_{12}^{(n)})^{-1}(\phi_{13}^{(n)})_{n \in \mathbb{N}}$ are bounded. Assume without loss of generality that $\phi_{12}^{(n)}(\phi_{13}^{(n)})^{-1} \xrightarrow[n \to \infty]{} c \in \mathbb{C}$. Now, if $cw_o + 1 \neq 0$ we can deduce that

$$g_n k_n = [\frac{\pm z_n + \phi_{12}^{(n)} w_n}{\phi_{13}^{(n)}} + 1; \frac{\phi_{22}^{(n)} w_n + \phi_{23}^{(n)}}{\phi_{13}^{(n)}}; \frac{1}{\phi_{13}^{(n)}}] \xrightarrow[n \to \infty]{} [cw_0 + 1; 0; 0].$$

Since this is not the case, we conclude that $c \neq 0$ and $\phi_{13}^{(n)}(\phi_{12}^{(n)})^{-1} \xrightarrow[n \to \infty]{} -w_0$. Thus $\Im(\phi_{13}^{(n)}(\phi_{12}^{(n)})^{-1}) < 0$ for n large, which is a contradiction since, for all $n \in \mathbb{N}$, $\Im(\phi_{13}^{(n)}(\phi_{12}^{(n)})^{-1}) = 0$.

Possibility 2. $(\phi_{12}^{(n)})_{n\in\mathbb{N}}$ is bounded and $(\phi_{13}^{(n)})_{n\in\mathbb{N}}$ is unbounded. We may assume that $\phi_{12}^{(n)} \xrightarrow[n\to\infty]{} \phi_{12} \in \mathbb{C}$ and $\phi_{13}^{(n)} \xrightarrow[n\to\infty]{} \infty$. Then

$$g_n k_n = [\frac{z_n + \phi_{12}^{(n)} w_n}{\phi_{13}^{(n)}} + 1; \frac{\phi_{22}^{(n)} w_n + \phi_{23}^{(n)}}{\phi_{13}^{(n)}}; \frac{1}{\phi_{13}^{(n)}}] \xrightarrow[n\to\infty]{} [1; 0; 0].$$

This is not possible. Hence we can deduce $\phi_{12}^{(n)} \xrightarrow[n\to\infty]{} \phi_{12} \in \mathbb{C}$ and $\phi_{13}^{(n)} \xrightarrow[n\to\infty]{} \phi_{13} \in \mathbb{C}$. This implies that

$$g_n[\mp z - \phi_{13} - i\phi_{12}; i; 1]$$
$$= [z + i(\phi_{12}^{(n)} - \phi_{12}) + \phi_{13}^{(n)} - \phi_{13}; \phi_{22}^{(n)} i + \phi_{33}^{(n)}; 1] \xrightarrow[n\to\infty]{} [z; r; 1].$$

That is $[z; r; 1] \in L_0(\Gamma)$.

3. Assume now $\Gamma_0 \subset \mathrm{Sol}_1'^4$. By Proposition 9.1 of [231] we know there is a nontrivial $\gamma \in \mathrm{Ker}(\Pi\,|_{\Gamma_0})$. Now one has, see [41], that γ has a lift $\tilde\gamma$ given by

$$\tilde\gamma = \begin{pmatrix} 1 & 0 & \gamma_{13} \\ 0 & 1 & 0 \\ 0 & 0 & 1 \end{pmatrix}.$$

Let $(\gamma_n)_{n\in\mathbb{N}} \subset \Gamma_0$, where each element has a given lift $(\gamma_{ij}^{(n)})_{i,j=1}^3$, be a sequence such that $(\Pi(\gamma_n))_{n\in\mathbb{N}}$ is a sequence of distinct elements and $\Pi(\gamma_n) \xrightarrow[n\to\infty]{} \mathrm{Id}$. From here $\gamma_{22}^{(n)} \xrightarrow[n\to\infty]{} 1$, $\gamma_{23}^{(n)} \xrightarrow[n\to\infty]{} 0$ and $\Im(\gamma_{13}^{(n)}) \xrightarrow[n\to\infty]{} 0$. Thus we can assume that there is $(l_n)_{\mathbb{Z}} \in \mathbb{Z}$ such that $\gamma_{13}^{(n)} + l_n\gamma_{13} \xrightarrow[n\to\infty]{} c \in \mathbb{C}$. Then one can check that

$$\gamma^{l_n} \circ \gamma_n[z; 0; 1] = [z + \gamma_{13}^{(n)} + l_n\gamma_{13}; \gamma_{23}^{(n)}; 1] \xrightarrow[n\to\infty]{} [z + c; 0; 1].$$

Thus $\overleftrightarrow{e_1, e_2} \subset L_0(\Gamma)$. To conclude observe that $g_n(\overleftrightarrow{e_1, e_2}) \xrightarrow[n\to\infty]{} \overleftrightarrow{e_1, [0; r; 1]}$.

Case 2. One has $\Omega_0 = \Omega \times \mathbb{C}^*$. Take Π, π and l as in the proof of the case $\Omega_0 = \Omega \times \mathbb{C}^*$ of Theorem 8.5.3. Then $\Pi(\Gamma)$ has the following properties:

Property 1. *The group* $\Pi(\Gamma)$ *is discrete.*

Otherwise we have that there is a sequence $(\gamma_n = [(\gamma_{i,j}^{(n)})_{i,j=1}^3]_2)_{n\in\mathbb{N}} \subset \Gamma$ such that $(\Pi(g_n))_{n\in\mathbb{N}}$ is a sequence of distinct elements that verify $\Pi(g_n) \xrightarrow[n\to\infty]{} \mathrm{Id}$. This implies that

$$\sqrt{\gamma_{33}^{(n)}}\gamma_{11}^{(n)}, \sqrt{\gamma_{33}^{(n)}}\gamma_{22}^{(n)} \xrightarrow[n\to\infty]{} 1 \quad \text{and} \quad \sqrt{\gamma_{33}^{(n)}}\gamma_{12}^{(n)}, \sqrt{\gamma_{33}^{(n)}}\gamma_{21}^{(n)} \xrightarrow[n\to\infty]{} 0.$$

Now by Lemma 8.6.1 there is an element $\gamma = [(\gamma_{ij})_{i,j=1}^3]_2 \in \operatorname{Ker}(\Gamma)$ with infinite order. Thus we can assume there is $(l_n)_{n \in \mathbb{N}} \subset \mathbb{Z}$ such that $\gamma_{11}^{2l_n} \gamma_{33}^{(n)} \xrightarrow[n \to \infty]{} h^2 \in \mathbb{C}^*$. Hence

$$((\gamma_{ij})_{i,j=1}^3)^{-l_n} (\gamma_{ij}^{(}n))_{i,j=1}^3) \begin{pmatrix} \gamma_{11}^{-l_n} \gamma_{11}^{(n)} & \gamma_{11}^{-l_n} \gamma_{12}^{(n)} & 0 \\ \gamma_{11}^{-l_n} \gamma_{21}^{(n)} & \gamma_{11}^{-l_n} \gamma_{22}^{(n)} & 0 \\ 0 & 0 & \gamma_{11}^{2l_n} \gamma_{33}^{(n)} \end{pmatrix}$$

$$\xrightarrow[n \to \infty]{} \begin{pmatrix} h^{-1} & 0 & 0 \\ 0 & h^{-1} & 0 \\ 0 & 0 & h^2 \end{pmatrix}.$$

This is a contradiction since Γ is discrete.

Property 2. *The group* $\Pi(\Gamma_0)$ *acts discontinuously on* $\pi(\Omega_0)$.

This follows because $\partial(\pi(\Omega_0))$ is closed and $\Pi(\Gamma_0)$-invariant, so we have $\Lambda(\Pi(\Gamma_0)) \subset \partial \pi(\Omega_0)$.

Property 3. *The quotient* $\pi(\Omega_0)/\Pi(\Gamma_0)$ *is a compact orbifold.*

Let $R \subset \Omega_0$ be a fundamental domain for the action of Γ_0 on Ω_0. Then $\overline{\pi(R)} = \pi(\overline{R}) \subset \pi(\Omega_0)$ is compact and the assertion follows.

Property 4. *The set* $\bigcup_{w \in \Omega(\Pi(\Gamma))} \overleftrightarrow{w, e_3} - (\{e_3\} \cup \overleftrightarrow{e_1, e_2})$ *is the largest open set where* Γ *acts discontinuously.*

Without loss of generality we can assume that the points $[1; 1; 0]$, $[1; 0; 0]$ and $[0; 1; 0]$ are in $\Lambda(\Pi(\Gamma_0))$. We will show that $l = \overleftrightarrow{[1; 1; 0], e_3} \subset \mathbb{P}_\mathbb{C}^2 - \Omega$. Let $[1; 1; z] \in l$; since $\operatorname{Ker}(\Pi)$ is infinite we can assume that $z \neq 0$. Let $(g_n)_{n \in \mathbb{N}} \subset \Gamma_0$, where each element has the lift $(a_{ij}^{(n)})_{i,j=1}^3$, be such that $(\Pi(g_n))_{n \in \mathbb{N}}$ is a sequence of distinct elements with $\Pi(g_n) \xrightarrow[n \to \infty]{} [1; 1; 0]$ uniformly on compact sets of $\pi(\Omega_0)$. Set $[z_0; 1; 1] \in \pi(\Omega_0)$. Then we can assume that there is $l_n \in \mathbb{Z}$ such that

$$\frac{a_{33}^{(n)} \gamma_{11}^{3l_n}}{a_{21}^{(n)} z_0 + a_{22}^{(n)}} \xrightarrow[n \to \infty]{} c \in \mathbb{C}^*.$$

Finally, let $\gamma \in \operatorname{Ker}(\Gamma) = [(\gamma_{ij})_{i,j=1}^3]_2$. Then we have the convergence

$$\gamma^{-l(n)} \gamma_n [z_0; 1; zc^{-1}] = [\pi g_n(z_0); 1; \frac{a_{33}^{(n)} \gamma_{11}^{3l(n)} c^{-1} z}{a_{21}^{(n)} z_0 + a_{22}^{(n)}}] \xrightarrow[n \to \infty]{} [1; 1; z],$$

and Proposition 3.3.3 completes the proof. $\qquad \square$

Remark 8.7.2. Observe that from the above proof we deduce that for the fundamental groups $\pi_1(M)$ of Inoue surfaces that satisfy $\pi_1(M) \subset \operatorname{Sol}_0^4$, we have $\operatorname{Eq}(\pi_1(M)) = \emptyset$.

8.8   Concluding remarks

We now summarise the information we have about quasi-cocompact subgroups of $\mathrm{PSL}(3,\mathbb{C})$, and we discuss some lines of further research.

8.8.1   Summary of results for quasi-cocompact groups

Theorem 8.4.12 says that there are four types of **compact orbifolds of complex dimension 2** having a $(\mathbb{P}^2_{\mathbb{C}}, \mathrm{PSL}(3,\mathbb{C}))$-structure. These are either:

- quotients of $\mathbb{P}^2_{\mathbb{C}}$ by a finite group;

- complex hyperbolic;

- elementary (i.e. the holonomy leaves invariant finitely many lines);

- affine.

Then we looked (in Section 8.5) at **quasi-cocompact actions** on $\mathbb{P}^2_{\mathbb{C}}$. In this setting we have a group $\Gamma \subset \mathrm{PSL}(3,\mathbb{C})$ acting on $\mathbb{P}^2_{\mathbb{C}}$ so that there is an open invariant set Ω where the action is discontinuous and the quotient Ω/Γ is a compact $(\mathbb{P}^2_{\mathbb{C}}, \mathrm{PSL}(3,\mathbb{C}))$-orbifold. Theorem 8.5.3 says all such actions are either **complex hyperbolic, affine or elementary.**

We then studied **the elementary case** in Section 8.6. We proved that there are essentially four possible cases, corresponding to the nature of the limit set, which can be:

- a line;

- two lines;

- three lines,

- a line and a point;

and we gave examples of each type. In several of these cases the corresponding group is actually affine or complex hyperbolic, but there are examples which are of none of these types, as for instance the kissing-Schottky groups in Chapter 9.

We then looked (in Section 8.7) at **the nonelementary affine case**; we saw that the Kulkarni discontinuity set $\Omega_{\mathrm{Kul}}(\Gamma)$ is the largest open set on which Γ acts discontinuously. There are essentially two different possibilities for quasi-cocompact groups:

- either the orbit space is an Inoue surface (or a union of two such surfaces); or

- the group Γ is controllable and every compact connected component of $\Omega_{\mathrm{Kul}}(\Gamma)/\Gamma$ is a finite covering of an elliptic affine surface.

If Γ is **quasi-cocompact** and there is an open invariant set Ω such that Ω/Γ contains a connected component which is compact and **complex hyperbolic**, then by Theorem 7.3.1 we have that, up to conjugation, $\mathbb{H}_{\mathbb{C}}^2$ is the maximal region of discontinuity of Γ, and the Chen-Greenberg limit set $\Lambda_{CG}(\Gamma)$ is the whole sphere at infinity, $\mathbb{S}_{\infty}^3$. This is the setting we envisaged in Chapter 7. The Kulkarni limit set $\Lambda_{\mathrm{Kul}}(\Gamma)$ is now the union of all complex projective lines in $\mathbb{P}_{\mathbb{C}}^2$ which are tangent to $\mathbb{S}_{\infty}^3$ at points in $\Lambda_{CG}(\Gamma)$. Moreover, the orbit of each of these lines is dense in $\Lambda_{\mathrm{Kul}}(\Gamma)$, and its complement $\Omega_{\mathrm{Kul}}(\Gamma) = \mathbb{P}_{\mathbb{C}}^2 \setminus \Lambda_{\mathrm{Kul}}(\Gamma)$ coincides with the region of equicontinuity and it is the maximal open set where the action is discontinuous (Theorem 7.7.1).

It is worth noting that if $\Gamma \subset \mathrm{PU}(2,1)$ is such that its limit set $\Lambda_{CG}(\Gamma)$ is not all of $\mathbb{S}_{\infty}^3$, then something interesting happens, noticed in [162]: there is a connected component Ω of $\Omega_{\mathrm{Kul}}(\Gamma)$ which contains $\mathbb{H}_{\mathbb{C}}^2$, and it is a Γ-invariant complete Kobayashi-hyperbolic metric space.

8.8.2 Comments and open questions

If we now drop the condition of "quasi-cocompactness" we know for instance that there are kissing-Schottky groups in $\mathrm{PSL}(3,\mathbb{C})$ which are neither elementary, nor affine nor hyperbolic: This type of groups will be studied in Chapter 9. Can we describe all types of discrete subgroups of $\mathrm{PSL}(3,\mathbb{C})$ in some comprehensible way? In other words:

Problem 1. *Study the "zoo" of Kleinian groups in dimension 2.*

Of course this includes, among other things, understanding the discrete subgroups of $\mathrm{PU}(2,1)$ and this is already a rich enough problem in itself, with lots of interesting ongoing research and questions. For instance, much of the interesting results concerning complex hyperbolic Kleinian groups on $\mathbb{H}_{\mathbb{C}}^2$ start by looking at a discrete subgroup Γ of $\mathrm{PSL}(2,\mathbb{R})$, which is a 3-dimensional Lie group, isomorphic to $\mathrm{PU}(1,1)$. Then one observes that this group embeds as a subgroup of $\mathrm{PU}(2,1)$, the group of hyperbolic motions of complex hyperbolic 2-space $\mathbb{H}_{\mathbb{C}}^2$. Since $\mathrm{PU}(2,1)$ has real dimension 8, we have a lot of "freedom" to deform Γ in $\mathrm{PU}(2,1)$, and one gets very interesting phenomena. Now we observe that $\mathrm{PU}(2,1)$ is naturally a subgroup of $\mathrm{PSL}(3,\mathbb{C})$, which has complex dimension 15, and we may as well deform Γ in $\mathrm{PSL}(3,\mathbb{C})$ where we have lots more space!. This brings us to the next problem:

Problem 2. *Study deformations in $\mathrm{PSL}(3,\mathbb{C})$ of discrete subgroups of $\mathrm{PSL}(2,\mathbb{R})$.*

Of course this line of research is in some sense analogous to that followed by M. Burger and A. Lozzi, studying representations of surface groups in $\mathrm{PU}(n,k)$.

It would also be interesting to compare the deformations of the group in $\mathrm{PSL}(3,\mathbb{C})$ with the corresponding deformations of the manifold obtained as quotient of its Kulkarni region of discontinuity by the group action. For instance, in the following chapter we study a type of Schottky groups in $\mathrm{PSL}(2n+1,\mathbb{C})$ that produce compact manifolds as quotient. They have a rich deformation theory and their infinitesimal deformations can be regarded as deformations of the embedding of the group in $\mathrm{PSL}(2n+1, \mathbb{C})$. Yet, this type of groups do not appear in dimension 2, as we see in Chapter 9.

Chapter 9

Complex Schottky Groups

Classical Schottky groups in $\mathrm{PSL}(2, \mathbb{C})$ play a key role in both complex geometry and holomorphic dynamics. On one hand, Köbe's retrosection theorem says that every compact Riemann surface can be obtained as the quotient of an open set in the Riemann sphere $\mathbb{S}^2$ which is invariant under the action of a Schottky group. On the other hand, the limit sets of Schottky groups have rich and fascinating geometry and dynamics, which has inspired much of the current knowledge we have about fractal sets and 1-dimensional holomorphic dynamics.

In this chapter we study generalisations to higher dimensions of the classical Schottky groups.

Firstly we study a construction of Schottky groups from [203], which is similar to a previous construction in [164] aimed at giving a method for constructing compact complex manifolds with a rich geometry. In dimension 3 the manifolds one gets are reminiscent of the *Pretzel Twistor Spaces* of [176]. We then give a general abstract definition, from [38], of complex Schottky groups, which captures the essence of this type of groups and generalises a characterisation given by Maskitt for Schottky subgroups of $\mathrm{PSL}(2, \mathbb{C})$.

In Section 9.3 we focus on the Schottky groups from [203]. We study the limit sets of these actions on $\mathbb{P}_{\mathbb{C}}^{2n+1}$, which turn out to be solenoids of the form $\mathcal{C} \times \mathbb{P}_{\mathbb{C}}^n$, where $\mathcal{C}$ is a Cantor set. We also look at the compact complex manifolds obtained as the quotient of the region of discontinuity, divided by the action. We determine their topology and the dimension of the space of their infinitesimal deformations. This shows that for $n > 2$, every such deformation arises from a deformation of the embedding of the group in question in $\mathrm{PSL}(2n+2, \mathbb{C})$: reminiscent of the classical Teichmüller theory.

The Schottky groups of [164] and [203] are all groups of automorphims of odd-dimensional complex projective spaces, and their construction extends easily to real projective spaces of odd dimension (see for instance [92]). In Section 9.4 we follow [38] and prove, using the results of Section 9.2, that Schottky groups do not exist in even dimensions (neither over the complex nor the real numbers).

The proof requires having a certain understanding of the dynamics of projective transformations in general, so we begin Section 9.4 by doing so.

Finally, in Section 9.5 we show that if we relax slightly the conditions imposed for defining Schottky groups, we get *kissing-Schottky* groups that exist in all dimensions. The examples of kissing-Schottky groups that we give in this section are also interesting because they are not elementary (see the text for the definition), nor affine, nor complex hyperbolic, a fact which shows that the hypothesis in Chapter 8 of the group being *quasi-cocompact* is necessary to get the conclusions of that chapter.

9.1   Examples of Schottky groups

Recall that classical Schottky groups are obtained by considering pairs of pairwise disjoint closed n-discs $D_1, \ldots, D_{2r}$ in $\mathbb{S}^n$, and for each pair of discs D_{2i-1}, D_{2i} one has a Möbius transformation T_i of $\mathbb{S}^n$ such that $T_i(D_{2i-1}) = \overline{\mathbb{S}^n - D_{2i}}$. The Schottky group is defined to be the group of conformal maps generated by these transformations. The main purpose of this chapter is to study generalisations of Schottky groups to higher-dimensional complex projective spaces, and we begin with some examples:

9.1.1   The Seade-Verjovsky complex Schottky groups

Consider the subspaces of $\mathbb{C}^{2n+2} = \mathbb{C}^{n+1} \times \mathbb{C}^{n+1}$ defined by $L_0 := \{(a,0) \in \mathbb{C}^{2n+2}\}$ and $M_0 := \{(0,b) \in \mathbb{C}^{2n+2}\}$. Let S be the involution of $\mathbb{C}^{2n+2}$ defined by $S(a,b) = (b,a)$, which clearly interchanges L_0 and M_0.

Lemma 9.1.1. *Let $\Phi \colon \mathbb{C}^{2n+2} \to \mathbb{R}$ be given by $\Phi(a,b) = |a|^2 - |b|^2$. Then:*

(i) *The set $E_S := \Phi^{-1}(0)$ is a real algebraic hypersurface in $\mathbb{C}^{2n+2}$ with an isolated singularity at the origin 0. It is embedded in $\mathbb{C}^{2n+2}$ as a (real) cone over $\mathbb{S}^{2n+1} \times \mathbb{S}^{2n+1}$, with vertex at $0 \in \mathbb{C}^{2n+2}$.*

(ii) *This set E_S is invariant under multiplication by $\lambda \in \mathbb{C}$, so it is in fact a complex cone that separates $\mathbb{C}^{2n+2} \setminus \{(0,0)\}$ in two diffeomorphic connected components U and V, which contain respectively $L_0 \setminus \{(0,0)\}$ and $M_0 \setminus \{(0,0)\}$. These two components are interchanged by the involution S, for which E_S is an invariant set.*

(iii) *Every linear subspace K of $\mathbb{C}^{2n+2}$ of dimension $n+2$ containing L_0 meets transversally E_S and M_0. Therefore a tubular neighbourhood V of $M_0 \setminus \{(0,0)\}$ in $\mathbb{P}^{2n+1}_{\mathbb{C}}$ is obtained, whose normal disc fibres are of the form $K \cap V$, with K as above.*

Proof. The first part of statement (i) is clear because Φ is a quadratic form with $0 \in \mathbb{C}^{2n+2}$ as unique critical point. That $E_S \cap \mathbb{S}^{4n+3} = \mathbb{S}^{2n+1} \times \mathbb{S}^{2n+1} \subset \mathbb{C}^{2n+2}$ is because this intersection consists of all pairs (x,y) so that $|x| = |y| = \frac{1}{\sqrt{2}}$. That S

leaves E_S invariant is obvious, and so is that S interchanges the two components of $\mathbb{C}^{2n+2} \setminus \{(0,0)\}$ determined by E_S, which must be diffeomorphic because S is an automorphism.

Now let K be a subspace as in (iii), then K meets transversally E_S because through every point in E_S there exists an affine line in K which is transverse to E_S. $\qquad\square$

Since E_S is a complex cone, we have that $[E_S \setminus \{0\}]_{2n+1}$ is a codimension 1 real submanifold of $\mathbb{P}^{2n+1}_{\mathbb{C}}$, that we denote simply by E_S.

Corollary 9.1.2. (i) E_S *is an invariant set of* $[S]_{2n+1}$.

(ii) E_S *is an* $\mathbb{S}^{2n+1}$*-bundle over* $\mathbb{P}^n_{\mathbb{C}}$*; in fact* E_S *is the sphere bundle associated to the holomorphic bundle* $(n+1)\mathcal{O}_{\mathbb{P}^n_{\mathbb{C}}}$*, which is the normal bundle of* $\mathbb{P}^n_{\mathbb{C}}$ *in* $\mathbb{P}^{2n+1}_{\mathbb{C}}$.

(iii) E_S *separates* $\mathbb{P}^{2n+1}_{\mathbb{C}}$ *in two connected components which are interchanged by* $[S]_{2n+1}$ *and each one is diffeomorphic to a tubular neighbourhood of the canonical* $\mathbb{P}^n_{\mathbb{C}}$ *in* $\mathbb{P}^{2n+1}_{\mathbb{C}}$.

Definition 9.1.3. We call E_S the *canonical mirror* and $[S]_{2n+1}$ the *canonical involution*.

It is an exercise to show that Lemma 9.1.1 holds in the following more general setting. Of course one has the equivalent of Corollary 9.1.2 too.

Lemma 9.1.4. *Let* λ *be a positive real number and consider the involution*

$$S_\lambda : \mathbb{C}^{n+1} \times \mathbb{C}^{n+1} \to \mathbb{C}^{n+1} \times \mathbb{C}^{n+1} \,,$$

given by $S_\lambda(a,b) = (\lambda b, \lambda^{-1} a)$. *Then* S_λ *also interchanges* L_0 *and* M_0, *and the set*

$$E_\lambda = \{(a,b) : |a|^2 = \lambda^2 |b|^2\}$$

satisfies, with respect to S_λ, *the analogous properties* (i)–(iii) *of Lemma 9.1.1 above.*

We notice that as λ tends to ∞, the manifold E_λ gets thinner and approaches the L_0-axes.

Consider now two arbitrary disjoint projective subspaces L and M of dimension n in $\mathbb{P}^{2n+1}_{\mathbb{C}}$. As before, we denote by $[\ ]_{2n+1}$ the natural projection $\mathbb{C}^{2n+2} \to \mathbb{P}^{2n+1}_{\mathbb{C}}$. It is clear that $\mathbb{C}^{2n+2}$ splits as a direct sum $\mathbb{C}^{2n+2} = [L]^{-1}_{2n+1} \oplus [M]^{-1}_{2n+1}$ and there is a linear automorphism H of $\mathbb{C}^{2n+2}$ taking $[L]^{-1}_{2n+1}$ to L_0 and $[L]^{-1}_{2n+1}$ to M_0. For every $\lambda \in \mathbb{R}_+$ the automorphism $[H^{-1} \circ S_\lambda \circ H]_{2n+1}$ is an involution of $\mathbb{P}^{2n+1}_{\mathbb{C}}$ that interchanges L and M. Conversely one has:

Lemma 9.1.5. *Let* T *be a linear projective involution of* $\mathbb{P}^{2n+1}_{\mathbb{C}}$, *with a lifting* $\tilde{T}$ *that interchanges* L *and* M. *Then* T *is conjugate in* $\mathrm{PSL}(2n+2, \mathbb{C})$ *to the canonical involution* S.

Proof. Let $\{l_1, \ldots, l_{n+1}\}$ be a basis of $[L]_{2n+1}^{-1}$. Then

$$\{l_1, \ldots, l_{n+1}, \tilde{T}(l_1), \ldots, \tilde{T}(l_{n+1})\}$$

is a basis of $\mathbb{C}^{2n+2}$. The linear transformation that sends the canonical basis of $\mathbb{C}^{2n+2} = \mathbb{C}^{n+1} \oplus \mathbb{C}^{n+1}$ to this basis induces a projective transformation which realises the required conjugation. $\square$

Definition 9.1.6. A *mirror* in $\mathbb{P}_{\mathbb{C}}^{2n+1}$ means the image of the canonical mirror E_S under an element of $\mathrm{PSL}(2n+2, \mathbb{C})$.

Observe that a mirror is the boundary of a tubular neighbourhood of a $\mathbb{P}_{\mathbb{C}}^{n}$ in $\mathbb{P}_{\mathbb{C}}^{2n+1}$, so it is an $\mathbb{S}^{2n+1}$-bundle over $\mathbb{P}_{\mathbb{C}}^{n}$. For $n = 0$ mirrors are just circles in $\mathbb{P}_{\mathbb{C}}^{1} \cong \mathbb{S}^2$; one has that for $n = 1$ (and only in these dimensions) this bundle is trivial, so in $\mathbb{P}_{\mathbb{C}}^{3}$ mirrors are copies of $\mathbb{S}^3 \times \mathbb{S}^2$.

One has:

Lemma 9.1.7. *Let L and M be as above. Given an arbitrary constant λ, $0 < \lambda < 1$, we can find an involution T interchanging L and M, with a mirror E such that if U^* is the open component of $\mathbb{P}_{\mathbb{C}}^{2n+1} \setminus E$ which contains M and $x \in U^*$, then $d(T(x), L) < \lambda d(x, M)$, where the distance d is induced by the Fubini-Study metric.*

Proof. The involution $T_\lambda := H^{-1} \circ S_\lambda \circ H$, with H and S_λ as above, satisfies the lemma. $\square$

We notice that the parameter λ in Lemma 9.1.7 gives control upon the degree of expansion and contraction of the generators of the group, so one can estimate bounds on the Hausdorff dimension of the limit set (see Subsection 9.3.2 below).

The previous discussion can be summarised in the following theorem:

Theorem 9.1.8. *Let $L_1, \ldots, L_r$ be disjoint projective subspaces of dimension n; $r > 1$, then:*

(i) *There exist involutions $T_1, \ldots, T_r$ of $\mathbb{P}_{\mathbb{C}}^{2n+1}$ with mirrors E_{T_j} such that if N_j denotes the connected component of $\mathbb{P}^{2n+1} \setminus E_{T_j}$ that contains L_j, then $\{N_j \cup E_{T_j}\}$ is a closed family of pairwise disjoint sets.*

(ii) *$\Gamma = \langle T_1, \ldots, T_r \rangle$ is a discrete group with a nonempty region of discontinuity.*

(iii) *Given a constant $C > 0$, we can choose the T_j's so that if $T := T_{j_1} \cdots T_{j_k}$ is a reduced word of length $k > 0$ (i.e., $j_1 \neq j_2 \neq \cdots \neq j_{k-1} \neq j_k$), then $T(N_i)$ is a tubular neighbourhood of the projective subspace $T(L_i)$ which becomes very thin as k increases: $d(x, T(L_i)) < C\lambda^k$ for all $x \in T(N_i)$.*

A group as in Theorem 9.1.8 was called *Complex Schottky* in [203].

9.1.2 Nori's construction of complex Schottky groups

In [164] there is the following construction of higher-dimensional analogues of the classical Schottky groups: let $n = 2k + 1$, $k > 1$ and $g \geq 1$. Choose $2g$ mutually disjoint projective subspaces $L_1, \ldots, L_{2g}$ of dimension k in $\mathbb{P}^n_{\mathbb{C}}$ and $0 < \alpha < \frac{1}{2}$. For every integer $1 \leq j \leq g$ choose a basis of $\mathbb{C}^{n+1}$ so that $L_j = [\{z_0, \ldots, z_k = 0\} \backslash \{0\}]_n$ and $L_{g+j} = [\{z_{k+1}, \ldots, z_n = 0\} \backslash \{0\}]_n$. Define $\phi_j : \mathbb{P}^n_{\mathbb{C}} \to \mathbb{R}$ by the formula

$$\phi_j[z_0, \ldots, z_n] = \frac{|z_0|^2 + \ldots + |z_k|^2}{|z_0|^2 + \ldots + |z_n|^2},$$

and consider the open neighbourhoods

$$V_j = \{x \in \mathbb{P}^n_{\mathbb{C}} : \phi_j(x) < \alpha\}, \quad V_{g+j} = \{x \in \mathbb{P}^n_{\mathbb{C}} : \phi_j(x) > \alpha\},$$

of L_j and L_{g+j} respectively. Consider the projective transformation $\gamma_j \in \mathrm{PSL}(n+1, \mathbb{C})$ given by

$$\gamma_j[z_0, \ldots, z_n] = [\lambda z_0, \ldots, \lambda z_k, z_{k+1}, \ldots, z_n],$$

where $\lambda \in \mathbb{C}$ and $|\lambda| = \frac{1}{\alpha} - 1$. Then $\gamma_j(V_j) = \mathbb{P}^n_{\mathbb{C}} \backslash \overline{V_{g+j}}$. Moreover, for all α small the group Γ generated by $\gamma_1, \ldots, \gamma_g$ is a Schottky group.

9.2 Schottky groups: definition and basic facts

The previous section leads naturally to the following abstract definition of Schottky groups given in [38]:

Definition 9.2.1. A subgroup $\Gamma \subset \mathrm{PSL}(n+1, \mathbb{C})$ is called a *Schottky group* if:

(i) There are $2g$, $g \geq 2$, open sets $R_1, \ldots, R_g, S_1, \ldots, S_g$ in $\mathbb{P}^n_{\mathbb{C}}$ with the property that:

    (a) each of these open sets is the interior of its closure; and

    (b) the closures of the $2g$ open sets are pairwise disjoint.

(ii) Γ has a generating set $\mathrm{Gen}(\Gamma) = \{\gamma_1, \ldots, \gamma_g\}$ such that for all $1 \leq j \leq g$ one has that

$$\gamma(R_j) = \mathbb{P}^n_{\mathbb{C}} \backslash \overline{S_j},$$

where the bar means topological closure.

From now on $\mathrm{Int}(A)$ will denote the topological interior and $\partial(A)$ the topological boundary of the set A, and for each $1 \leq j \leq g$, R_j and S_j will be denoted by $R^*_{\gamma_j}$ and $S^*_{\gamma_j}$ respectively, to emphasise their relation with the corresponding generator of the Schottky group.

Remark 9.2.2.   (i) By the characterisation of Schottky groups acting on the Riemann sphere given in [143] (see Chapter 1), one has that a subgroup of $\mathrm{PSL}(2, \mathbb{C})$ is a Schottky group in the sense of Definition 9.2.1 (see [143], [144] and [139]) if and only if it is Schottky in the sense of Definition 9.2.1.

(ii) Let Γ be a complex Schottky group generated by $T_1, \ldots, T_g$ and let $\tilde{\Gamma}$ be the index 2 subgroup consisting of elements of Γ which can be written as reduced words of even length in the generators (recall that $w = z_n^{\varepsilon_n} \cdots z_2^{\varepsilon_2} z_1^{\varepsilon_1} \in \Gamma$ is a reduced word of length n if $z_\ell \in \{T_1, \ldots T_g\}$; $\varepsilon_\ell \in \{-1, +1\}$ and if $z_j = z_{j+1}$, then $\varepsilon_j = \varepsilon_{j+1}$). For $g > 2$ one has that $\tilde{\Gamma}$ is also a Schottky group.

Definition 9.2.3. For a Schottky subgroup $\Gamma \subset \mathrm{PSL}(n, \mathbb{C})$, as in Definition 9.2.1, we define:

(i) A *fundamental domain* $F(\Gamma) = \mathbb{P}_{\mathbb{C}}^n \setminus \left(\bigcup_{\gamma \in \mathrm{Gen}(\Gamma)} R_\gamma^* \cup S_\gamma^* \right)$.

(ii) The *region of discontinuity* $\Omega_S(\Gamma) = \bigcup_{\gamma \in \Gamma} \gamma(F(\Gamma))$.

(iii) The *limit set* $\Lambda := \mathbb{P}_{\mathbb{C}}^n \setminus \Omega_S(\Gamma)$.

In Proposition 9.2.8 we prove that the set $F(\Gamma)$ indeed is a fundamental region for the action of Γ on $\Omega_S(\Gamma)$.

The subscript S in these definitions stands for the fact that they are made ad-hoc for Schottky groups. An open question is to decide whether they coincide in general with the corresponding Kulkarni limit set and discontinuity region.

Remark 9.2.4. When $n = 1$ the set $\Omega_S(\Gamma)$ is the discontinuity domain of Γ as defined in Chapter 1. Its complement $\Lambda(\Gamma) = \mathbb{P}_{\mathbb{C}}^1 \setminus \Omega_S(\Gamma)$ is the usual limit set, which is now a Cantor set.

Definition 9.2.5. Let $\Gamma \subset \mathrm{PSL}(n, \mathbb{C})$ be a subgroup. For an infinite subset $H \subset \Gamma$ and a nonempty, Γ-invariant open set $\Omega_S \subset \mathbb{P}_{\mathbb{C}}^n$, we define $Ac(H, \Omega_S)$ to be the closure of the set of cluster points of HK, where K runs over all the compact subsets of Ω_S. Recall that p is a *cluster point* of HK if there is a sequence $(g_n)_{n \in \mathbb{N}} \subset H$ of different elements and $(x_n)_{n \in \mathbb{N}} \subset K$ such that $g_n(x_n) \xrightarrow[n \to \infty]{} p$.

Lemma 9.2.6. *Let* $\Gamma \subset \mathrm{PSL}(n, \mathbb{C})$ *be a subgroup satisfying Definition* 9.2.1. *Then:*

(i) *For each reduced word* $w = z_n^{\varepsilon_n} \cdots z_2^{\varepsilon_2} z_1^{\varepsilon_1} \in \Gamma$ *one has:*

    (a) *If* $\varepsilon_n = 1$, *then* $w(\mathrm{Int}(F(\Gamma))) \subset S_{z_n}^*$.

    (b) *If* $\varepsilon_n = -1$, *then* $w(\mathrm{Int}(F(\Gamma))) \subset R_{z_n}^*$.

(ii) *Let* $\gamma \in \mathrm{Gen}(\Gamma)$. *Then*

$$R(\gamma) = \bigcap_{k \in \mathbb{N} \cup \{0\}} \gamma^{-k}(R_\gamma^*) \quad and \quad S(\gamma) = \bigcap_{k \in \mathbb{N} \cup \{0\}} \gamma^k(S_\gamma^*)$$

are closed disjoint sets contained in $\mathbb{P}_{\mathbb{C}}^n \setminus \Omega_S(\Gamma)$.

(iii) *Let* $\Gamma_k = \{\gamma \in \Gamma : \gamma$ *is a reduced word of length at most* $k\}$ *and* $F_k(\Gamma) = \Gamma_k(F(\Gamma))$. *Then* $F(\Gamma) \subset F_1(\Gamma) \subset \cdots \subset F_k(\Gamma) \subset \cdots$ *and*

$$\Omega_S(\Gamma) = \bigcup_{k \in \mathbb{N} \cup \{0\}} \mathrm{Int}(F_k(\Gamma)).$$

(iv) *For each $\gamma \in \text{Gen}(\Gamma)$ one has that $\emptyset \neq Ac(\{\gamma^n\}_{n\in\mathbb{N}}, \Omega_S(\Gamma)) \subset S(\gamma)$ and $\emptyset \neq Ac(\{\gamma^{-n}\}_{n\in\mathbb{N}}, \Omega_S(\Gamma)) \subset R(\gamma)$.*

Proof. For (i) we proceed by induction on the length of the reduced words. Clearly the case $k = 1$ follows from the definition of Schottky group. Now assume we have proven the statement for $j = k$. Let $w = z_{k+1}^{\epsilon_{k+1}} \cdots z_1^{\epsilon_1}$ be a reduced word and $x \in \text{Int}(F(\Gamma))$. By the induction hypothesis we deduce that $z_{k+1}^{-\epsilon_{k+1}} w(x) \in \mathbb{P}_{\mathbb{C}}^n \setminus \overline{R_{z_{k+1}}^*}$ if $\epsilon_{k+1} = 1$ and $z_{k+1}^{-\epsilon_{k+1}} w(x) \in \mathbb{P}_{\mathbb{C}}^n - \overline{S_{z_{k+1}}^*}$ if $\epsilon_{k+1} = -1$. Now the proof follows by the definition of Schottky group.

Let us prove (ii). Let $\gamma \in \text{Gen}(\Gamma)$. Since $\gamma^m(\overline{S_\gamma^*}) \subset \gamma^{m-1}(S_\gamma^*)$, we deduce that

$$\bigcap_{m\in\mathbb{N}} \gamma^m(\overline{S_\gamma^*}) \subset \bigcap_{m\in\mathbb{N}} \gamma^{m-1}(S_\gamma^*) = S(\gamma).$$

To finish observe that

$$S(\gamma) = \bigcap_{m\in\mathbb{N}} \gamma^{m-1}(S_\gamma^*) \subset \bigcap_{m\in\mathbb{N}} \gamma^{m-1}(\overline{S_\gamma^*}) \subset \bigcap_{m\in\mathbb{N}} \gamma^m(\overline{S_\gamma^*}).$$

For (iii) we will prove that that $F(\Gamma) \subset \text{Int}(F_1(\Gamma))$. Let $x \in \partial(F(\Gamma))$, then there is $\gamma_0 \in \text{Gen}(\Gamma)$ such that $x \in \partial S_{\gamma_0}^* \cup \partial R_{\gamma_0}^*$. For simplicity we assume $x \in \partial S_{\gamma_0}^*$. Define:

$$r_1 = \min\{d(x, \gamma_0(\overline{S_\gamma^*})) : \gamma \in \text{Gen}(\Gamma)\};$$
$$r_2 = \min\{d(x, \overline{R_\gamma^*}) : \gamma \in \text{Gen}(\Gamma)\};$$
$$r_3 = \min\{d(x, \gamma_0(\overline{R_\gamma^*})) : \gamma \in \text{Gen}(\Gamma) - \{\gamma_0\}\};$$
$$r_4 = \min\{d(x, \overline{S_\gamma^*}) : \gamma \in \text{Gen}(\Gamma) - \{\gamma_0\}\};$$
$$r = \min\{r_1, r_2, r_3, r_4\}$$

where d denotes the Fubini-Study metric. Clearly $r > 0$. Now let $y \in B_{r/4}(x) \cap \overline{S_{\gamma_0}^*}$, then by the definition of r we have that $y \in F(\Gamma) \cup \gamma(F(\Gamma))$. If $y \in B_{r/2}(x) \cap \mathbb{P}_{\mathbb{C}}^n \setminus \overline{S_{\gamma_0}^*}$, then by definition of r we deduce $y \in F(\Gamma)$. In other words, we have shown that $F(\Gamma) \subset \text{Int}(F_1(\Gamma))$. Therefore $F_k(\Gamma) \subset \Gamma_k(\text{Int}(F_1)) \subset \Gamma_k(F_1(\Gamma)) \subset F_{k+1}(\Gamma)$, i.e., $F_k(\Gamma) \subset \text{Int}(F_{k+1}(\Gamma))$. To finish the proof observe that

$$\Omega_S(\Gamma) = \bigcup_{k\in\mathbb{N}\cup\{0\}} F_k(\Gamma) \subset \bigcup_{k\in\mathbb{N}\cup\{0\}} \text{Int}(F_{k+1}(\Gamma)) \subset \bigcup_{k\in\mathbb{N}\cup\{0\}} \text{Int}(F_k(\Gamma)).$$

We now prove (iv). Let $K \subset \Omega_S(\Gamma)$ be a compact set and x a cluster point of $\{\gamma^m(K)\}_{m\in\mathbb{N}}$. Then there is a subsequence $(n_m)_{m\in\mathbb{N}} \subset (m)_{m\in\mathbb{N}}$ and a sequence $(x_m)_{m\in\mathbb{N}} \subset K$ such that $\gamma^{n_m}(x_m) \xrightarrow[m\to\infty]{} x$. If $x \notin S(\gamma)$, then there is $k_0 \in \mathbb{N}$ such that $x \notin \gamma^{k_0}(\overline{S_\gamma^*})$. Taking $r = d(x, \gamma^{k_0}(\overline{S_\gamma^*}))$ we have that

$$B_{r/2}(x) \cap \gamma^{k_0}(\overline{S_\gamma^*}) = \emptyset. \tag{9.2.7}$$

On the other hand, observe that since K is compact, by part (iii) of the present lemma there is $l_0 \in \mathbb{N}$ such that $K \subset F_{l_0}(\Gamma)$; also observe that since $(n_m)_{m \in \mathbb{N}}$ is a strictly increasing sequence, there is $k_1 \in \mathbb{N}$ such that $n_m > l_0 + 1 + k_0$ for $m > k_1$. It follows that we have $\gamma^{l_0+1}(K) \subset \overline{S_\gamma^*}$ and therefore

$$\gamma^{n_m}(x_m) \in \gamma^{n_m - l_0 - 1}(\overline{S_\gamma^*}) \subset \gamma^{k_0}(\overline{S_\gamma^*}) \text{ for } m > k_1.$$

Hence $x \in \gamma^{k_0}(\overline{S_\gamma^*})$, which contradicts (9.2.7). Thus $\emptyset \neq Ac(\{\gamma^n\}_{n \in \mathbb{N}}, \Omega_S(\Gamma)) \subset S(\gamma)$. Observe that similar arguments prove also $\emptyset \neq Ac(\{\gamma^{-n}\}_{n \in \mathbb{N}}, \Omega_S(\Gamma)) \subset R(\gamma)$. $\qquad\square$

Proposition 9.2.8. *If Γ is a Schottky group, then:*

(i) Γ *is a free group generated by* $\mathrm{Gen}(\Gamma)$.

(ii) $\Omega_S(\Gamma)/\Gamma$ *is a compact complex n-manifold and* $\mathrm{Int}(F(\Gamma))$ *is a fundamental domain for the action of* Γ.

Proof. (i) Assume there is a reduced word h with length > 0 such that $h = \mathrm{Id}$. Now let $x \in \mathrm{Int}(F(\Gamma))$. Then by part (i) of Lemma 9.2.6 we have $x = h(x) \in \bigcup_{\gamma \in \mathrm{Gen}(\Gamma)}(R_\gamma^* \cup S_\gamma^*)$, which contradicts the choice of x. Therefore Γ is free.

(ii) Let $K \subset \Omega_S(\Gamma)$ be a compact set, then there is a $k \in \mathbb{N}$ such that $K \subset F_k(\Gamma)$. Assume there is a word w with length $\geq 2k + 2$ such that $w(F_k(\Gamma)) \cap F_k(\Gamma) \neq \emptyset$. Then there are $x_1, x_2 \in F(\Gamma)$ and a word w_2 of length at most k such that $x_1 = w_1^{-1} w^{-1} w_2 x_2$. On the other hand $w_1^{-1} w^{-1} w_2$ is a word with length ≥ 2. By (i) of Lemma 9.2.6, $x_1 = w_1^{-1} w^{-1}(w_2(x_2)) \in \bigcup_{g \in \mathrm{Gen}(\Gamma)} S_j^* \cup R_j^*$, but this contradicts the choice of x_1. Therefore Γ acts properly discontinuously and freely on $\Omega_S(\Gamma)$. $\qquad\square$

Remark 9.2.9. Notice that the results in this section remain valid if we replace $\mathbb{P}_{\mathbb{C}}^n$ by $\mathbb{P}_{\mathbb{R}}^n$.

Definition 9.2.10. A discrete subgroup Γ of $\mathrm{PSL}(n, \mathbb{C})$ is an *ideal complex Schottky group* if it contains a finite index subgroup which is Schottky.

Example 9.2.11. Let Γ be a complex Kleinian group in $\mathbb{P}_{\mathbb{C}}^{2n+1}$ generated by involutions $\{T_1, \ldots, T_r\}$, $n \geq 1$, $r > 1$ as in Theorem 9.1.8 above. Let $\check{\Gamma}$ be the index 2 subgroup of Γ consisting of the elements that can be written as reduced words of even length in the generators. It is an exercise to show that $\check{\Gamma}$ is Schottky, and therefore Γ is an ideal Schottky group.

This is analogous to what happens in conformal geometry, where the groups one gets by taking as generators the inversions on pairwise disjoint circles is not Schottky, but its subgroup of words of even length is Schottky. Even so, these groups of inversions are also called Schottky in the literature. Similarly, for simplicity we will say that the groups obtained as in Theorem 9.1.8 are Schottky groups.

9.3 On the limit set and the discontinuity region

This section is based on [203]. We look at the corresponding complex Schottky groups and determine their limit set and the region of discontinuity. We also study the topology of the space of orbits in the region of discontinuity, which is a compact complex manifold, and the infinitesimal deformations of their complex structure.

Theorem 9.3.1. *Let Γ be a complex Schottky group in $\mathbb{P}^{2n+1}_{\mathbb{C}}$, generated by involutions $\{T_1, \ldots, T_r\}$, $n \geq 1$, $r > 1$, as in Theorem 9.1.8 above. Let $\Omega_S(\Gamma)$ be the region of discontinuity of Γ and let $\Lambda_S(\Gamma) = \mathbb{P}^{2n+1}_{\mathbb{C}} \setminus \Omega_S(\Gamma)$ be the limit set. Then one has:*

(i) *Let $W = \mathbb{P}^{2n+1}_{\mathbb{C}} \setminus \cup_{i=1}^r \text{Int}(N_i)$, where $\text{Int}(N_i)$ is the interior of the tubular neighbourhood N_i as in Theorem 9.1.8. Then W is a compact fundamental domain for the action of Γ on $\Omega_S(\Gamma)$. One has: $\Omega_S(\Gamma) = \bigcup_{\gamma \in \Gamma} \gamma(W)$, and the action on Ω_S is properly discontinuous.*

(ii) *If $r = 2$, then $\Gamma \cong \mathbb{Z}/2\mathbb{Z} * \mathbb{Z}/2\mathbb{Z}$, the infinite dihedral group, and $\Lambda_S(\Gamma)$ is the union of two disjoint projective subspaces L and M of dimension n. In this case Γ is elementary,.*

(iii) *If $r > 2$, then $\Lambda_S(\Gamma)$ is a complex solenoid (lamination), homeomorphic to $\mathbb{P}^n_{\mathbb{C}} \times \mathcal{C}$, where $\mathcal{C}$ is a Cantor set. Γ acts minimally on the set of projective subspaces in $\Lambda_S(\Gamma)$ considered as a closed subset of the Grassmannian $G_{2n+1,n}$.*

(iv) *If $r > 2$, let $\check{\Gamma} \subset \Gamma$ be the index 2 subgroup consisting of the elements which are reduced words of even length. Then $\check{W} = W \cup T_1(W)$ is a fundamental domain for the action of $\check{\Gamma}$ on $\Omega_S(\Gamma)$.*

(v) *Each element $\gamma \in \check{\Gamma}$ leaves invariant two copies, P_1 and P_2, of $\mathbb{P}^n_{\mathbb{C}}$ in $\Lambda_S(\Gamma)$. For every $L \subset \Lambda_S(\Gamma)$, $\gamma^i(L)$ converges to P_1 (or to P_2) as $i \to \infty$ (or $i \to -\infty$).*

In fact one has that if $r > 2$, then Γ acts on a graph whose vertices have all valence either 2 or r. This graph is actually a tree, which can be compactified by adding its "ends". These form a Cantor set and the action of Γ can be extended to this compactification. The limit set $\Lambda_S(\Gamma)$ corresponds to the uncountable set of ends of this tree. We use this to prove statement $v)$ above.

Proof. Notice that (i) follows by a slight modification of Proposition 9.2.8. For (ii), observe we have two involutions, T and S, and two neighbourhoods, N_T and N_S, whose boundaries are the mirrors of T and S, respectively. The limit set is the disjoint union $A \cup B$, where $A := \bigcap_{\gamma \in \Gamma'} \gamma(N_S)$, $B := \bigcap_{\gamma \in \Gamma''} \gamma(N_T)$, Γ' is the set of elements in Γ which are words ending in T and Γ'' is the set of elements which are words ending in S. By Theorem 9.1.8, A and B are each the intersection of a nested sequence of tubular neighbourhoods of projective subspaces of dimension n, whose intersection is a projective subspace of dimension n. Hence A and B

are both projective subspaces of dimension n, and they are disjoint. Two reduced words ending in T and S act differently on N_T (or N_S). Hence Γ is the free product of the groups generated T and S, proving (ii).

For (iii), let $L \subset \mathbb{P}_{\mathbb{C}}^{2n+1}$ be a subspace of dimension n and let N be a closed tubular neighbourhood of L as above. Let D be a closed disc which is an intersection of the form $\widehat{L} \cap N$, where $\widehat{L}$ is a subspace of complex dimension $n+1$, transversal to L. If M is a subspace of dimension n contained in the interior of N, then M is transverse to D, otherwise the intersection of M with L would contain a complex line and M would not be contained in N. From part (i) of Theorem 9.3.1 and Theorem 9.1.8 we know that $\Lambda_S(\Gamma)$ is the disjoint union of uncountable subspaces of dimension n. Let $x \in \Lambda_S(\Gamma)$ and let $L \subset \Lambda_S(\Gamma)$ be a projective subspace with $x \in L$. Let N be a tubular neighbourhood of L and D a transverse disc as above. Then $\Lambda_S(\Gamma) \cap D$ is obtained as the intersection of families of discs of decreasing diameters, exactly as in the construction of Cantor sets. Therefore $\Lambda_S(\Gamma) \cap D$ is a Cantor set and $\Lambda_S(\Gamma)$ is a solenoid (or lamination) by projective subspaces which is transversally Cantor. It follows that $\Lambda_S(\Gamma)$ is a fibre bundle over $\mathbb{P}_{\mathbb{C}}^n$, with fibre a Cantor set $\mathcal{C}$. Since $\mathbb{P}_{\mathbb{C}}^n$ is simply connected and $\mathcal{C}$ is totally disconnected, this fibre bundle must be trivial, hence the limit set is a product $\mathbb{P}_{\mathbb{C}}^n \times \mathcal{C}$, as stated.

There is another way to describe the above construction: Γ acts, via the differential, on the Grassmannian $G_{2n+1,n}$ of projective subspaces of dimension n of $\mathbb{P}_{\mathbb{C}}^{2n+1}$. This action also has a region of discontinuity and contains a Cantor set which is invariant. This Cantor set corresponds to the closed family of disjoint projective subspaces in $\Lambda_S(\Gamma)$. It is clear that the action on the Grassmannian is minimal on this Cantor set.

We now prove (iv). Choose a point x_0 in the interior of W. Let Γ_{x_0} be the Γ-orbit of x_0. We construct a graph $\check{\mathcal{G}}$ as follows: to each $\gamma(x_0) \in \Gamma_{x_0}$ we assign a vertex v_γ. Two vertices $v_\gamma, v_{\gamma'}$ are joined by an edge if $\gamma(W)$ and $\gamma'(W)$ have a common boundary component, which corresponds to a mirror E_i. This means that γ' is γ followed by an involution T_i or vice-versa. This graph can be realised geometrically by joining the corresponding points $\gamma(x_0), \gamma'(x_0) \in \Omega_S(\Gamma)$ by an arc $\alpha_{\gamma,\gamma'}$ in $\Omega_S(\Gamma)$, which is chosen to be transversal to the corresponding boundary component of $\gamma(W)$; we also choose these arcs so that no two of them intersect but at the extreme points. Clearly $\check{\mathcal{G}}$ is a tree and each vertex has valence r. To construct a graph $\mathcal{G}$ with an appropriate Γ-action we introduce more vertices in $\check{\mathcal{G}}$: we put one vertex at the middle point of each edge in $\check{\mathcal{G}}$; these new vertices correspond to the points where the above arcs intersect the boundary components of $\gamma(W)$. Then we have an obvious simplicial action of Γ on $\mathcal{G}$. Let $\check{\Gamma}$ be as in the statement of the theorem. It is clear that $\check{W} = W \cup T_1(W)$ is a fundamental domain for $\check{\Gamma}$ in $\Omega_S(\Gamma)$. Hence this group acts freely on the vertices of $\check{\mathcal{G}}$, and it is a free group of rank $r - 1$. The tree $\check{\mathcal{G}}$ can be compactified by its ends by adding a Cantor set on which $\check{\Gamma}$ acts minimally; this corresponds to the fact that Γ acts minimally on the set of projective subspaces which constitute $\Lambda_S(\Gamma)$.

(v) By Theorem 9.1.8, if $\gamma \in \check{\Gamma}$, then either $\gamma(N_1)$ is contained in N_1 or $\gamma^{-1}(N_1)$ is contained in N_1; say $\gamma(N_1)$ is contained in N_1. Thus $\{\gamma^i(N_1)\}$, $i > 0$, is a nested sequence of tubular neighbourhoods of projective subspaces whose intersection is a projective subspace P_1 of dimension n; $\{\gamma^i(N_1)\}$, $i < 0$, is also nested sequence of tubular neighbourhoods of projective subspaces whose intersection is a projective subspace P_2 of dimension n. For every $L \subset \Lambda_S(\Gamma)$, $\gamma^i(L)$ converges to P_1 and P_2 as $i \to \infty$ or $i \to -\infty$, respectively, and both P_1 and P_2 are invariant under γ, as claimed. $\qquad\qquad\square$

Remark 9.3.2. (i) The action of $\check{\Gamma}$ in the Cantor set of projective subspaces is analogous to the action of a classical Fuchsian group of the second kind on its Cantor limit set. We also observe that, since each involution T_i is conjugate to the canonical involution defined in Lemma 9.1.1, the laminations obtained in Theorem 9.3.1 are transversally *Projectively self-similar*. Hence one could try to apply results analogous to the results for (conformally) self-similar sets (for instance the formula in [31]) to estimate the transverse Hausdorff dimension of the laminations obtained. Here by *transverse Hausdorff dimension* we mean the Hausdorff dimension of the Cantor set $\mathcal{C}$ of projective subspaces of $G_{2n+1,n}$ which conform to the limit set. If $\widetilde{T}_i$, $i = 1, \ldots, r$, denote the maps induced in the Grassmannian $G_{2n+1,n}$ by the linear projective transformations T_i, then $\mathcal{C}$ is dynamically-defined by the group generated by the set $\{\widetilde{T}_i\}$.

(ii) The construction of Kleinian groups given in Theorem 9.3.1 actually provides families of Kleinian groups, obtained by varying the size of the mirrors that bound tubular neighbourhoods around the $L'_i s$. In Subsection 9.3.2 below we will look at these families .

(iii) The above construction of complex Kleinian groups, using involutions and mirrors, can be adapted to produce discrete groups of automorphisms of quaternionic projective spaces of odd (quaternionic) dimension. Every "quaternionic Kleinian group" on $P_{\mathcal{H}}^{2n+1}$ lifts canonically to a complex Kleinian group on $\mathbb{P}_{\mathbb{C}}^{4n+3}$.

9.3.1 Quotient spaces of the region of discontinuity

We now discuss the nature of the quotients $\Omega_S(\Gamma)/\Gamma$ and $\Omega_S(\Gamma)/\check{\Gamma}$, for the groups of Subsection 9.1.1. The proof of Proposition 9.3.3 is straightforward and is left to the reader. Notice that this is similar to the construction of a "control group" done in Chapter 5.

Proposition 9.3.3. *Let L be a copy of the projective space $\mathbb{P}_{\mathbb{C}}^n$ in $\mathbb{P}_{\mathbb{C}}^{2n+1}$ and let x be a point in $\mathbb{P}_{\mathbb{C}}^{2n+1} \setminus L$. Let $K_x \subset \mathbb{P}_{\mathbb{C}}^{2n+1}$ be the unique copy of the projective space $\mathbb{P}_{\mathbb{C}}^{n+1}$ in $\mathbb{P}_{\mathbb{C}}^{2n+1}$ that contains L and x. Then K_x intersects transversally every other copy of $\mathbb{P}_{\mathbb{C}}^n$ embedded in $\mathbb{P}_{\mathbb{C}}^{2n+1} \setminus L$, and this intersection consists of one single point.*

Thus, given two disjoint copies L and M of $\mathbb{P}_{\mathbb{C}}^n$ in $\mathbb{P}_{\mathbb{C}}^{2n+1}$, there is a canonical projection map

$$\pi := \pi_L : \mathbb{P}_{\mathbb{C}}^{2n+1} \setminus L \to M,$$

which is a (holomorphic) submersion. Each fibre $\pi_i^{-1}(x)$ is diffeomorphic to $\mathbb{R}^{2n+2}$.

Theorem 9.3.4. *Let Γ be a complex Schottky group as in Theorem 9.3.1, with $r > 2$. Then:*

(i) *The fundamental domain W of Γ is (the total space of) a locally trivial differentiable fibre bundle over $\mathbb{P}_{\mathbb{C}}^n$ with fibre $\mathbb{S}^{2n+2} \setminus \mathrm{Int}(D_1) \cup \cdots \cup \mathrm{Int}(D_r)$, where each $\mathrm{Int}(D_i)$ is the interior of a smooth closed $(2n+2)$-disc D_i in $\mathbb{S}^{2n+2}$ and the D_i's are pairwise disjoint.*

(ii) *$\Omega_S(\Gamma)$ fibres differentially over $\mathbb{P}_{\mathbb{C}}^n$ with fibre $\mathbb{S}^{2n+2}$ minus a Cantor set.*

(iii) *If $\check{\Gamma}$ is the subgroup of index 2 as in Theorem 9.3.1, which acts freely on $\Omega_S(\Gamma)$, then $\Omega_S(\Gamma)/\check{\Gamma}$ is a compact complex manifold that fibres differentiably over $\mathbb{P}_{\mathbb{C}}^n$ with fibre $(\mathbb{S}^{2n+1} \times \mathbb{S}^1) \# \cdots \# (\mathbb{S}^{2n+1} \times \mathbb{S}^1)$, the connected sum of $r-1$ copies of $\mathbb{S}^{2n+1} \times \mathbb{S}^1$.*

Proof. For (i), let $P_1, P_2 \subset \Lambda_S(\Gamma)$ be two disjoint projective subspaces of dimension n contained in $\Lambda_S(\Gamma) \subset \mathbb{P}_{\mathbb{C}}^{2n+1}$. Since $\Omega_S(\Gamma)$ is open in P^{2n+1}, the restriction to $\Omega_S(\Gamma)$ of the map π given by Proposition 9.3.3, using P_1 as L and P_2 as M, is a holomorphic submersion. We know, by part (iii) of Theorem 9.3.1, that $\Lambda_S(\Gamma)$ is a compact set which is a *disjoint* union of projective subspaces of dimension n and which is a transversally Cantor lamination. By Proposition 9.3.3, for each $y \in P_2$, K_y meets transversally each of these projective subspaces (in other words, K_y is transverse to the lamination $\Lambda_S(\Gamma)$, outside P_1). Hence, by Theorem 9.3.1, for each $y \in P_2$, K_y intersects $\Lambda_S(\Gamma) \setminus P_1$ in a Cantor set minus one point (this point corresponds to P_1). The family of subspaces K_y of dimension $n+1$ are all transverse to P_2.

Let us now choose P_1 and P_2 as in part (iii) of Theorem 9.3.1, so they are invariant sets for some $\gamma \in \check{\Gamma}$, and $\gamma^j(L)$ converges to P_2 as $j \to \infty$ for every projective n-subspace $L \subset \Lambda_S(\Gamma) \setminus P_1$. We see that *every* mirror E_i, $i \in \{1, \ldots, r\}$ is transverse to all K_y. Hence the restriction

$$\pi_1 := \pi_{P_1}|_W : W \to P_2 \cong \mathbb{P}_{\mathbb{C}}^n$$

of π to W, is a submersion which restricted to each component of the boundary is also a submersion. For each $y \in P_2$ one has $\pi_1^{-1}(\{y\}) = K_y \cap W$, so $\pi_1^{-1}(\{y\})$ is compact. Thus π_1 is the projection of a locally trivial fibre bundle with fibres $K_y \cap W$, $y \in P_2$, by the fibration lemma in [55]. On the other hand for a fixed $y_0 \in P_2$, $K_{y_0} \cap W$ is a closed $(2n+2)$-disc with $r-1$ smooth closed $(2n+2)$-discs removed from its interior. This is true because P_1 is contained in exactly one of the N_i's, say N_1, the tubular neighbourhood of P_1, and K_{y_0} intersects each N_j, $j \neq 1$, in a smooth closed $(2n+2)$-disc. This proves (i).

(ii) The above arguments show that for each $\bar{\gamma} \in \Gamma$, the image $\bar{\gamma}(E_i)$ of a mirror E_i is transverse to K_y for all $y \in P_2$ and $i \in \{1, \dots, r\}$. Hence the restriction $\pi_1^k := \pi_{P_1}|_{W_k}$, where W_k is as above, is a submersion whose restriction to each boundary component of W_k is also a submersion. Thus π_1^k is a locally trivial fibration. Since $\Omega_S(\Gamma) = \bigcup_{k \geq 0} W_k$, we finish the proof of the first of part (ii) by applying the slight generalisation of Ehresmann's fibration lemma; we leave the proof to the reader.

Lemma 9.3.5. *Let $\mathcal{M} = \bigcup_{i=1}^{\infty} \mathcal{N}_i$ be a smooth manifold which is the union of compact manifolds with boundary $\mathcal{N}_i$, so that each $\mathcal{N}_i$ is contained in the interior of $\mathcal{N}_{i+1}$. Let $\mathcal{L}$ be a smooth manifold and $f : \mathcal{M} \to \mathcal{L}$ a submersion whose restriction to each boundary component of $\mathcal{N}_i$, for each i, is also a submersion. Then f is a locally trivial fibration.*

Thus $\pi_{P_1} : \Omega_S(\Gamma) \to P_2 \cong \mathbb{P}_{\mathbb{C}}^n$ is a holomorphic submersion which is a locally trivial differentiable fibration. To finish the proof of (ii) we only need to show that the fibres of π_{P_1} are $\mathbb{S}^{2n+2}$ minus a Cantor set. Just as above, one shows that $K_y \cap W_k$ is diffeomorphic to the sphere $\mathbb{S}^{2n+2}$ minus the interior of $r(r-1)^k$ disjoint $(2n+2)$-discs. Therefore the fibre of π_{P_1} at y, which is $K_y \cap \Omega_S(\Gamma)$, is the intersection of $\mathbb{S}^{2n+2}$ minus a nested union of discs, which gives a Cantor set as claimed in (ii).

For (iii) we recall that by part (iv) of Theorem 9.3.1, the fundamental domain of $\check{\Gamma}$ is the manifold $\check{W} = W \cup T_1(W)$. Then, as above, the restriction of π to $\check{W}$ is a submersion which is also a submersion in each connected component of the boundary:

$$\partial \check{W} = \left(\bigcup_{j \neq 1} T_1(E_j) \right) \bigcup_{j \neq 1} E_j,$$

which is the disjoint union of the $r-1$ mirrors E_j, $j \neq 1$, together with the mirrors $E_{1j} := T_1(E_j)$, $j \neq 1$.

The mirror E_j is identified with E_{1j}, $j \neq 1$, by T_1, and $\Omega_S(\Gamma)/\check{\Gamma}$ is obtained through these identifications. Let $\check{\pi} : \check{W} \to P_2$ be the restriction of π to $\check{W}$. By the proof of (i), $\check{\pi}^{-1}(y) = K_y \cap \check{W}$, $y \in P_2$, is diffeomorphic to $\mathbb{S}^{2n+2}$ minus the interior of $2(r-1)$ disjoint $(2n+2)$-discs. The restriction of π to each E_j and E_{1j} determines fibrations $\check{\pi}_{i_j} : E_j \to P_2$ and $\check{\pi}_{1j} : E_{1j} \to P_2$, respectively, whose fibres are $\mathbb{S}^{2n+1}$. Set $\hat{\pi}_j := \check{\pi}_{1j} \circ (T_1|_{E_j})$. If we had that $\hat{\pi}_j = \check{\pi}_j$ for all $j = 2, \dots, r$, then we would have a fibration from $\check{W}/\check{\Gamma}$ to P_2, because we would have compatibility of the projections on the boundary. In fact we only need that $\hat{\pi}_j$ and $\check{\pi}_j$ be homotopic through a smooth family of fibrations $\pi_t : E_{1j} \to P_2$; $\pi_1 = \hat{\pi}_j$, $\pi_0 = \check{\pi}_j$, $t \in [0, 1]$.

Actually, to be able to glue well the fibrations at the boundary we need that $\pi_t = \check{\pi}_j$ for t in a neighbourhood of 0 and $\pi_t = \hat{\pi}_j$ for t in a neighbourhood of 1. But this is almost trivial: $\check{\pi}_j : E_{1j} \to P_2$ is the projection of E_{1j} onto P_2 from P_1 and $\hat{\pi} - j$ is the projection of E_{1j} from $T(P_1)$ onto P_2.

The n-dimensional subspaces P_1 and $T(P_1)$ are disjoint from P_2, so there exists a smooth family of n-dimensional subspaces P_t, $t \in [0, 1]$, such that the family is disjoint from P_2 and $P_t = P_1$ for t in a neighbourhood of 0 and $P_t = T(P_1)$ for t in a neighbourhood of 1. We can choose the family so that for each $t \in [0, 1]$, the set of $(n + 1)$-dimensional subspaces which contain P_t meet transversally E_{1j}. To achieve this we only need to take an appropriate curve in the Grassmannian of projective n-planes in $\mathbb{P}_{\mathbb{C}}^{2n+1}$, consisting of a family P_t which is transverse to all K_y; this is possible by Proposition 9.3.3 and the fact that the set of n-dimensional subspaces which are *not* transverse to the $K'_y s$, is a proper algebraic variety of $\mathbb{P}_{\mathbb{C}}^{2n+1}$. In this way we obtain the desired homotopy.

It follows that $\check{W}$ fibres over $P_2 \cong \mathbb{P}_{\mathbb{C}}^n$; the fibre is obtained from $\mathbb{S}^{2n+2}$ minus the interior of $2(r-1)$ disjoint $(2n+2)$-discs whose boundaries are diffeomorphic to $\mathbb{S}^{2n+1}$ and are identified by pairs by diffeomorphisms which are *isotopic* to the identity (using a fixed diffeomorphism to $\mathbb{S}^{2n+1}$). Hence the fibre is *diffeomorphic* to $(\mathbb{S}^{2n+1} \times \mathbb{S}^1) \# \cdots \# (\mathbb{S}^{2n+1} \times \mathbb{S}^1)$, the connected sum of $r-1$ copies of $\mathbb{S}^{2n+1} \times \mathbb{S}^1$. This proves (iii). $\qquad\square$

Theorem 9.3.6. *Let M_Γ be the compact complex orbifold $M_\Gamma := \Omega_S(\Gamma)/\Gamma$, which has complex dimension $(2n + 1)$. Then:*

(i) *The singular set of M_Γ, $\mathrm{Sing}(M_\Gamma)$, is the disjoint union of r submanifolds analytically equivalent to $\mathbb{P}_{\mathbb{C}}^n$, one contained in (the image in M_Γ of) each mirror E_i of Γ.*

(ii) *Each component of $\mathrm{Sing}(M_\Gamma)$ has a neighbourhood homeomorphic to the normal bundle of $\mathbb{P}_{\mathbb{C}}^n$ in $\mathbb{P}_{\mathbb{C}}^{2n+1}$ modulo the involution $v \mapsto -v$, for v a normal vector.*

(iii) *M_Γ fibres over $\mathbb{P}_{\mathbb{C}}^n$ with fibre a real analytic orbifold with r singular points, each having a neighbourhood (in the fibre) homeomorphic to the cone over the real projective space $P_{\mathbb{R}}^{2n+1}$.*

Proof. We notice that M_Γ is obtained from the fundamental domain W after an identification on the boundary E_j by the action of T_j. The singular set of M_Γ is the union of the images, under the canonical projection $p : \Omega_S(\Gamma) \to \Omega_S(\Gamma)/\Gamma$, of the fixed point sets of the r involutions T_j. Now, T_j is conjugate to the canonical involution S of Corollary 9.1.2. The lifting of S to $\mathbb{C}^{2n+2} = \mathbb{C}^{n+1} \times \mathbb{C}^{n+1}$ has as fixed point set the $(n + 1)$-subspace $\{(a, a) : a \in \mathbb{C}^{n+1}\}$. This projectivises to an n-dimensional projective subspace.

Since we can assume, *for a fixed j*, that T_j is an isometry, we obtain the local structure of a neighbourhood of each component of the singular set. The same arguments as in part (iii) of Theorem 9.3.4 prove that $\Omega_S(\Gamma)/\Gamma$ fibres over $\mathbb{P}_{\mathbb{C}}^n$ and that the fibre has r singular points, corresponding to the r components of $\mathrm{Sing}(M_\Gamma)$, and each of these r points has a neighbourhood (in the fibre) homeomorphic to the cone over $P_{\mathbb{R}}^{2n+1}$. $\qquad\square$

Remark 9.3.7. (i) The map π in the part (ii) of Theorem 9.3.4 is holomorphic, but the fibration is not holomorphically locally trivial, because the complex structure on the fibres may change.

(ii) The Kleinian groups of Theorem 9.3.4 provide a method for constructing complex manifolds which is likely to produce interesting examples (c.f., [107], [109], [108], [110], [164], [176], [205]). These are never Kähler, because the fibration $\pi : \Omega_S(\Gamma)/\check{\Gamma} \to \mathbb{P}^n_{\mathbb{C}}$ has a section, by dimensional reasons, so there can not exist a 2-cocycle with a power which is the fundamental class of $\Omega_S(\Gamma)/\check{\Gamma}$. The bundle $(n+1)\mathcal{O}_{\mathbb{P}^n_{\mathbb{C}}}$ is nontrivial as a real bundle, because it has nonvanishing Pontryagin classes (except for $n=1$), hence π is a nontrivial fibration.

We notice that the fundamental group of a compact Riemann surface of genus greater than zero is never a free group; similarly, by Kodaira's classification, the only compact complex surface with nontrivial free fundamental group is the Hopf surface $\mathbb{S}^3 \times \mathbb{S}^1$. The examples above give compact complex manifolds with free fundamental groups (of arbitrarily high rank) in all odd dimensions greater than 1. Multiplying these examples by $\mathbb{P}^1_{\mathbb{C}}$, one obtains similar examples in all even dimensions.

As pointed out in [203], it would be interesting to know if there are other examples which are minimal, i.e., they are not obtained by blowing up along a smooth subvariety of the examples above. It is natural to conjecture that these examples in odd dimensions are the only ones which have a projective structure and free fundamental group of rank greater than 1.

(iii) The manifolds obtained by resolving the singularities of the orbifolds in Theorem 9.3.6 have very interesting topology. We recall that the orbifold M_Γ is singular along r disjoint copies of $\mathbb{P}^n_{\mathbb{C}}$: $S_1, \ldots, S_r$. The resolution $\widetilde{M}_\Gamma$ of M_Γ is obtained by a monoidal transformation along each S_i, and it replaces each point $x \in S_i$, $1 \le i \le r$ by a projective space $\mathbb{P}^n_{\mathbb{C}}$. Hence, if $\mathcal{P} : \widetilde{M} \to M$ denotes the resolution map, then $\mathcal{P}^{-1}(S_i)$ is a nonsingular divisor in $\widetilde{M}$, which fibres holomorphically over $\mathbb{P}^n_{\mathbb{C}}$ with fibre $\mathbb{P}^n_{\mathbb{C}}$, $1 \le i \le r$.

9.3.2 Hausdorff dimension and moduli spaces

Let $\mathcal{L} := \{L_1, \ldots, L_r\}$ be r pairwise disjoint projective subspaces of dimension n in $\mathbb{P}^{2n+1}_{\mathbb{C}}$, $r > 2$; $\mathcal{L}$ will be called a *configuration*. Let Γ and Γ' be complex Schottky groups obtained from this same configuration, i.e., they are generated by sets $\{T_1, \ldots, T_r\}$ and $\{T'_1, \ldots, T'_r\}$ of holomorphic involutions of $\mathbb{P}^{2n+1}_{\mathbb{C}}$ such that $T'_j \circ T_j^{-1}$ preserves the subspaces L_j.

It is an exercise to see that the subgroup of $\mathrm{PSL}(n+2, \mathbb{C})$ of transformations that preserve these subspaces is the projectivisation of a copy of $\mathrm{GL}(n+1, \mathbb{C}) \times \mathrm{GL}(n+1\mathbb{C}) \subset \mathrm{GL}(2n+2, \mathbb{C})$. Therefore we can always find an analytic family $\{\Gamma_t\}$, $0 \le t \le 1$, of complex Schottky groups, with configuration $\mathcal{L}$, such that $\{\Gamma_0\} = \Gamma$ and $\{\Gamma_1\} = \Gamma'$. Furthermore, let $\mathcal{L} := \{L_1, \ldots, L_r\}$ and $\mathcal{L}' := \{L_1', \ldots, L_r'\}$ be two configurations of $\mathbb{P}_{\mathbb{C}}^n$'s in $\mathbb{P}_{\mathbb{C}}^{2n+1}$ as before.

Due to dimensional reasons, we can always move these configurations to obtain a differentiable family of pairs of disjoint n-dimensional subspaces $\{L_{1,t}, \ldots, L_{r,t}\}$, with $0 \le t \le 1$, providing an isotopy between $\mathcal{L}$ and $\mathcal{L}'$. Thus one has a differentiable family Γ_t of complex Kleinian groups, where $\Gamma_0 = \Gamma$ and $\Gamma_1 = \Gamma'$. The same statements hold if we replace Γ and Γ' by their subgroups $\check{\Gamma}$ and $\check{\Gamma}'$, consisting of words of even length. So one has a differentiable family $\check{\Gamma}_t$ of Kleinian groups, where $\check{\Gamma}_0 = \check{\Gamma}$ and $\check{\Gamma}_1 = \check{\Gamma}'$. Hence the manifolds $\Omega_S(\Gamma_t)/\check{\Gamma}_t$ are all diffeomorphic. By Subsection 9.3.1, these manifolds are (in general nontrivial) fibre bundles over $\mathbb{P}_{\mathbb{C}}^n$ with fibre $\#^{(r-1)}(\mathbb{S}^{2n+1} \times \mathbb{S}^1)$, a connected sum of $(r-1)$-copies of $\mathbb{S}^{2n+1} \times \mathbb{S}^1$.

If $n = 1$, given any configuration of r pairwise disjoint lines in $\mathbb{P}_{\mathbb{C}}^3$, there exist an isotopy of $\mathbb{P}_{\mathbb{C}}^3$ which carries the configuration into a family of r twistor lines. Hence $\mathbb{P}_{\mathbb{C}}^3$ minus this configuration is diffeomorphic to the Cartesian product of $\mathbb{S}^4$ minus r points with $\mathbb{P}_{\mathbb{C}}^1$. Moreover, the attaching functions that we use to glue the boundary components of W, the fundamental domain of Γ, are all isotopic to the identity, because they live in $\mathrm{PSL}(4, \mathbb{C})$, which is connected. Thus, if $n = 1$, then $\Omega_S(\Gamma_t)/\check{\Gamma}_t$ is diffeomorphic to a product $\mathbb{P}_{\mathbb{C}}^1 \times \#^{(r-1)}(\mathbb{S}^3 \times \mathbb{S}^1)$. Hence we have:

Proposition 9.3.8. *The differentiable type of the compact (complex) manifold $\Omega_S(\Gamma_t)/\check{\Gamma}_t$ is independent of the choice of configuration. It is a manifold of real dimension $(4n+2)$, which is a fibre bundle over $\mathbb{P}_{\mathbb{C}}^n$ with fibre $\#^{(r-1)}(\mathbb{S}^{2n+1} \times \mathbb{S}^1)$; moreover, this bundle is trivial if $n = 1$. We denote the corresponding manifold by M_r^n.*

The fact that the bundle is trivial when $n = 1$ is interesting because when the configuration $\mathcal{L}$ consists of twistor lines in $\mathbb{P}_{\mathbb{C}}^3$ (see Chapter 10), the quotient $\Omega_S(\Gamma)/\check{\Gamma}$ is the twistor space of the conformally flat manifold $p(\Omega_S(\Gamma))/p(\check{\Gamma})$, which is a connected sum of the form $\#^{(r-1)}(\mathbb{S}^3 \times \mathbb{S}^1)$. Hence, in this case the natural fibration goes the other way round, i.e., it is a fibre bundle over $\#^{(r-1)}(\mathbb{S}^3 \times \mathbb{S}^1)$ with fibre $\mathbb{P}_{\mathbb{C}}^1$.

Given a configuration $\mathcal{L}$ as above, let us denote by $[\mathcal{L}]_G$ its orbit under the action of the group $G = \mathrm{PSL}(2n + 2, \mathbb{C})$. These orbits are equivalence classes of such configurations. Let us denote by $\mathcal{C}^n$ the set of equivalence classes of configurations of $\mathbb{P}_{\mathbb{C}}^n$'s as above. Then $\mathcal{C}^n$ is a Zariski open set of the moduli space $\mathfrak{M}^n$, of configurations of unordered projective subspaces of dimension n in $\mathbb{P}_{\mathbb{C}}^{2n+1}$, which is obtained as the Mumford quotient [157] of the action of G on such configurations.

By [157], $\mathcal{C}^n$ is a complex algebraic variety: the *moduli space of configurations* of n-planes $\mathbb{P}_{\mathbb{C}}^n$ in $\mathbb{P}_{\mathbb{C}}^{2n+1}$. Similarly, we denote by $\mathfrak{G}^n$ the equivalence classes, or moduli space, of the corresponding Schottky groups, where two such groups are equivalent if they are conjugate by an element in $\mathrm{PSL}(n + 2, (\mathbb{C}))$.

Given $\mathcal{L} := \{L_1, \ldots, L_r\}$, and r-tuples of involutions $(T_1, \ldots, T_r)$, $(S_1, \ldots, S_r)$ as above, i.e., interchanging $T_j \circ S_j$ leaves L_j invariant for all $j = 1, \ldots, r$, and having pairwise disjoint mirrors, we say that these r-tuples are *equivalent* if there exists $h \in G$ such that $hT_i h^{-1} = S_i$ for all i. Let $\mathfrak{T}_{\mathcal{L}}$ denote the set of equivalence classes of such r-tuples of involutions.

It is clear that a conjugation h as above must leave $\mathcal{L}$ invariant. Hence, if r is big enough with respect to n, then h must actually be the identity, so the equivalence classes in fact consist of a single element.

Theorem 9.3.9. *There exists a holomorphic surjective map* $\pi \colon \mathfrak{G}_r^n \to \mathcal{C}_r^n$ *which is a* C^∞ *locally trivial fibration with fibre* $\mathfrak{T}_{\mathcal{L}}$. *Furthermore, let* Γ, Γ' *be complex Schottky groups as above and let* $\Omega_S(\Gamma), \Omega_S(\Gamma')$ *be their regions of discontinuity. Then the complex orbifolds* $M_\Gamma := \Omega_S(\Gamma)/\Gamma$ *and* $M_{\Gamma'} := \Omega_S(\Gamma')/\Gamma'$ *are biholomorphically equivalent if and only if* Γ *and* Γ' *are projectively conjugate, i.e., they represent the same element in* $\mathfrak{G}_r^n$. *Similarly, if* $\check{\Gamma}, \check{\Gamma}'$ *are the corresponding index 2 subgroups, consisting of the elements which are words of even length, then the manifolds* $M_{\check{\Gamma}} := \Omega_S(\Gamma)/\check{\Gamma}$ *and* $M_{\check{\Gamma}'} := \Omega_S(\Gamma')/\check{\Gamma}'$, *are biholomorphically equivalent if and only if* $\check{\Gamma}$ *and* $\check{\Gamma}'$ *are projectively conjugate.*

Proof. The first statement in Theorem 9.3.9 is obvious, i.e., that we have a holomorphic surjection $\pi \colon \mathfrak{G}_r^n \to \mathcal{C}_r^n$ with kernel $\mathfrak{T}_{\mathcal{L}}$. The other statements are immediate consequences of Lemma 9.3.10, proved by Sergei Ivashkovich. The proof below, taken from [203], is a variation of Ivashkovich's proof. $\square$

Lemma 9.3.10. *Let* U *be a connected open set in* $\mathbb{P}_{\mathbb{C}}^{2n+1}$ *that contains a subspace* $L \subset \mathbb{P}_{\mathbb{C}}^{2n+1}$ *of dimension* n, *and let* $h \colon U \to V$ *be a biholomorphism onto an open set* $V \subset \mathbb{P}_{\mathbb{C}}^{2n+1}$. *Suppose that* V *also contains a subspace* M *of dimension* n. *Then* h *extends uniquely to an element in* $\mathrm{PSL}(2n+2, \mathbb{C})$.

Proof. Let $f : U \to \mathbb{P}_{\mathbb{C}}^n$ be a holomorphic map. Then f is defined by n meromorphic functions $f_1, \ldots, f_n$ from U to $\mathbb{P}_{\mathbb{C}}^1$ (see [101]), i.e., holomorphic functions which are defined outside of an analytic subset of U (the indeterminacy set).

Consider the set of all subspaces of $\mathbb{P}_{\mathbb{C}}^{2n+1}$ of dimension $n + 1$ which contain L. Then, if N is such a subspace, one has a neighbourhood U_N of L in N which is the complement of a round ball in the affine part, $\mathbb{C}^{n+1}$, of N. Since the boundary of such a ball is a round sphere S_N and, hence, it is pseudo-convex, it follows from E. Levi's extension theorem, applied to each f_i, that the restriction, f_N, of f to $U \cap N$ extends to all of N as a meromorphic function.

The union of all subspaces N is $\mathbb{P}_{\mathbb{C}}^{2n+1}$ and they all meet in L. The functions f_N depend holomorphically on N as is shown in [101]. One direct way to prove this is by considering the Henkin-Ramirez reproducing kernel defined on each round sphere S_N, [88], [183]. One can choose the spheres S_N in such a way that the kernel depends holomorphically on N by considering a tubular neighbourhood of L in N whose radius is independent of N. Hence the extended functions to all $N's$ define a meromorphic function in all of $\mathbb{P}_{\mathbb{C}}^{2n+1}$, which extends f.

Now let h be as in the statement of Lemma 9.3.10 and let $\tilde{h}$ be its meromorphic extension. Then, since by hypothesis h is a biholomorphism from the open set $U \subset \mathbb{P}^n_\mathbb{C}$ onto the open set $V := h(U) \subset \mathbb{P}^n_\mathbb{C}$, one can apply the above arguments to $h^{-1} : V \to U$. Let $g : \mathbb{P}^n_\mathbb{C} \to \mathbb{P}^n_\mathbb{C}$ be the meromorphic extension of h^{-1}. Then, outside of their sets of indeterminacy, one has $\tilde{h}g = g\tilde{h} = \mathrm{Id}$. Hence the indeterminacy sets are empty and both $\tilde{h}$ and g are biholomorphisms of $\mathbb{P}^n_\mathbb{C}$.

In fact, in [101] it is shown that if f is as in the statement of Lemma 9.3.10 and if f is required only to be locally injective, then f extends as a holomorphic function. $\qquad\square$

Notice that if $n = 1$, then Lemma 9.3.10 becomes Lemma 3.2 in [107].

Corollary 9.3.11. *For $r > 2$ sufficiently large, the manifold $\Omega_S(\Gamma)/\check{\Gamma}$ has nontrivial moduli.*

In fact, if the manifolds $\Omega_S(\Gamma)/\check{\Gamma}$ and $\Omega_S(\Gamma')/\check{\Gamma}'$ are complex analytically equivalent, then $\check{\Gamma}$ is conjugate to $\check{\Gamma}'$ in $\mathrm{PSL}(2n+2, \mathbb{C})$, by Theorem 9.3.9, and the corresponding configurations $\mathcal{L}$ and $\mathcal{L}'$ are projectively equivalent. Now it is sufficient to choose r big enough to have two such configurations which are not projectively equivalent. This is possible because the action induced by the projective linear group G on the Grassmannian $G_{2n+1,n}$ is obtained from the projectivisation of the action of $\mathrm{SL}(2n+2, \mathbb{C})$ acting on the Grassmann algebra Λ^{n+1}_S, of $(n+1)$-vectors of $\mathbb{C}^{2n+2}$, restricted to the set of decomposable $(n+1)$-vectors $\mathcal{D}^{n+1}$. The set $\mathcal{D}^{n+1}$ generates the Grassmann algebra and $G_{2n+1,n} = (\mathcal{D}^{n+1} \setminus \{0\})/\sim$, where $\sim$ is the equivalence relation of projectivisation.

If r is small with respect to n, then $\mathcal{C}^n$ consists of one point, because any two such configurations are in the same $\mathrm{PSL}(2n+2, \mathbb{C})$-orbit. Therefore, in this case $\mathfrak{T}_{\mathcal{L}}$ coincides with $\mathfrak{S}^n$. That is, to change the complex structure of M^n we need to change the corresponding involutions into a family of involutions, with the same configuration (up to conjugation), which is not conjugate to the given one.

The following result is a generalisation of Theorem 1.2 in [107]. This can be regarded as a restriction for a complex orbifold (or manifold) to be of the form $\Omega_S(\Gamma)/\Gamma$ (or $\Omega_S(\Gamma)/\check{\Gamma}$).

Proposition 9.3.12. *If $r > 2$, then the compact complex manifolds and orbifolds $\Omega_S(\Gamma)/\check{\Gamma}$ and $\Omega_S(\Gamma)/\Gamma$, obtained in Theorem 9.3.4, have no nonconstant meromorphic functions.*

Proof. Let f be a meromorphic function on one of these manifolds (or orbifolds). Then f lifts to a meromorphic function $\tilde{f}$ on $\Omega_S(\Gamma) \subset \mathbb{P}^{2n+1}_\mathbb{C}$, which is $\check{\Gamma}$-invariant. By lemma 9.3.13 below, f extends to a meromorphic function on all of $\mathbb{P}^{2n+1}_\mathbb{C}$. Hence $\tilde{f}$ must be constant, because $\check{\Gamma}$ is an infinite group. $\qquad\square$

Lemma 9.3.13 (see [101]). *Let $U \subset \mathbb{P}^{2n+1}_\mathbb{C}$, $n \geq 1$, be an open set that contains a projective subspace $\mathbb{P}^n_\mathbb{C}$. Let $f : U \to \mathbb{P}^1_\mathbb{C}$ be a meromorphic function. Then f can be extended to a meromorphic function $\tilde{f} : U \to \mathbb{P}^1_\mathbb{C}$.*

We refer to [101] for the proof of Lemma 9.3.13.

The following proposition estimates an upper bound for the Hausdorff dimension of the limit set of some Schottky groups.

Proposition 9.3.14. *Let $r > 2$, $0 < \lambda < (r-1)^{-1}$ and let Γ and $\check{\Gamma}$ be as in Theorem 9.1.8. Then, for every $\delta > 0$, the Hausdorff dimension of $\Lambda_S(\Gamma) = \Lambda_S(\check{\Gamma})$ is less than $2n+1+\delta$, i.e., the transverse Hausdorff dimension of $\Lambda_S(\Gamma) = \Lambda_S(\check{\Gamma})$ is less than $1 + \delta$.*

Proof. By part (i) of Theorem 9.3.1 one has that $\Lambda_S(\Gamma) = \bigcap_{k=0}^{\infty} F_k$, where F_k is the disjoint union of the $r(r-1)^k$ closed tubular neighbourhoods $\gamma(N_i)$, $i \in \{1,\dots,r\}$, where $\gamma \in \Gamma$ is an element which can be represented as a reduced word of length k in terms of the generators. $\gamma(N_i)$ is a closed tubular neighbourhood of $\gamma(L_i)$, as in Theorem 9.1.8, and the "width" of each $\gamma(N_i)$, $w_{(\gamma,i)} := d(\gamma(E_i), L_i)$, satisfies $w_{(\gamma,i)} \leq C\lambda^k$, as was shown in Lemma 9.1.7 and Theorem 9.1.8. Hence:

$$w(k) := \sum_{\substack{i \in \{1,\dots,r\} \\ l(\gamma)=k}} w_{(\gamma,i)}^{1+\delta} \leq Cr(r-1)^k \lambda^{k(1+\delta)} < C\, r(r-1)^{-\delta k}.$$

Thus $\lim_{k\to\infty} w(k) = 0$. Hence, just as in the proof of the theorem of Marstrand [142], the Hausdorff dimension of $\Lambda_S(\Gamma)$ can not exceed $2n+1+\delta$. $\qquad \square$

Next we will apply the previous estimates to compute the versal deformations of manifolds obtained from complex Schottky groups as in Proposition 9.3.14, whose limit sets have small Hausdorff dimension.

We first recall ([122]) that given a compact complex manifold X, a deformation of X consists of a triple $(\mathcal{X}, \mathcal{B}, \omega)$, where $\mathcal{X}$ and $\mathcal{B}$ are complex analytic spaces and $\omega : \mathcal{X} \to \mathcal{B}$ is a surjective holomorphic map such that $\omega^{-1}(t)$ is a complex manifold for all $t \in \mathcal{B}$ and $\omega^{-1}(t_0) = X$ for some t_0, which is called the reference point. It is known [134] that given X, there is always a deformation $(\mathcal{X}, \mathfrak{K}_X, \omega)$ which is *universal*, in the sense that every other deformation is induced from it (see also [123, 122]).

The space $\mathfrak{K}_X$ is the Kuranishi space of *versal* deformations of X [134]. If we let $\Theta := \Theta_X$ be the sheaf of germs of local holomorphic vector fields on X, then every deformation of X determines, via differentiation, an element in $H^1(X, \Theta)$, so $H^1(X, \Theta)$ is called the *space of infinitesimal deformations* of X ([122, Ch. 4]). Furthermore ([123] or [122, Th. 5.6]), if $H^2(X, \Theta) = 0$, then the Kuranishi space $\mathfrak{K}_X$ is smooth at the reference point t_0 and its tangent space at t_0 is canonically identified with $H^1(X, \Theta)$. In particular, in this case *every infinitesimal deformation* of X comes from an actual deformation, and vice-versa, every deformation of the complex structure on X, which is near the original complex structure, comes from an infinitesimal deformation.

The following lemma is an immediate application of Proposition 9.3.14 and Harvey's Theorem 1 in [86], which generalises the results of [191].

Lemma 9.3.15. *Let $r > 2$, $0 < \lambda < (r-1)^{-1}$, let $\check{\Gamma}$ be as in Proposition 9.3.14 and let $\Omega_S := \Omega_S(\Gamma) \subset \mathbb{P}_{\mathbb{C}}^{2n+1}$ be its region of discontinuity . Then one has*

$$H^j(\Omega_S, i^*(\Theta_{\mathbb{P}_{\mathbb{C}}^{2n+1}})) \cong H^j(\mathbb{P}_{\mathbb{C}}^{2n+1}, \Theta_{\mathbb{P}_{\mathbb{C}}^{2n+1}}) , \ for \ 0 \le j < n ,$$

where i is the inclusion of Ω_S in $\mathbb{P}_{\mathbb{C}}^{2n+1}$. Hence, if $n > 1$, then one has

$$H^0(\Omega_S, i^*(\Theta_{\mathbb{P}_{\mathbb{C}}^{2n+1}})) \cong \mathfrak{sl}_{2n+2}(\mathbb{C}) \ and \ H^j(\Omega_S, i^*(\Theta_{\mathbb{P}_{\mathbb{C}}^{2n+1}})) \cong 0 ,$$

for all $0 < j < n$, where $\mathfrak{sl}_{2n+2}(\mathbb{C})$ is the Lie algebra of $\mathrm{PSL}(2n+2, \mathbb{C})$, and it is being considered throughout this section as an additive group.

Proof. By Proposition 9.3.14 we have that the Hausdorff dimension d of the limit set $\Lambda_S(\check{\Gamma})$ satisfies $d < 2n+1+\delta$ for every $\delta > 0$. Therefore the Hausdorff measure of $\Lambda_S(\Gamma)$ of dimension s, $\mathcal{H}_s(\Lambda_S(\Gamma))$, is zero for every $s > 2n + 1$. Hence the first isomorphism in Lemma 9.3.15 follows from part (ii) of Theorem 1 in [86], because the sheaf Θ is locally free. The second statement in Lemma 9.3.15 is now immediate, because $H^0(\mathbb{P}_{\mathbb{C}}^{2n+1}, \Theta_{\mathbb{P}_{\mathbb{C}}^{2n+1}}) \cong \mathfrak{sl}_{2n+2}(\mathbb{C})$ and $H^j(\mathbb{P}_{\mathbb{C}}^{2n+1}, \Theta_{\mathbb{P}_{\mathbb{C}}^{2n+1}}) \cong 0$ for $j > 0$, a fact which follows immediately by applying the long exact sequence in cohomology derived from the short exact sequence

$$0 \to \mathcal{O} \to [\mathcal{O}(1)]^{n+1} \to \Theta_{\mathbb{P}_{\mathbb{C}}^{2n+1}} \to 0 ,$$

where $\mathcal{O}$ is the structural sheaf of $\mathbb{P}_{\mathbb{C}}^{2n+1}$ and $[\mathcal{O}(1)]^{n+1}$ is the direct sum of $n+1$ copies of $\mathcal{O}_{\mathbb{P}_{\mathbb{C}}^{2n+1}}(1)$, the sheaf of germs of holomorphic sections of the holomorphic line bundle over $\mathbb{P}_{\mathbb{C}}^{2n+1}$ with Chern class 1. See [87], Example 8.20.1, page 182. □

We let $M := \Omega_S/\check{\Gamma}$, where $\check{\Gamma}$ is as above. We notice that Ω_S is simply connected when $n > 0$, so that Ω_S is the universal covering $\widetilde{M}$ of M. Let $p : \widetilde{M} \to M$ be the covering projection; since $\check{\Gamma}$ acts freely on Ω_S, this projection is actually given by the group action. Let Θ_M be the sheaf of germs of local holomorphic vector fields on M and let $\widetilde{\Theta}$ be the pull-back of Θ to $\widetilde{M}$ under the covering p; $\widetilde{\Theta}$ is the sheaf $i^*(\Theta_{\mathbb{P}_{\mathbb{C}}^{2n+1}})$ on $\widetilde{M} = \Omega_S$.

Lemma 9.3.16. *If $n > 2$, then for $0 \le j \le 2$ we have*

$$H^j(M, \Theta_M) \cong H^j_\rho(\check{\Gamma}, \mathfrak{sl}_{2n+2}(\mathbb{C})) ,$$

where $\mathfrak{sl}_{2n+2}(\mathbb{C})$ is considered as a $\check{\Gamma}$-left module via the representation

$$\rho : \check{\Gamma} \to \mathrm{Aut}\,(\mathfrak{sl}_{2n+2}(\mathbb{C}))$$

given by

$$\rho(\gamma)(v) = dT_g \circ v \circ T_g^{-1}, \ v \in \mathfrak{sl}_{2n+2}(\mathbb{C}) ,$$

where T_g is the action of $g \in \check{\Gamma}$ on $\mathbb{P}_{\mathbb{C}}^{2n+1}$.

Proof. If $n > 2$, then Lemma 9.3.15 and [156, formula (c) in page 23] (see also [82, Chapter V]) imply that there exists an isomorphism

$$\Phi : H^j_\rho(\check{\Gamma}, H^0(\Omega_S, \widetilde{\Theta})) \to H^j(M, \Theta_M),$$

for $0 \le j \le 2$, where $H^0(\Omega_S, \widetilde{\Theta})$ is the vector space of holomorphic vector fields on the universal covering $\widetilde{M} = \Omega_S \subset \mathbb{P}^{2n+1}_{\mathbb{C}}$ of M.

Now, by part (i) of Theorem 1 in [86], every holomorphic vector field in $\Omega_S(\Gamma)$, extends to a holomorphic vector field defined in all of $\mathbb{P}^{2n+1}_{\mathbb{C}}$. Therefore,

$$H^0(\Omega_S, \widetilde{\Theta}) = H^0(\mathbb{P}^{2n+1}_{\mathbb{C}}, \Theta_{\mathbb{P}^{2n+1}_{\mathbb{C}}}) = \mathfrak{sl}_{2n+2}(\mathbb{C}).$$

$\square$

We recall that $\check{\Gamma}$ is a free group of rank $r - 1$; let $g_1, \ldots, g_{r-1}$ be generators of $\check{\Gamma}$. By [89], page 195, Corollary 5.2, applied to $\check{\Gamma}$, we obtain

$$H^1_\rho(\check{\Gamma}, \mathfrak{sl}_{2n+2}(\mathbb{C})) \cong \mathfrak{sl}_{2n+2}(\mathbb{C}) \times \cdots \times \mathfrak{sl}_{2n+2}(\mathbb{C})/\Im(\psi),$$

where

$$\psi : \mathfrak{sl}_{2n+2}(\mathbb{C}) \to \mathfrak{sl}_{2n+2}(\mathbb{C}) \times \cdots \times \mathfrak{sl}_{2n+2}(\mathbb{C})$$

is given by $\psi(v) = (g_1(v) - v, \ldots, g_{r-1}(v) - v)$. We claim that ψ is injective. Indeed, if v is a linear vector field in $\mathbb{P}^{2n+1}_{\mathbb{C}}$ which is invariant by $g_1, \ldots, g_{r-1}$, then, by Jordan's theorem, this vector field is tangent to a hyperplane Π which is $\check{\Gamma}$-invariant. This can not happen. In fact, if L is an n-dimensional projective subspace contained in $\Lambda_S(\check{\Gamma})$, then L must intersect Π transversally in a subspace of dimension $n-2$, for otherwise Π would contain the whole limit set $\Lambda_S(\check{\Gamma})$, which is a disjoint union of projective subspaces of dimension n. Hence, there exists $L \subset \Pi$, a projective n-subspace such that $L \cap \Lambda_S(\check{\Gamma}) = \emptyset$. Then, as proved before, there exists a sequence $\{\gamma_i\}_{i \in \mathbb{N}}$ such that $\lim_{i \to \infty}(\gamma_i(L)) = L_1$, where $L_1 \subset \Lambda_S(\check{\Gamma})$, where L_1 is not contained in Π. This is a contradiction to the invariance of Π.

Therefore,

$$\dim_{\mathbb{C}} H^1(\Omega_S, \widetilde{\Theta}) = \dim_{\mathbb{C}} [\mathfrak{sl}_{2n+2}(\mathbb{C})^{r-2}] = (r-2)\left((2n+2)^2 - 1\right).$$

By [89], page 197, Corollary 5.6 we have $H^2_\rho(\check{\Gamma}, \mathfrak{sl}_{2n+2}(\mathbb{C})) = 0$. Hence, by Lemma 9.3.16 above, one obtains that

$$H^2(M, \Theta_M) \cong H^2_\rho(\check{\Gamma}, \mathfrak{sl}_{2n+2}(\mathbb{C})) = 0.$$

Thus we arrive at the following theorem from [203]:

Theorem 9.3.17. *Let $n, r > 2$ and let λ be an arbitrary scalar such that $0 < \lambda < (r-1)^{-1}$. Let Γ be a Schottky group as in part (iii) of Theorem 9.1.8, so that the (Fubini-Study) distance from $\gamma(x)$ to the limit set Λ_S decreases faster than $C\lambda^k$ for every point $x \in \mathbb{P}^{2n+1}_{\mathbb{C}}$ and any $\gamma \in \Gamma$ of word-length k (where C is some* *positive constant). Let $\check{\Gamma}$ be the index 2 subgroup of Γ consisting of words of even length. Let Ω_S be the region of discontinuity of Γ, $M := \Omega_S/\check{\Gamma}$, and let $\mathfrak{K}_r^n$ denote the Kuranishi space of versal deformations of M, with reference point $t_0 \in \mathfrak{K}_r^n$ corresponding to M. Then, we have*

$$H^1(M, \Theta_M) \cong H^1_\rho(\check{\Gamma}, \mathfrak{sl}_{2n+2}(\mathbb{C})) \cong \mathbb{C}^{(r-2)((2n+2)^2-1)} ,$$

and

$$H^2(M, \Theta_M) = 0 .$$

Hence $\mathfrak{K}_r^n$ is nonsingular at t_0, of complex dimension $(r-2)((2n+2)^2-1)$, and every small deformation of M is obtained by a small deformation of $\check{\Gamma}$ as a subgroup of $\mathrm{PSL}(2n+2, \mathbb{C})$, unique up to conjugation.

Remark 9.3.18. Although we only considered $n > 2$ above, the last theorem remains valid for $n = 0, 1$. In fact, if $n = 0$ and $r > 2$, we have the classical Schottky groups. The manifold $\Omega_S/\check{\Gamma}$ is a compact Riemann surface of genus $r - 1$. It is well-known that in this case the moduli space has dimension $3(r - 1) - 3 = 3(r - 2)$, which, of course, coincides with the formula above. When $n = 1$ and $r > 2$ the manifolds $\Omega_S/\check{\Gamma}$ are Pretzel twistor spaces of genus $g = r - 1$, in the sense of [176, 206]. The theorem above gives that the dimension of the moduli space of this manifold is $15g - 15$, which coincides with Penrose's calculations in page 251 of [176].

9.4   Schottky groups do not exist in even dimensions

This section is taken from [38]. The aim is to show that there are no Schottky groups acting on even-dimensional projective spaces. For this we need a better understanding of the dynamics of projective transformations in general.

9.4.1   On the dynamics of projective transformations

The purpose now is to describe the closure of the set of accumulation points under the action of cyclic groups $\langle \gamma \rangle$ in $\mathrm{PSL}(n, \mathbb{C})$. In Kulkarni's terms this means describing the set $L_0(\langle \gamma \rangle) \cup L_1(\langle \gamma \rangle)$. For this we need:

Definition 9.4.1. Let V be a $\mathbb{C}$-linear space with $\dim_\mathbb{C}(V) = n$ and let $T : V \to V$ be a $\mathbb{C}$-linear transformation. We define

$$\mathrm{Eve}(T) = \langle\langle\{v \in V : v \text{ is an eigenvector of } T\}\rangle\rangle .$$

This is the linear subspace generated by the eigenvectors of T.

Remark 9.4.2. If M is an $n \times n$ matrix with coefficients in $\mathbb{C}$ we have that $\mathrm{Eve}(M)$ is the smaller complex linear subspace which contains the eigenvectors of M. For example if M is diagonalisable and 0 is not an eigenvalue, then $\mathrm{Eve}(M)$ is $\mathbb{C}^n$.

Definition 9.4.3. Let $\gamma \in \mathrm{PSL}(n, \mathbb{C})$ be an element of infinite order and let $\tilde{\gamma}$ be a lifting of γ. Then we define:

(i) The set $\mathrm{Eva}(\gamma)$ of nonnegative real numbers defined by:

$$\mathrm{Eva}(\gamma) = \{r \in \mathbb{R} : \text{ there is an eigenvalue } \lambda \text{ of } \tilde{\gamma} \text{ such that } r = |\lambda|\}.$$

(ii) The set $\mathrm{Lat}_r(\gamma) = \langle [\{v \in \mathbb{C}^n : v \text{ is an eigenvector of } \tilde{\gamma} \text{ and } |\tilde{\gamma}(v)| = r|v|\}]_n \rangle$.

(iii) The set $\Lambda(\gamma)$ is the closure of the accumulation points of orbits $\{\gamma^m(z)\}_{m \in \mathbb{Z}}$ where $z \in \mathbb{P}^n_{\mathbb{C}}$.

Notice that $\Lambda(\gamma)$ is the union of the Kulkarni sets $L_0(\langle \gamma \rangle)$ and $L_1(\langle \gamma \rangle)$.

Remark 9.4.4. (i) In the one-dimensional case we can show that if γ is an elliptic Möbius transformation then

$$\mathrm{Eva}(\gamma) = \{1\}, \ \mathrm{Lat}_1(\gamma) = \Lambda(\gamma) = \mathbb{P}^1_{\mathbb{C}},$$

where $\Lambda(\gamma)$ is the usual limit set of $\langle \gamma \rangle$. If γ is parabolic, then

$$\mathrm{Eva}(\gamma) = \{1\}, \ \mathrm{Lat}_1(\gamma) = \Lambda(\gamma) = \mathrm{Fix}(\gamma).$$

Finally, if γ is loxodromic and λ^2, λ^{-2} are the corresponding multipliers at the fixed points, then $\mathrm{Eva}(\gamma) = \{|\lambda|, |\lambda^{-1}|\}$ and

$$\mathrm{Lat}_\lambda(\gamma) \cup \mathrm{Lat}_{\lambda^{-1}}(\gamma) = \Lambda(\gamma) = \mathrm{Fix}(\gamma).$$

(ii) It is possible to check that $\Lambda(\gamma) = L_0(\langle \gamma \rangle) \cup L_1(\langle \gamma \rangle)$.

(iii) Parts 1 and 2 of this definition do not depend on the choice of lifting $\tilde{\gamma}$.

The following is the main result of this subsection and its proof takes the rest of it.

Proposition 9.4.5. *Let $\gamma \in \mathrm{PSL}_{n+1}(\mathbb{C})$ be an element of infinite order, then*

$$\Lambda(\gamma) = \bigcup_{r \in \mathrm{Eva}(\gamma)} \mathrm{Lat}_r(\gamma).$$

To prove this result we need several lemmas:

Lemma 9.4.6. *Let V be a $\mathbb{C}$-linear space with $\dim_{\mathbb{C}}(V) = n$ and $T : V \to V$ an invertible linear transformation such that $|\lambda| < 1$ for every eigenvalue λ of T. For every $l \in \mathbb{N}$ we have uniform convergence $\binom{m}{l} T^m \xrightarrow[m \to \infty]{} 0$ on compact subsets of V.*

Here $\begin{pmatrix} m \\ l \end{pmatrix}$ denotes the number of sets with l elements from a set with m elements.

Proof. We decompose T into one or more Jordan blocks according to Jordan's Normal Form Theorem. This reduces the problem to the case where there is $0 < |\lambda| < 1$ and an ordered basis $\beta = \{v_1, \ldots, v_n\}$, $n \geq 2$, such that the matrix of T with respect to β (in symbols $[T]_\beta$) satisfies

$$[T]_\beta = \begin{pmatrix} \lambda & 1 & 0 & \cdots & 0 \\ 0 & \lambda & 1 & \cdots & 0 \\ \cdots & \cdots & \cdots & \cdots & \cdots \\ 0 & 0 & 0 & \cdots & \lambda \end{pmatrix}.$$

An induction argument shows that, for all $m > n$,

$$[T^m]_\beta = \begin{pmatrix} \lambda^m & \begin{pmatrix} m \\ 1 \end{pmatrix} \lambda^{m-1} & \begin{pmatrix} m \\ 2 \end{pmatrix} \lambda^{m-2} & \cdots & \begin{pmatrix} m \\ n-1 \end{pmatrix} \alpha^{m+1-n} \\ 0 & \lambda^m & \begin{Bmatrix} m \\ 1 \end{Bmatrix} \lambda^{m-1} & \cdots & \begin{Bmatrix} m \\ n-2 \end{Bmatrix} \lambda^{m+2-n} \\ \cdots & \cdots & \cdots & \cdots & \cdots \\ 0 & 0 & 0 & \cdots & \lambda^m \end{pmatrix}.$$

$$(9.4.7)$$

Given a compact subset $K \subset V$ we set

$$\sigma(K) = \sup\{\sum_{j=1}^n |\alpha_j| : \sum_{j=1}^n \alpha_j v_j \in K\}.$$

Let $z \in K$, $z = \sum_{j=1}^n \alpha_j v_j$, then by equation (9.4.7) we deduce that

$$|T^m(z)| \leq \sigma(K) \max\{|v_j| : 1 \leq j \leq n\} \sum_{j=1}^n \sum_{k=0}^{j-1} \begin{pmatrix} m \\ k \end{pmatrix} |\alpha^{m-k}|.$$

Hence it is sufficient to observe that

$$\left| \begin{pmatrix} m \\ l \end{pmatrix} \begin{pmatrix} m \\ k \end{pmatrix} \alpha^{m-k} \right| \leq m^{2\max\{k,l\}} |\alpha|^{m-k} \xrightarrow[m \to \infty]{} 0,$$

and the lemma follows. $\square$

Lemma 9.4.8. *Let V be a $\mathbb{C}$-linear space with $\dim_{\mathbb{C}}(V) = n > 1$ and let $T : V \to V$ be an invertible linear transformation such that there are $\lambda \in \mathbb{C}$ with $|\lambda| = 1$, and an ordered basis $\beta = \{v_1, \ldots, v_n\}$ for which*

$$[T]_\beta = \begin{pmatrix} \lambda & 1 & 0 & \cdots & 0 \\ 0 & \lambda & 1 & \cdots & 0 \\ \cdots & \cdots & \cdots & \cdots & \cdots \\ 0 & 0 & 0 & \cdots & \lambda \end{pmatrix}.$$

That is, $[T]_\beta$ is an $n \times n$-Jordan block. Then for every $v \in V - \{0\}$ there is a unique
$k(v, T) \in \mathbb{N} \cup \{0\}$ *such that the set of cluster points of* $\left\{ \begin{pmatrix} m \\ k(v, T) \end{pmatrix}^{-1} T^m(v) \right\}_{m \in \mathbb{N}}$
lies in $\langle\langle v_1 \rangle\rangle \setminus \{0\}$.

Proof. Let $z = \sum_{j=0}^{n} \alpha_j v_j$ and define

$$k(z, T) = \max\{1 \le j \le n : \alpha_j \ne 0\} - 1,$$

then we have that

$$\begin{pmatrix} m \\ k(v, T) \end{pmatrix}^{-1} T^m(z) = \sum_{j=1}^{n} (\sum_{k=0}^{n-j} \begin{pmatrix} m \\ k \end{pmatrix} \begin{pmatrix} m \\ k(v, T) \end{pmatrix}^{-1} \lambda^{m-k} \alpha_{k+j}) v_j .$$

The result now follows easily. $\qquad\square$

Corollary 9.4.9. *Let V be a $\mathbb{C}$-linear space with $\dim_\mathbb{C}(V) = n$ and let $T : V \to V$ be a linear transformation which is diagonalisable and all its eigenvalues are unitary complex numbers. Then for every $v \in V \setminus \{0\}$, the number $k(v, T) = 0$ is the unique positive integer for which the set of cluster points of* $\left\{ \begin{pmatrix} m \\ k(v, T) \end{pmatrix}^{-1} T^m(v) \right\}_{m \in \mathbb{N}}$
lies in $\mathrm{Eve}(T) \setminus \{0\}$.

Corollary 9.4.10. *Let V be a $\mathbb{C}$-linear space with $\dim_\mathbb{C}(V) = n$ and let $T : V \to V$ be an invertible linear transformation such that each of its eigenvalues is a unitary complex number. Then for every $v \in V \setminus \{0\}$ there is a unique $k(v, T) \in \mathbb{N} \cup \{0\}$ for which the set of cluster points of* $\left\{ \begin{pmatrix} m \\ k(v, T) \end{pmatrix}^{-1} T^m(v) \right\}_{m \in \mathbb{N}}$ *lies in $\mathrm{Eve}(T) \setminus \{0\}$.*

Proof. By the Jordan Normal Form Theorem there are $k \in \mathbb{N}$; $V_1, \ldots, V_k \subset V$ linear subspaces and $T_i : V_i \to V_i$, $1 \le i \le k$ such that:

(i) $\bigoplus_{j=1}^{k} V_j = V$.

(ii) For each $1 \le i \le k$, T_i is a nonzero $\mathbb{C}$-linear map whose eigenvalues are unitary complex numbers.

(iii) $\bigoplus_{j=1}^{k} T_j = T$.

(iv) For each $1 \le i \le k$, T_i is either diagonalisable or V_i contains an ordered basis β_i for which $[T]_\beta$ is an $n_i \times n_i$-Jordan block.

Let $v \in V \setminus \{0\}$, then there is a nonempty finite set $W \subset \bigcup_{j=1}^{k} V_j \setminus \{0\}$ such that $v = \sum_{w \in W} w$. Take $i : W \to \mathbb{N}$ where $i(w)$ is the unique element in $\{1, \ldots, k\}$ such that $w \in V_{i(w)}$. Define

$$k(v, T) = \max\{k(w, T_{i(w)}) : w \in W\},$$

$$W_1 = \{w \in W : k(w, T_{i(w)}) < k(v, T)\},$$
$$W_2 = W \setminus W_1,$$

then

$$\frac{T^m(v)}{\binom{m}{k(v,T)}} = \sum_{w \in W_1} \frac{\binom{m}{k(w,T_{i(w)})}}{\binom{m}{k(v,T)}} \frac{T^m_{i(w)}(w)}{\binom{m}{k(w,T_{i(w)})}} + \sum_{w \in W_2} \frac{T^m_{i(w)}(w)}{\binom{m}{k(w,T_{i(w)})}}.$$
$$(9.4.10)$$

The result now follows from equation (9.4.10), Lemma 9.4.8 and Corollary 9.4.9. □

Proof of Proposition 9.4.5. Let the set $\Lambda(\Gamma)$, $\text{Eva}(\Gamma)$ and $\text{Lat}_r(\gamma)$ be as in Definition 9.4.3. Since $\bigcup_{r \in \text{Eva}(\Gamma)} \text{Lat}_r(\gamma) \subset \Lambda(\Gamma)$, thence it is enough to show that $\Lambda(\gamma) \subset \bigcup_{r \in \text{Eva}(\Gamma)} \text{Lat}_r(\gamma)$.

Let $\tilde{\gamma}$ be a lifting of γ, then by the Jordan Normal Form Theorem there are $k \in \mathbb{N}$; $V_1, \ldots, V_k \subset \mathbb{C}^{n+1}$ linear subspaces; $\gamma_i : V_i \to V_i$, $1 \le i \le k$ and $r_1, \ldots, r_k \in \mathbb{R}$ which satisfy:

(i) $\bigoplus_{j=1}^k V_j = \mathbb{C}^{n+1}$.

(ii) For each $1 \le i \le k$, γ_i is a nonzero $\mathbb{C}$-linear map whose eigenvalues are unitary complex numbers.

(iii) $0 < r_1 < r_2 < \cdots < r_k$.

(iv) $\bigoplus_{j=1}^k r_j \gamma_j = \tilde{\gamma}$.

In what follows $(\tilde{\gamma}, k, \{V_i\}_{i=1}^k, \{\gamma_i\}_{i=1}^k, \{r_i\}_{i=1}^k)$ will be called a decomposition for γ. Now let $[v]_n \in \mathbb{P}_{\mathbb{C}}^n$, thus $v = \sum_{j=1}^k v_j$ where $v_j \in V_j$. Set $j_0 = \max\{1 \le j \le k : v_j \ne 0\}$. One has

$$\binom{m}{k(v_{j_0}, T_{j_0})}^{-1} \frac{\tilde{\gamma}^m(v)}{r_{j_0}^m} = \sum_{j=1}^k \binom{m}{k(v_{j_0}, T_{j_0})}^{-1} \frac{r_j^m \gamma_j^m(v_j)}{r_{j_0}^m}. \qquad (9.4.10)$$

By Lemma 9.4.6 and Corollary 9.4.10 we conclude that the set of cluster points of $\{\gamma^m(v)\}_{m \in \mathbb{Z}}$ lies in $[\text{Eve}(\gamma_{j_0}) \setminus 0]_n = L_{r_{j_0}}(\gamma)$. □

9.4.2 Nonrealizability of Schottky groups in $\text{PSL}(2n+1, \mathbb{C})$

As we know from Chapter 3, for every group $G \subset \text{PSL}(3, \mathbb{C})$ and every region Ω where the group acts discontinuously, there is a projective line contained in $\mathbb{P}_{\mathbb{C}}^2 \setminus \Omega$. If we had a Schottky group in $\mathbb{P}_{\mathbb{C}}^2$, this would mean that the generating sets R_i, S_i would contain each at least one complex line, and these lines cannot meet since the generating sets are pairwise disjoint. This is not possible since in $\mathbb{P}_{\mathbb{C}}^2$ there are no pairwise disjoint complex lines. In higher dimensions we will show that the existence of a Schottky group acting on $\mathbb{P}_{\mathbb{C}}^{2n}$ would imply that the set $\Lambda(\gamma)$ is contained in a single connected component of the complement of the discontinuity region, and this leads to a contradiction.

Lemma 9.4.11. *If* $\Gamma \subset \mathrm{PSL}(2n+1, \mathbb{C})$ *is a Schottky group, then* $\mathbb{P}_{\mathbb{C}}^{2n} \setminus \Omega_S(\Gamma)$ *does not contain a projective subspace* $\mathcal{V}$ *with* $\dim_{\mathbb{C}}(\mathcal{V}) \geq n$.

Proof. If $\mathcal{V} \subset \mathbb{P}_{\mathbb{C}}^{2n} \setminus \Omega_S(\Gamma)$ is a projective subspace with $\dim_{\mathbb{C}}(\mathcal{V}) \geq n$, then

$$\mathcal{V} \subset \mathbb{P}_{\mathbb{C}}^{2n} \setminus \Omega_S(\Gamma) = \mathbb{P}_{\mathbb{C}}^{2n} \setminus \bigcup_{\gamma \in \Gamma} \gamma(F(\Gamma)) \subset \mathbb{P}_{\mathbb{C}}^{2n} \setminus F(\Gamma) = \bigcup_{g \in \mathrm{Gen}(\Gamma)} R_\gamma^* \cup S_\gamma^*.$$

Since $\mathcal{V}$ is connected and $(\mathcal{V} \cap \bigcup_{\gamma \in \mathrm{Gen}(\Gamma)} R_\gamma^*, \mathcal{V} \cap \bigcup_{\gamma \in \mathrm{Gen}(\Gamma)} S_\gamma^*)$ is a disconnection for $\mathcal{V}$ we deduce that $\mathcal{V} \subset \bigcup_{\gamma \in \mathrm{Gen}(\Gamma)} R_\gamma^*$ or $\mathcal{V} \subset \bigcup_{\gamma \in \mathrm{Gen}(\Gamma)} S_\gamma^*$. Moreover by an induction argument we deduce that there is $\gamma_0 \in \mathrm{Gen}(\Gamma)$ such that $\mathcal{V} \subset S_{\gamma_0}^*$ or $\mathcal{V} \subset R_{\gamma_0}^*$. For simplicity let us assume that $\mathcal{V} \subset S_{\gamma_0}^*$. Taking $\sigma \in \mathrm{Gen}(\Gamma) \setminus \{\gamma_0\}$ we have

$$\sigma^{-1}(\mathcal{V}) \subset \sigma^{-1}(S_{\gamma_0}^*) \subset \sigma^{-1}(\mathbb{P}_{\mathbb{C}}^{2n} - \overline{S}_\sigma^*) = R_\sigma^*. \tag{9.4.12}$$

Observe that $\mathcal{V}$ and $\sigma^{-1}\mathcal{V}$ are projective subspaces with $\dim_{\mathbb{C}}(\mathcal{V}) + \dim_{\mathbb{C}}(\sigma^{-1}\mathcal{V}) \geq 2n$. Thence $\mathcal{V} \cap \sigma^{-1}(\mathcal{V}) \neq \emptyset$, which is a contradiction since by equation (9.4.12) we have $\mathcal{V} \cap \sigma^{-1}\mathcal{V} \subset R_\sigma^* \cap S_{\gamma_0}^* = \emptyset$. $\square$

Lemma 9.4.13. *Let* $Ac(\{\gamma\}$ *be as in Definition 9.2.5 and* $L(\gamma)$ *be the closure of the set of accumulation points of the orbits* $\gamma^m(z)_{m \in \mathbb{Z}}$ *with* $z \in \mathbb{P}_{\mathbb{C}}^n$. *Let* $\Gamma \subset \mathrm{PSL}(2n+1, \mathbb{C})$ *be a group and* Ω_S *a nonempty,* Γ-invariant open set where Γ acts properly discontinuously. Assume further that whenever l is a projective subspace contained in $\mathbb{P}_{\mathbb{C}}^{2n} \setminus \Omega_S$, then $\dim_{\mathbb{C}}(l) < n$. Then for every $\gamma \in \Gamma$ with infinite order there is a connected set $\mathcal{L}(\gamma) \subset Ac(\{\gamma\}_{m \in \mathbb{Z}}, \Omega_S) \cup L(\gamma)$, such that $L(\gamma) \subset \mathcal{L}(\gamma)$.

Proof. Let $\gamma \in \Gamma$ be an element with infinite order, and choose a decomposition $(\tilde{\gamma}, k, \{V_i\}_{i=1}^k, \{\gamma\}_{i=1}^k, \{r_i\}_{i=1}^k)$ for γ. Take $j_0 = \min\{1 \leq j \leq k : \sum_{i=1}^j \dim_C(V_i) \geq n+1\}$. From Proposition 9.4.5 we can assume that $k \geq 2$. For the moment we assume that $j_0 \neq 1, k$. Observe that, since $\sum_{i=1}^{j_0} \dim_C(V_i) \geq n+1$, we know there is a nonzero $w = \sum_{i=1}^{j_0} w_j \in \bigoplus_{j=1}^{j_0} V_j$, where $w_i \in V_i$, such that $[w]_{2n} \in \Omega_S$. Since Ω_S is open we can assume that w_{j_0} is nonzero. Now, let $z \in \bigoplus_{j>j_0} V_j \setminus \{0\}$, then by Lemma 9.4.6 we have

$$w_m(z) = \left[w + \binom{m}{k(w, \gamma_{j_0})} \sum_{j>j_0} \left(\frac{r_{j_0}}{r_j} \right)^m \gamma_j^{-m}(z_j) \right]_{2n} \xrightarrow[m \to \infty]{} [w]_{2n}.$$

Thus for $m(z)$ large one has $(w_m(z))_{m \geq m(z)} \subset \Omega_S$. On the other hand, by Corollary 9.4.10 there is a strictly increasing sequence $(n_m)_{m \in \mathbb{N}} \subset \mathbb{N}$ and $w_0 \in \mathrm{Eve}(\gamma_{j_0}) \setminus \{0\}$

such that

$$\left(\begin{array}{c} n_m \\ k(w_{j_0}, \gamma_{j_0}) \end{array}\right)^{-1} \gamma_{j_0}^{n_m}(w_0) \xrightarrow[m \to \infty]{} w_0.$$

From this and Lemma 9.4.6 we deduce that

$$\gamma^{n_m}(w_{n_m}) = \left[\left(\begin{array}{c} n_m \\ k(w_{j_0}, \gamma_{j_0}) \end{array}\right)^{-1} \sum_{j \leq j_0} (\frac{r_j}{r_{j_0}})^{n_m} \gamma_j^{n_m}(w_j) + z\right]_{2n} \xrightarrow[m \to \infty]{} [w_0 + z]_{2n}.$$

Hence

$$\bigcup_{j > j_0} L_{r_j}(\gamma) \subset \langle [w_0]_{2n}, [\bigoplus_{j > j_0} V_j \setminus \{0\}]_{2n} \rangle \subset Ac(\{\gamma^m\}_{m \in \mathbb{Z}}, \Omega_S) \cup L(\gamma).$$

Now consider the following observations:

(i) If $j_0 = 1, k$, then we have that $L(\gamma)$ contains a single connected component and therefore $\mathcal{L}(\gamma)$ can be taken as $L(\Gamma)$.

(ii) In the case $j_0 = 1$ one has that there is a $w_1 \in L(r_1)$ such that

$$\bigcup_{j > 1} L(r_j) \subset \langle w_1, [\bigoplus_{j > 1} V_i \setminus \{0\}]_{2n} \rangle \subset Ac(\{\gamma^m\}_{m \in \mathbb{Z}}, \Omega_S) \cup L(\gamma).$$

So in this case, to finish the proof of Lemma 9.4.13 it is enough to take

$$\mathcal{L}(\gamma) = \langle w_1, [\bigoplus_{j > 1} V_j \setminus \{0\}]_{2n} \rangle \cup L_{r_1}(\gamma).$$

(iii) To obtain the result in the case $j = k$ it is enough to apply to γ^{-1} the same argument used in the first observation above.

(iv) In the case $j_0 \neq 1, k$, applying the same arguments to γ^{-1} we get that there is $v \in L(r_{j_0})$ such that

$$\bigcup_{j < j_0} L(r_j) \subset \langle v, [\bigoplus_{j < j_0} V_i \setminus \{0\}]_{2n} \rangle \subset Ac(\{\gamma^m\}_{m \in \mathbb{Z}}, \Omega_S).$$

Therefore to finish the proof of Lemma 9.4.13 in this case it is enough to take

$$\mathcal{L}(\gamma) = \langle v, [\bigoplus_{j < j_0} V_j \setminus \{0\}]_{2n} \rangle \cup \langle [w_0]_{2n}, [\bigoplus_{j > j_0} V_j \setminus \{0\}]_{2n} \rangle \cup L_{r_{j_0}}(\gamma). \qquad \square$$

Theorem 9.4.14. *If $\Gamma \subset \mathrm{PSL}(2n+1, \mathbb{C})$ is a discrete subgroup, then Γ cannot be a Schottky group acting on $\mathbb{P}^{2n}_{\mathbb{C}}$.*

Proof. Assume Γ is a Schottky group acting on $\mathbb{P}_{\mathbb{C}}^{2n}$, and let γ be one of its generators. Since the group Γ is free, γ must have infinite order. Now, since the action is on a projective space of even dimension, the previous Lemmas 9.4.11 and 9.4.13 insure that the set $A = L_0(\langle\gamma\rangle)\cup L_1(\langle\gamma\rangle)$ is contained in a connected component of the limit set as defined in Definition 3.3.9. On the other hand, since the dynamics of the group is of the "ping-pong" type, one has that the set A must be contained in at least two connected components of the limit set, which is a contradiction, and the theorem follows. $\qquad\square$

Remark 9.4.15. Notice that Theorem 9.4.14 and its proof remain valid if we replace $\mathbb{C}$ by $\mathbb{R}$.

9.5 Complex kissing-Schottky groups

We recall that in Chapter 1 we looked at *kissing-Schottky subgroups* of conformal maps of $\mathbb{P}_{\mathbb{C}}^1$. These are groups obtained by reflections on circles which were not all of them disjoint, but we allowed them to touch their neighbours slightly. That is, each circle could *kiss* their neighbours, touching them tangentially. In this section we construct a family of complex Kleinian groups acting on the 2-dimensional projective space with similar features. The idea is to start with sets R_j and S_j in $\mathbb{P}_{\mathbb{C}}^1$ with the properties of Definition 9.2.1, but allowing them to touch others in a "little part", thus obtaining examples of free kissing-Schottky groups as in [158]; then from these we get complex kissing-Schottky subgroups of PSL$(3,\mathbb{C})$.

The precise definition of complex kissing-Schottky goes exactly as in Definition 9.2.1, replacing the condition on the sets $\{\overline{R_i}, \overline{S_j} : j = 1, \dots, g\}$ being pairwise disjoint by the condition $\bigcup_{j=1}^{g} \overline{R_j \cup S_j} \neq \mathbb{P}_{\mathbb{C}}^n$. In this way we can guarantee that the corresponding group is free in g generators and it has a nonempty region of discontinuity, with noncompact quotient.

We recall that the construction of Schottky groups in [203], explained before, is based on having mirrors that play in $\mathbb{P}_{\mathbb{C}}^{2n+1}$ the same role that circles play in $\mathbb{P}_{\mathbb{C}}^1 \cong \mathbb{S}^2$. Then we proved in the previous section that there are no Schottky groups in even dimensions, but we did not say that there are no mirrors. In fact the arguments used there show that the existence of such mirrors implies that these sets necessarily intersect each other.

So the question now is how to create mirrors with "small" intersections, in such a way that they enable us to construct complex Kleinian groups?

For this, let us first construct mirrors in $\mathbb{P}_{\mathbb{C}}^2$. A way to do so is to consider the complex line ℓ in $\mathbb{P}_{\mathbb{C}}^2$ given by $[z_1 : z_2 : 0]$ and the point $p = [0 : 0 : 1]$. Now consider a Euclidean circle C on $\ell = \mathbb{P}_{\mathbb{C}}^1$ and the complex inversion ι_C on it; this is the composition of the Euclidean inversion defined in Chapter 1, followed by a reflection on a line (actually a circle) through its centre. Let $\mathcal{C}(C, p)$ be the union of all the complex lines which join points in C with p, so it is a complex cone, and it is a 3-dimensional subvariety of $\mathbb{P}_{\mathbb{C}}^2$ with an isolated singularity at the vertex.

Lemma 9.5.1. *The involution $\imath_C$ extends canonically to an involution $\tilde{\imath}_C$ of $\mathbb{P}^2_{\mathbb{C}}$ which leaves the cone $\mathcal{C}(C,p)$ invariant, and this cone $\mathcal{C}(C,p)$ splits $\mathbb{P}^2_{\mathbb{C}}$ in two diffeomorphic halves which are permuted by $\tilde{\imath}_C$.*

Proof. Let $m \in \mathrm{SL}(2,\mathbb{C})$ be a lift of $\imath_C$, then $\tilde{\imath}_C$ is the projectivisation of the matrix

$$\begin{pmatrix} m & 0 \\ 0 & 1 \end{pmatrix}.$$

It is clear that this map has the properties stated in the lemma. $\square$

In other words the cone $\mathcal{C}(C,p)$, together with the map $\imath_C$, play the role of a mirror. More generally,

Definition 9.5.2. A *mirror* in $\mathbb{P}^2_{\mathbb{C}}$ is the image of $\mathcal{C}(C,p)$ by a projective automorphism.

So we can now start playing ping-pong as before. Consider a disjoint family of circles in ℓ with the corresponding inversions, and apply the procedure described before. Then we generate a Kleinian group. But notice that the mirrors intersect at the vertex of the cone, so the group is not Schottky but is often kissing-Schottky.

As an example, consider the Möbius transformations given by

$$m_1(z) = \frac{(1+i)z - i}{iz + 1 - i} \;;\; m_2(z) = \frac{(1-i)z - i}{iz + 1 + i} \;;\; m_3(z) = \frac{3iz + 10i}{iz + 3i} \;.$$

We have:

$$m_1(\mathbb{D} + 1 + i) = \mathbb{P}^1_{\mathbb{C}} \setminus \overline{\mathbb{D} + 1 - i}\,;$$
$$m_2(\mathbb{D} - 1 + i) = \mathbb{P}^1_{\mathbb{C}} \setminus \overline{\mathbb{D} - 1 - i}\,;$$
$$m_3(\mathbb{D} - 3) = \mathbb{P}^1_{\mathbb{C}} \setminus \overline{\mathbb{D} + 3}\,.$$

Thus $\Sigma_s = \langle m_1, m_2, m_3 \rangle$ is a kissing-Schottky group.

Now let $\Gamma_\epsilon = \langle M_1, M_2, M_\epsilon \rangle$ where $\epsilon \in \mathbb{C}^*$ and

$$M_1 = \begin{pmatrix} -1-i & i & 0 \\ -i & -1+i & 0 \\ 0 & 0 & 1 \end{pmatrix}\,;\; M_2 = \begin{pmatrix} 1-i & -i & 0 \\ i & 1+i & 0 \\ 0 & 0 & 1 \end{pmatrix}\,;\; M_\epsilon = \begin{pmatrix} 3i\epsilon & 10i\epsilon & 0 \\ i\epsilon & 3i\epsilon & 0 \\ 0 & 0 & \epsilon^{-2} \end{pmatrix}\,.$$

Proposition 9.5.3. *The group Γ_ϵ is a complex kissing-Schottky subgroup of $\mathrm{PSL}(3,\mathbb{C})$ with three generators. The discontinuity region in the sense of Kulkarni is the largest open set where Γ_ϵ acts properly discontinuously on $\mathbb{P}^2_{\mathbb{C}}$ and its complement is given by $\Lambda_{\mathrm{Kul}}(\Gamma_\epsilon) = \bigcup_{p \in \Lambda(\Sigma_s)} \overleftrightarrow{p, e_3}$, where $\Lambda(\Sigma_s)$ is the usual limit set of a subgroup in $\mathrm{PSL}(2,\mathbb{C})$.*

Proof. Let $r\mathbb{D} + p$ be the Euclidean disc on $\overleftrightarrow{e_1, e_2}$ with radius r and centre p. Consider the following disjoint family of open sets

$$R_1 = \bigcup_{q \in \mathbb{D}+1+i} \overleftrightarrow{q, e_3}\,;\qquad\qquad S_1 = \bigcup_{q \in \mathbb{D}+1-i} \overleftrightarrow{q, e_3}\,;$$

$$R_2 = \bigcup_{q \in \mathbb{D}-1+i} \overleftrightarrow{q, e_3}; \qquad\qquad S_2 = \bigcup_{q \in \mathbb{D}-1-i} \overleftrightarrow{q, e_3};$$

$$R_3 = \bigcup_{q \in \mathbb{D}-4} \overleftrightarrow{q, e_3}; \qquad\qquad S_3 = \bigcup_{q \in \mathbb{D}+4} \overleftrightarrow{q, e_3}.$$

One deduces that

$$M_1(R_1) = \mathbb{P}^2_{\mathbb{C}} \setminus \overline{S_1};$$
$$M_2(R_2) = \mathbb{P}^2_{\mathbb{C}} \setminus \overline{S_2};$$
$$M_{3\epsilon}(R_3) = \mathbb{P}^2_{\mathbb{C}} \setminus \overline{S_3};$$
$$\bigcup_{i=1}^{3} \overline{R_i \cup S_i} \neq \mathbb{P}^2_{\mathbb{C}}.$$

Therefore Γ_ϵ is a Kissing-Schottky group with three generators. Finally, since $tr^2(m_2) = 4$ and $\det(M_2 + \mathrm{Id}) = 8$, we deduce that M_2 has the following normal Jordan form:

$$\begin{pmatrix} 1 & 1 & 0 \\ 0 & 1 & 0 \\ 0 & 0 & 1 \end{pmatrix}.$$

Then there is a complex line ℓ such that $e_3 \in l = \mathrm{Fix}(M_2)$. Hence $\ell \cap \overleftrightarrow{e_1, e_2} = \mathrm{Fix}(m_2)$ and $\overline{\Gamma_\epsilon \ell} = \bigcup_{q \in \Lambda_S(\Sigma_s)} \overleftrightarrow{q, e_3}$. Thus $\bigcup_{q \in \Lambda_S(\Sigma_s)} \overleftrightarrow{q, e_3} \subset L_0(\Gamma_\epsilon)$. Now, by Proposition 3.3.6 we conclude that $\Lambda_{\mathrm{Kul}}(\Gamma_\epsilon) = \bigcup_{q \in \Lambda_S(\Sigma_s)} \overleftrightarrow{q, e_3}$ is the largest open set on which Γ_ϵ acts discontinuously. $\qquad\square$

9.6 The "zoo" in dimension 2

We know from Chapter 8 that if a subgroup $\Gamma \subset \mathrm{PSL}(3, \mathbb{C})$ has an invariant region $\Omega \subset \mathbb{P}^2_{\mathbb{C}}$ where it acts properly discontinuously and with compact quotient, then the group must be either complex hyperbolic, affine or elementary. So it is natural to ask whether there are any other type of groups when we relax the compactness condition.

The following is an example of a kissing-Schottky group which is not elementary and it is not topologically conjugate to either an affine or a complex hyperbolic group.

Example 9.6.1. Now let us consider the following modification of the previous example. Let $\Gamma_\epsilon = \langle M_1, M_2, M_\epsilon \rangle$ where $\epsilon = (\epsilon_1, \epsilon_2, \epsilon_3) \in \mathbb{C}^* \times \mathbb{C}^2$ and

$$M_1 = \begin{pmatrix} -1-i & i & 0 \\ -i & -1+i & 0 \\ 0 & 0 & 1 \end{pmatrix}; \quad M_2 = \begin{pmatrix} 1-i & -i & 0 \\ i & 1+i & 0 \\ 0 & 0 & 1 \end{pmatrix}; \quad M_\epsilon = \begin{pmatrix} 3i\epsilon_1 & 10i\epsilon_1 & 0 \\ i\epsilon_1 & 3i\epsilon_1 & 0 \\ \epsilon_2 & \epsilon_3 & \epsilon_1^{-2} \end{pmatrix}.$$

The proof of the following lemma is a straightforward computation.

Lemma 9.6.2. *Let $P_\epsilon(\lambda)$ denote the characteristic polynomial of M_ϵ. Then*

$$P_\epsilon(\lambda) = -(\lambda - \epsilon_1^{-2})(\lambda - i\epsilon_1(3 - \sqrt{10}))(\lambda - i\epsilon_1(\sqrt{10} + 3)).$$

Thus, if we take $\epsilon_1 = -(3 + \sqrt{10})^{1/3}e^{-i\pi(1+4\vartheta)/6}$ with $\vartheta \in \mathbb{R} - \mathbb{Q}$, we deduce:

(i) *The set* $\beta = \left\{ p_1 = \begin{pmatrix} -\sqrt{10} \\ 1 \\ k_\epsilon^- \end{pmatrix}, \; p_2 = \begin{pmatrix} \sqrt{10} \\ 1 \\ k_\epsilon^+ \end{pmatrix}, \; \begin{pmatrix} 0 \\ 0 \\ 1 \end{pmatrix} \right\}$ *is an ordered*

   basis of eigenvectors, where

$$k_\epsilon^\pm = \frac{i(\pm\sqrt{10}\epsilon_2 + \epsilon_3)e^{i\pi(1+4\vartheta)/6}}{(3 + \sqrt{10})^{1/3}(3(1 - e^{2i\pi\vartheta}) - \sqrt{10}(\mp 1 - e^{2i\pi\vartheta}))}.$$

   The corresponding eigenvalues are $\{\alpha_-, \alpha_+, e^{2\pi i\vartheta}\alpha_-\}$, where

$$\alpha_\pm = \frac{-i(3 \pm \sqrt{10})(3 + \sqrt{10})^{1/3}}{e^{i\pi(1+4\vartheta)/6}}.$$

(ii) *For every point $x \in \mathbb{P}^2_\mathbb{C} \setminus (\overleftrightarrow{p_1, p_2} \cup \overleftrightarrow{p_1, e_3} \cup \overleftrightarrow{e_3, p_2})$ the set of cluster points of $\{M_\epsilon^{-n}(x)\}_{n\in\mathbb{N}}$ is contained in $\overleftrightarrow{p_1, e_3}$ and is diffeomorphic to $\mathbb{S}^1$.*

Recall that a kissing-Schottky group is like a Schottky group but allowing the corresponding mirrors to touch each other tangentially (compare with [158]).

Proposition 9.6.3. *The group Γ_ϵ is a complex kissing-Schottky subgroup of $PSL(3, \mathbb{C})$ with three generators. The Kulkarni discontinuity region is the largest open set where Γ_ϵ acts properly discontinuously on $\mathbb{P}^2_\mathbb{C}$ and its complement is given by $\Lambda_{\mathrm{Kul}}(\Gamma_\epsilon) = \bigcup_{p\in\Lambda_S(\Sigma_s)} \overleftrightarrow{p, e_3}$.*

Moreover, if we assume $\epsilon_1 = -(3 + \sqrt{10})^{1/3}e^{-i\pi(1+4\vartheta)/6}$, where $\vartheta = \mathbb{R} - \mathbb{Q}$, and $k_\epsilon \neq k_\epsilon^-$, then these kissing-Schottky groups Γ_ϵ are not topologically conjugate to either an elementary, an affine or a complex hyperbolic group.

Proof. Let us first take $p = e_3$, $l = \overleftrightarrow{e_2, e_1}$. Let $\pi = \pi_{p,\ell} : \mathbb{P}^2_\mathbb{C} \setminus \{p\} \longrightarrow \ell$, be the canonical projection and $\Pi = \Pi_{p,\ell} : \Gamma \longrightarrow \mathrm{Bihol}(\ell) \cong PSL(2, \mathbb{C})$, given by $\Pi(g)(x) = \pi(g(x))$, the group morphism associated to π. Then $\Pi(M_1) = m_1$, $\Pi(M_2) = m_2$, $\Pi(M_\epsilon) = m_3$.

Let us prove that Γ_ϵ is a kissing-Schottky group. Consider the disjoint family of open sets

$$\begin{aligned} R_1 &= \pi^{-1}(\mathbb{D} + 1 + i); & S_1 &= \pi^{-1}(\mathbb{D} + 1 - i); \\ R_2 &= \pi^{-1}(\mathbb{D} - 1 + i); & S_2 &= \pi^{-1}(\mathbb{D} - 1 - i); \\ R_3 &= \pi^{-1}(\mathbb{D} - 4); & S_3 &= \pi^{-1}(\mathbb{D} + 4). \end{aligned}$$

One has:

$$M_1(R_1) = \mathbb{P}^2_\mathbb{C} \setminus \overline{S_1};$$
$$M_2(R_2) = \mathbb{P}^2_\mathbb{C} \setminus \overline{S_2};$$

$$M_{3\epsilon}(R_3) = \mathbb{P}^2_{\mathbb{C}} \setminus \overline{S_3};$$

$$\bigcup_{i=1}^{3} \overline{R_i \cup S_i} \neq \mathbb{P}^2_{\mathbb{C}}.$$

Therefore Γ_ϵ is a Kissing-Schottky group with three generators.

Now consider the following properties of Γ_ϵ:

Property 1. Since $tr^2(m_2) = 4$ and $\det(M_2 + \mathrm{Id}) = 8$ we deduce that M_2 has the normal Jordan form

$$\begin{pmatrix} 1 & 1 & 0 \\ 0 & 1 & 0 \\ 0 & 0 & 1 \end{pmatrix}.$$

Then there is a complex line ℓ such that $e_3 \in l = \mathrm{Fix}(M_2)$. Thus $\pi(l \setminus \{e_3\}) = \mathrm{Fix}(m_2)$ and therefore $\overline{\Gamma_\epsilon l} = \bigcup_{q \in \Lambda_S(\Sigma_s)} \overleftrightarrow{q, e_3}$. Hence $\bigcup_{q \in \Lambda_S(\Sigma_s)} \overleftrightarrow{q, e_3} \subset L_0(\Gamma_\epsilon)$.

Now, by Theorem 5.8.2, it follows that Γ_ϵ acts discontinuously on $\mathbb{P}^2_{\mathbb{C}} \setminus \bigcup_{q \in \Lambda_S(\Sigma_s)} \overleftrightarrow{q, e_3}$. By Proposition 3.3.6 we conclude that $\Lambda_{\mathrm{Kul}}(\Gamma_\epsilon) = \bigcup_{q \in \Lambda_S(\Sigma_s)} \overleftrightarrow{q, e_3}$.

Property 2. *The group Γ_ϵ is not topologically conjugate to a complex hyperbolic group.* In fact, it is easy to see that $\Lambda_{\mathrm{Kul}}(M_\epsilon) = \overleftrightarrow{p_1, e_3} \cup \{p_2\}$. On the other hand, by Theorem 7.3.1 we have that M_ϵ cannot be topologically conjugate to an element of $\mathrm{PU}(2, 1)$.

Property 3. *The group Γ_ϵ is not topologically conjugate to an affine group.* In fact, assume on the contrary that there is a homeomorphism $\phi : \mathbb{P}^2_{\mathbb{C}} \longrightarrow \mathbb{P}^2_{\mathbb{C}}$ such that $\phi^{-1}\Gamma_\epsilon\phi(\ell) = \ell$ for some complex line ℓ. Then $\phi(\ell)$ is a Γ_ϵ-invariant 2-sphere. If there is a point $q \in \phi(\ell)$ such that $q \notin \overleftrightarrow{p_1, p_2} \cup \overleftrightarrow{p_2, e_3} \cup \overleftrightarrow{p_1, e_3}$, then by (ii) of Lemma 9.6.2 the set of cluster points of $\{M_\epsilon^{-n} q\}_{n \in \mathbb{N}}$ is contained in $\overleftrightarrow{p_1, e_3}$ and is diffeomorphic to $\mathbb{S}^1$. Hence $|\phi^{-1}(\overleftrightarrow{p_1, e_3}) \cap \ell| > 2$. On the other hand, since $\overleftrightarrow{p_1, e_3} \subset L_0(M_\epsilon)$, it follows that $\phi^{-1}(\overleftrightarrow{p_1, e_3})$ is a complex line. Thus $\phi(\ell) = \overleftrightarrow{p_1, e_3}$, which is a contradiction. Therefore $\phi(\ell) \subset \overleftrightarrow{p_1, p_2} \cup \overleftrightarrow{p_2, e_3} \cup \overleftrightarrow{p_1, e_3}$. Now, since $\phi(\ell) \setminus \{p_1, p_2, e_3\}$ is connected we conclude that $\phi(\ell) = \overleftrightarrow{p_1, p_2}$ or $\phi(\ell) = \overleftrightarrow{p_2, e_3}$ or $\phi(\ell) = \overleftrightarrow{p_1, e_3}$. Since $\overline{\Gamma_\epsilon \overleftrightarrow{p_1, e_3}} = \overline{\Gamma_\epsilon \overleftrightarrow{p_2, e_3}} = \bigcup_{q \in \Lambda_S(\Sigma_s)} \overleftrightarrow{q, e_3}$ with $\mathrm{card}(\Lambda_S(\Sigma_s)) > 2$, we deduce $\phi(\ell) = \overleftrightarrow{p_1, p_2}$. On the other hand, one can easily check that $[1; 1; 0]$ and $[0; 0; 1]$ are the unique fixed points of M_1 and its normal Jordan form is

$$\begin{pmatrix} -1 & 1 & 0 \\ 0 & -1 & 0 \\ 0 & 0 & 1 \end{pmatrix}. \tag{9.6.3}$$

Now, by Corollary 4.3.4 we conclude that $\overleftrightarrow{e_1, e_2}$ and $\overleftrightarrow{[1; 1; 0], e_3}$ are the unique invariant complex lines for M_1. It follows that $\overleftrightarrow{p_1, p_2} = \overleftrightarrow{e_1, e_2}$. Thence there are $\alpha, \beta \in \mathbb{C}$ such that $|\alpha| + |\beta| \neq 0$ and

$$-\sqrt{10}\alpha + \beta\sqrt{10} = 1,$$

$$\alpha + \beta = 0,$$
$$k_\epsilon^- \alpha + k_\epsilon^+ \beta = 0,$$

which is a contradiction since $K_\epsilon^+ \neq K_\epsilon^-$.   □

Remark 9.6.4. By Chapter 4 the limit set $\Lambda_{\text{Kul}}(\langle [[M_\epsilon]]_2 \rangle)$ is not a subset of $\Lambda_{\text{Kul}}(\Gamma_\epsilon)$ for $|\epsilon_1| > (\sqrt{10} + 3)^{1/3}$. This shows that the limit set in the sense of Kulkarni is not monotone. In other words, having discrete subgroups $\Gamma_1 \subset \Gamma_2$ of $\text{PSL}(3, \mathbb{C})$ does not imply that the Kulkarni limit set of Γ_1 is contained in that of Γ_2.

9.7   Remarks on the uniformisation of projective 3-folds

In Chapter 8 we discussed projective structures on manifolds, and we spoke about the uniformisation problem for Riemann surfaces and for complex projective surfaces. Of course it is very interesting to study this problem in higher dimensions, and the constructions we give in this and in the following Chapter 10 point in that direction: we get subgroups of $\text{PSL}(n + 1, \mathbb{C})$ acting properly discontinuously on an open set $\Omega \subset \mathbb{P}_\mathbb{C}^n$ with compact quotient, and the projection map $\Omega \to \Omega/\Gamma$ can be regarded as a uniformisation.

When pursuing this line of ideas, it is important to mention the important contributions to the subject by M. Kato, particularly in dimension 3 (see references in the bibliography).

In [111] the author studies compact complex 3-manifolds whose universal covering is a "large" domain in the projective space $\mathbb{P}_\mathbb{C}^3$. A domain in a projective space $\mathbb{P}_\mathbb{C}^3$ is called large if it contains a projective line. So for instance, the discontinuity regions of Schottky groups envisaged in this chapter, are large domains. The author proves that if a large domain $\Omega \subset \mathbb{P}_\mathbb{C}^3$ is the universal covering of a smooth compact complex manifold M carrying nonconstant meromorphic functions, then Ω is dense in $\mathbb{P}_\mathbb{C}^3$ and its complement is contained in a finite union of hypersurfaces and a set with Hausdorff dimension ≤ 2. With some extra assumptions on the structure of $\mathbb{P}_\mathbb{C}^3 \backslash \Omega$, it is also shown that this complement is actually a disjoint union of two lines, or a line, or empty. This uses a previous result by the same author, saying that if Ω is large, then the fundamental group of M is a subgroup of $\text{PSL}(4, \mathbb{C})$, acting on Ω as deck transformations.

More recently, in [112] the author refined and improved the results from [111] studying properly discontinuous actions of discrete subgroups $\Gamma \subset \text{PSL}(4, \mathbb{C})$ on large open domains in $\Omega \subset \mathbb{P}^3(\mathbb{C})$, and the structure of their quotients. It is proved that every holomorphic automorphism of a large domain is actually a restriction of an element in $\text{PSL}(4, \mathbb{C})$, so the group $\text{Aut}(\Omega)$ of biholomorphisms of Ω is a subgroup of $\text{PSL}(4, \mathbb{C})$, the stabiliser of Ω. Using results of P. J. Myrberg about limit sets of normal sequences, the author shows that if Γ acts properly discontinuously on a large domain Ω, then Γ must be of the so-called *type L*, which means that there is an open subdomain $W \subset \mathbb{P}^3$, biholomorphic to $\{[z_0 : z_1 : z_2 : z_3] \in \mathbb{P}^3 \mid |z_0|^2 + |z_1|^2 < |z_2|^2 + |z_3|^2\}$, which satisfies $\gamma(W) \cap W = \emptyset$ for any $\gamma \in \Gamma \backslash \{id\}$.

This is of course reminiscent of the results from [202] explained in this chapter. And as before, one also has for $\Gamma \subset \mathrm{PSL}(4, \mathbb{C})$ of type L a corresponding concept of a *limit set* $\Lambda(\Gamma)$, which is a union of projective lines and a closed, nowhere dense Γ-invariant subset of $\mathbb{P}^3_{\mathbb{C}}$. The complement $\Omega(\Gamma) = \mathbb{P}^3_{\mathbb{C}} \backslash \Lambda(\Gamma)$ is a large domain and Γ acts properly discontinuously on it.

Groups of type L also have the nice property that given two such groups Γ_1 and Γ_2, one has a Klein combination $\Gamma_1 * \Gamma_2$, which is again a group of type L

The quotient $\Omega(\Gamma)/\Gamma$ may have infinitely many components, see *M. E. Kapovich* and *L. D. Potyagailo* [Sib. Math. J. 32, No.2, 227–237 (1991; Zbl 0741.30038)]. Let Ω be a large domain in $\mathbb{P}^3$ and assume that the quotient $X = \Omega/\Gamma$, by a properly discontinuous holomorphic action of $\Gamma \subset \mathrm{Aut}(\Omega)$ on Ω, is compact. Then Ω is a connected component of $\Omega(\Gamma)$. If moreover Γ acts fixed point free on Ω and the algebraic dimension $a(X)$ of X is positive, then $\Omega = \Omega(\Gamma)$, see [K], and the universal covering space of X is biholomorphic equivalent to the complement of one or two disjoint projective lines in $\mathbb{P}^3$. The latter statement is the main result of the paper and the extensive proof uses a broad spectrum of classical methods from algebraic geometry including deformation theory of Kodaira surfaces. The main difficulty occurs for $a(X) = 1$ and is solved by a thorough analysis of the possible singularities of the singular fibres of the algebraic reduction map of X, added by detailed considerations about free abelian group actions on $\mathbb{P}^2$ and $\mathbb{P}^3$.

Chapter 10

Kleinian Groups and Twistor Theory

Twistor theory is one of the jewels of mathematics in the 20th Century. A starting point of the celebrated "Penrose twistor programme" is that there is a rich interplay between the conformal geometry on even-dimensional spheres and the holomorphic on their twistor spaces. Here we follow [202] and explain how the relations between the geometry of a manifold and the geometry of its twistor space, can be carried forward to dynamics. In this way we get that the dynamics of conformal Kleinian groups embeds in the dynamics of complex Kleinian groups.

The prototype of a twistor fibration is $\mathbb{P}^3_{\mathbb{C}} \to \mathbb{S}^4$, also known as the Calabi-Penrose fibration, in which the fibre is the 2-sphere. In this case we get that every Kleinian subgroup of $\mathrm{Iso}_+(\mathbb{H}^5_{\mathbb{R}})$ embeds canonically in $\mathrm{PSL}(4, \mathbb{C})$ as a complex Kleinian group.

This chapter is based on [202]. In Section 10.1 we review the twistor space and the twistor fibration associated to Riemannian manifolds. We look in detail at the case of the 4-sphere, whose twistor space is $\mathbb{P}^3_{\mathbb{C}}$, where we define the twistor fibration in three equivalent ways. We then discuss the canonical lifting of the conformal group acting on the base, to a group of holomorphic automorphisms of the twistor space. We use this information to construct and study complex Kleinian groups, and finally we use the Patterson-Sullivan measure for conformal Kleinian groups, to define a measure for these "twistorial" Kleinian groups, and study their ergodicity and the minimality of the action on their Kulkarni limit set.

10.1 The twistor fibration

In this section we give a definition of the twistor fibration and we explain some of the basic ideas and results in this respect. We consider first the case of 4-manifolds, where this fibration was also studied, independently and from a different viewpoint in [35, 36]. Here we discuss both viewpoints, and we refer to [10], [12], [174] and [175] for more about twistor theory. See also [34], [189], [97] for clear accounts on the subject. We use also the spin representation of the orthogonal group, and we refer to [11] and [137] for clear accounts of spin geometry.

10.1.1   The twistor fibration in dimension 4

If M is a closed, oriented 4-manifold endowed with a Riemannian metric, its twistor space $3(M)$ is by definition the total space of the fibre bundle over M. The fibre of $3(M)$ at each $x \in M$ is the set $3_x(M)$ of all complex structures on $T_x M$ which are compatible with the metric and orientation on M.

It is clear that given one such complex structure J on $T_x M$, i.e., an element in $3_x(M)$, every element in $SO(4)$ transforms J into another complex structure on $T_x M$ compatible with the metric and the orientation. That is, $SO(4)$ acts on $3_x(M)$, and it is an exercise to show that this action is transitive, with isotropy $U(2)$. Hence the fibre is $SO(4)/U(2)$.

It is well known that $SO(4)/U(2)$ is diffeomorphic to the 2-sphere, i.e.,

$$SO(4)/U(2) \cong \mathbb{S}^2.$$

A way to prove this is by showing that $SO(4)$ acts transitively on $\mathbb{S}^2$ with fibre $U(2)$, and the best way to achieve this is via the spin representation. In fact, $SO(4)$ is isomorphic to $SU(2) \times SO(3)$ and its spin representation is the direct sum of two complex representations of dimension 2, the usual representation of $SU(2)$ as linear maps in $\mathbb{C}^2$. Its projectivisation provides a transitive action of $SO(4)$ on $\mathbb{P}^1_{\mathbb{C}} \cong \mathbb{S}^2$ with isotropy $U(2)$ and thence the fibre $3_x(M) = SO(4)/U(2)$ is actually $\mathbb{P}^1_{\mathbb{C}}$.

Thus one has a fibre bundle $p : 3(M) \to M$, with fibre $\mathbb{P}^1_{\mathbb{C}} \cong \mathbb{S}^2$:

Definition 10.1.1. The 2-sphere bundle,

$$p : 3(M) \to M,$$

is called the *Calabi-Penrose, or twistor*, fibration of the Riemannian manifold M; its fibres are called the *twistor lines*.

When $M = \mathbb{S}^4$ with its canonical metric, the twistor space $3 := 3(\mathbb{S}^4)$ turns out to be the complex projective space $\mathbb{P}^3_{\mathbb{C}}$. In fact, it is easy to see that one has

$$3(\mathbb{S}^4) \cong SO(5)/U(2) \cong SO(6)/U(3),$$

and the projectivisation of the spin representation of $SO(6)$ provides a diffeomorphism $SO(6)/U(3) \cong \mathbb{P}^3_{\mathbb{C}}$. One thus has the twistor fibration

$$3(\mathbb{S}^4) \cong \mathbb{P}^3_{\mathbb{C}} \longrightarrow \mathbb{S}^4,$$

with fibre $\mathbb{S}^2$.

Let us give an alternative (well-known) construction of this fibration. Let $\mathcal{H}$ be now the quaternionic line. We can think of it as being $\mathbb{R}^4$ equipped with three complex structures given by i, j, k, with $i^2 = j^2 = k^2 = -1$, which satisfy $ij = k$; $jk = i$; and $ki = j$.

We now consider the 2-dimensional quaternionic space $\mathcal{H}^2$. We leave it as an exercise to show that $\mathbb{S}^4$ is diffeomorphic to the space of right quaternionic lines in $\mathcal{H}^2$: $\mathbb{S}^4 \cong \mathbb{P}^1_{\mathcal{H}}$.

Multiplication in $\mathcal{H}^2$ on the right by i determines a complex structure on $\mathcal{H}^2 \cong \mathbb{R}^8$. In this way, each *right* quaternionic line L_q in $\mathcal{H}^2$ becomes a 2-dimensional complex space in $\mathbb{C}^2 \cong \mathcal{H}$. Moreover, given any $\alpha \in \mathcal{H}$, multiplication by α on the right preserves $L_q := \{q\lambda : \lambda \in \mathcal{H}, q \in \mathcal{H}^2 - \{(0,0)\}\}$, so each line L_q is covered by the complex lines $l_{q\alpha} := \{q\alpha\lambda : \lambda \in \mathbb{C}, \} \subset L_q$. If we identify each complex line $l_{q\alpha}$ to a point we obtain $\mathbb{P}^3_{\mathbb{C}} = 3(\mathbb{S}^4)$, and if we identify each quaternionic line L_q to a point we obtain $P^1_{\mathcal{H}} = \mathbb{S}^4$. This gives the 2-sphere bundle, the twistor fibration $p : \mathbb{P}^3_{\mathbb{C}} \to \mathbb{S}^4$, where each fibre is a projective line $\mathbb{P}^1_{\mathbb{C}}$.

We remark that there is yet another way for thinking of the fibres of this fibration $\mathbb{P}^3_{\mathbb{C}} \to \mathbb{S}^4$, which generalises immediately to all oriented, Riemannian 4-manifolds M. Or rather, there is a natural way for constructing elements in these fibres, that is complex structures on the tangent space at each point of M. This was Calabi's viewpoint. For this, notice that a choice of a metric and an orientation in $\mathbb{R}^2$ identifies this space with $\mathbb{C}$: multiplication by i being a rotation of $\pi/2$ degrees counterclockwise. Now consider $\mathbb{R}^4$ equipped with an orientation and its usual metric, and let P be an oriented 2-plane through the origin in $\mathbb{R}^4$. Then, by the previous remark, P is canonically a complex line. On the other hand, the metric on $\mathbb{R}^4$ determines a well-defined orthogonal complement $P^\perp$ of P. The orientation in $\mathbb{R}^4$ determines an orientation in $P^\perp$ compatible with that in P. Hence $P^\perp$ also has a canonical complex structure. Since $\mathbb{R}^4 \cong P \oplus P^\perp$, one thus has a complex structure on $\mathbb{R}^4$. Thence we have proved:

Lemma 10.1.2. *Every oriented 2-plane in $\mathbb{R}^4$ determines an isomorphism $\mathbb{R}^4 \cong \mathbb{C}^2$.*

In other words, given a point $x \in M^4$ and an oriented plane $P \subset T_x M$, one has a well-determined element in the twistor fibre $3_x(M) \cong \mathbb{S}^2$. Thus one has:

Lemma 10.1.3. *Let Σ be a closed, oriented, C^1 surface immersed in $\mathbb{S}^4$. Then Σ has a canonical lifting to a surface $\widetilde{\Sigma}$ in $\mathbb{P}^3_{\mathbb{C}}$.*

The problem studied by Calabi and others, was to determine geometric conditions on Σ to ensure that the lifting $\widetilde{\Sigma}$ was a complex submanifold of $\mathbb{P}^3_{\mathbb{C}}$. Calabi's work is particularly interesting in relation with minimal immersions of 2-spheres in $\mathbb{S}^4$, and that will be used later in this chapter. See [35], [36], and also [54], [34], [189].

10.1.2   The twistor fibration in higher dimensions

We now define the twistor fibration in higher dimensions. We consider the unitary group $\mathrm{U}(n) := \{A \in \mathrm{GL}\,(n, \mathbb{C}) : A^{-1} = \overline{A^t}\}$; the columns of each such matrix define linearly independent vectors in $\mathbb{C}^n$, so $\mathrm{U}(n)$ can be regarded as being the set of all unitary n-frames in $\mathbb{C}^n$. We also consider the special orthogonal group $\mathrm{SO}(n)$, which consists of all orthonormal, oriented n-frames in $\mathbb{R}^n$. These are both compact Lie groups of dimensions n^2 and $n(n-1)/2$, respectively. There are natural inclusions

$$\mathrm{U}(n) \hookrightarrow \mathrm{SO}(2n) \hookrightarrow \mathrm{SO}(2n+1) \qquad \text{and} \qquad \mathrm{U}(n) \hookrightarrow \mathrm{U}(n+1)\,.$$

This gives a map between symmetric spaces,

$$\mathrm{SO}(2n+1)/\mathrm{U}(n) \longrightarrow \mathrm{SO}(2n+2)/\mathrm{U}(n+1)\,, \quad n > 0\,,$$

which is easily seen to be a diffeomorphism. One also has a natural action of $\mathrm{SO}(2n+1)$ on the sphere $\mathbb{S}^{2n}$, with isotropy $\mathrm{SO}(2n)$, so that $\mathbb{S}^{2n} \cong \mathrm{SO}(2n+1)/\mathrm{SO}(2n)$. This induces a projection map

$$p: \mathfrak{Z}(\mathbb{S}^{2n}) := \mathrm{SO}(2n+1)/\mathrm{U}(n) \to \mathbb{S}^{2n}\,,$$

which is a submersion with fibre $\mathfrak{L}_x^{(2n)} := p^{-1}(x) \cong \mathrm{SO}(2n)/\mathrm{U}(n)\,,$ the set of all complex structures on the tangent space $T_x\mathbb{S}^{2n}$ which are compatible with the metric and orientation. For $n > 1$, the fibre $\mathfrak{L}_x^{(2n)}$ coincides with the space $\mathfrak{Z}(\mathbb{S}^{2n-2})$.

Definition 10.1.4. The manifold $\mathfrak{Z}(\mathbb{S}^{2n})$ is the *twistor space* of $\mathbb{S}^{2n}$, and $p: \mathfrak{Z}(\mathbb{S}^{2n}) \to \mathbb{S}^{2n}$ is its *twistor fibration*. The fibres of p are the *twistor fibres*.

More generally one has:

Definition 10.1.5. Let N be an oriented, Riemannian $2n$-manifold, $n > 0$. The *twistor space* of N is the total space of *the twistor fibration*, $p: \mathfrak{Z}(N) \to N$, whose fibre at $x \in N$ is the twistor space $\mathrm{SO}(2n)/\mathrm{U}(n)$ of $\mathbb{S}^{2n-2}$, i.e., the set of all complex structures on T_xN compatible with the metric and the orientation on N.

This generalises the twistor fibration studied above in dimension 4. We refer to Penrose's articles [174], [175], [176], and to the articles [95], [12], [52], [165] for more on the subject. See also to [34], [189] for clear accounts of twistor spaces.

The space $\mathfrak{Z} := \mathfrak{Z}(N)$ is always an almost complex manifold. In fact the Levi-Civita connexion ∇ on N gives rise to a splitting $T(\mathfrak{Z}) = H \oplus V$ of the tangent bundle $T(\mathfrak{Z})$, into the horizontal and vertical components, where V is the bundle tangent to the fibres of p and H is isomorphic to the pull-back $p^*(TN)$ of TN.

Each $\widetilde{x} \in \mathfrak{Z}(N)$ represents a point $x := p(\widetilde{x})$ in N, together with a complex structure on T_xN; since at each $\widetilde{x}$ in $\mathfrak{Z}(N)$ one has $T_{\widetilde{x}}(\mathfrak{Z}) = H_{\widetilde{x}} \oplus V_{\widetilde{x}}$, and $H_{\widetilde{x}}$ is naturally isomorphic to T_xN, one has a tautological complex structure on $H_{\widetilde{x}}$. Hence an almost complex structure on the fibre $3(\mathbb{S}^{2n-2}) = SO(2n)/U(n)$, determines an almost complex structure $\tilde{J}$ on $3(N)$, and an almost complex structure on the fibre is easily defined by induction: $3(\mathbb{S}^{2n-2})$ fibres over $\mathbb{S}^{2n-2}$ with fibre $3(\mathbb{S}^{2n-4})$, and so on; at each step $T(3(\mathbb{S}^{2i}))$ decomposes as above, into an horizontal and a vertical component, with the horizontal component having a tautological almost complex structure.

Hence, the complex structure on $\mathbb{P}^1_{\mathbb{C}} = 3(\mathbb{S}^2)$ determines an almost complex structure on $3(\mathbb{S}^4) = \mathbb{P}^3_{\mathbb{C}}$ and so on, till we get an almost complex structure $\tilde{J}$ on $3(N)$.

The question of the integrability of $\tilde{J}$ is very subtle: it is integrable if N is (locally) conformally flat (by [12] for $n = 2$, by [52], [165] for $n > 2$). In fact this condition is also necessary for $n > 2$, see [189, Th.3.3]. Hence, in particular, $3(\mathbb{S}^{2n})$ is always a complex manifold with the almost complex structure $\tilde{J}$. It has complex dimension $n(n + 1)/2$.

We summarise the previous discussion in the following well-known theorem.

Theorem 10.1.6. *Let N be a closed, oriented, Riemannian $2n$-manifold, $n > 1$, and let $p\colon 3(N) \to N$ be the twistor fibration of N. Then $3(N)$ has a (preferred) almost complex structure $\tilde{J}$, which is integrable whenever N is conformally flat. The twistor fibration $p\colon 3(N) \to N$, is a locally trivial fibre bundle with fibre $3(\mathbb{S}^{2n-2}) = SO(2n)/U(n)$. In particular,*

$$3(\mathbb{S}^{2n}) \cong SO(2n + 1)/U(n) \cong SO(2n + 2)/U(n + 1).$$

For $n = 2$, to have integrability of the almost complex structure $\tilde{J}$, one does not actually need to ask so much: the metric only needs to be anti self-dual.

Another important property of the twistor space $3(\mathbb{S}^{2n})$ is that it always embeds in a projective space $\mathbb{P}^N_{\mathbb{C}}$, for some N. We will return to this point later in the chapter.

10.2 The Canonical Lifting

A key-fact of Penrose's twistor programme is that every orientation preserving conformal automorphism of the $2n$-sphere lifts canonically to a holomorphic automorphism of its twistor space. Here we briefly explain these liftings, and prove that these are isometries on the twistor fibres, a fact which is essential for studying, later in this chapter, the dynamics of these liftings.

10.2.1 Lifting $\mathrm{Conf}_+(\mathbb{S}^4)$ to $\mathrm{PSL}(4, \mathbb{C})$

There are several ways for explaining the canonical lifting of (orientation preserving) conformal automorphisms of $\mathbb{S}^4$ to holomorphic automorphisms of $\mathbb{P}^3_{\mathbb{C}}$. Here we give one such description, following [202].

Let us give a different description of the group $\mathrm{Conf}_+(\mathbb{S}^4)$ which is appropriate for this work. We know already that $\mathbb{S}^4$ can be thought of as being the *projective quaternionic line* $\mathbb{P}^1_{\mathcal{H}} \cong \mathbb{S}^4$. This is the space of *right* quaternionic lines in $\mathcal{H}^2$, i.e., subspaces of the form

$$L_q := \{q\lambda \ : \ \lambda \in \mathcal{H}\}, \quad q \in \mathcal{H}^2 - \left\{ \begin{pmatrix} 0 \\ 0 \end{pmatrix} \right\},$$

where $\mathcal{H}$ is the space of quaternions and $\mathcal{H}^2 := \left\{ \begin{pmatrix} q_0 \\ q_1 \end{pmatrix} \ : \ q_0, q_1 \in \mathcal{H} \right\}$. Identify $\mathbb{S}^4$ with $\mathcal{H} \cup \{\infty\} := \widehat{\mathcal{H}}$ via the stereographic projection. Let $\mathrm{GL}\,(2, \mathcal{H}) := Gl_l(2, \mathcal{H})$ be the group of all invertible 2×2 quaternionic matrices $\begin{pmatrix} a & b \\ c & d \end{pmatrix}$ acting on $\mathcal{H}^2$ by the left in the obvious way. The space $\mathcal{H}^2$ is a right module over $\mathcal{H}$ and the action of $\mathrm{GL}\,(2, \mathcal{H})$ on $\mathcal{H}^2$ commutes with multiplication by the right: for every $\lambda \in \mathcal{H}$ and $A \in \mathrm{GL}\,(2, \mathcal{H})$ one has

$$A \circ R_\lambda = R_\lambda \circ A,$$

where R_λ is multiplication on the right by λ. Thus $\mathrm{GL}\,(2, \mathcal{H})$ carries right quaternionic lines into right quaternionic lines, so it defines an action of $\mathrm{GL}\,(2, \mathcal{H})$ on $P^1_{\mathcal{H}} = \mathbb{S}^4$. Now consider the map

$$\psi : \mathcal{H}^2 - \left\{ \begin{pmatrix} 0 \\ 0 \end{pmatrix} \right\} \to \mathbb{S}^4$$

given by: $\psi \begin{pmatrix} q_0 \\ q_1 \end{pmatrix} = q_0 q_1^{-1}$. For each $A = \begin{pmatrix} a & b \\ c & d \end{pmatrix} \in \mathrm{GL}\,(2, \mathcal{H})$ and a point (q_0, q_1) in $\mathcal{H}^2 - \left\{ \begin{pmatrix} 0 \\ 0 \end{pmatrix} \right\}$, one has: $T \circ \psi(q_0, q_1) = \psi(A \begin{pmatrix} q_0 \\ q_1 \end{pmatrix})$, where T is the Möbius transformation $T(q) = (aq + b)(cq + d)^{-1} \in \mathbb{S}^4$.

We refer to [4], [5], [10], [133] or [64] for proofs of the following theorem.

Theorem 10.2.1. *Let us denote by* $\mathrm{Möb}(2, \mathcal{H})$ *the set of all the quaternionic Möbius transformations in* $\widehat{\mathcal{H}} := \mathcal{H} \cup \infty = \mathbb{S}^4$ *of the form*

$$T(q) = (aq + b)(cq + d)^{-1}, \quad q \in \widehat{\mathcal{H}},$$

where a, b, c, d *are quaternions and the matrix* $\begin{pmatrix} a & b \\ c & d \end{pmatrix}$ *is in* $\mathrm{GL}\,(2, \mathcal{H})$. *We make the usual conventions about the point at infinity. Then* $\mathrm{Möb}(2, \mathcal{H})$ *is a group, and one has the group isomorphisms*

$$\mathrm{Möb}(2, \mathcal{H}) \cong \mathrm{PSL}(2, \mathcal{H}) \cong \mathrm{Conf}_+(\mathbb{S}^4) \cong \mathrm{SO}_0(5, 1),$$

where $\mathrm{PSL}(2, \mathcal{H}) = (\mathrm{SL}(8, \mathbb{R}) \cap \mathrm{GL}(2, \mathcal{H})) / \{\pm I\}$ *and* $\mathrm{SO}_0(5, 1)$ *is the connected component of the identity of* $\mathrm{SO}(5, 1)$.

As before, we endow $\mathbb{S}^4$ with its usual metric and we consider its twistor fibration

$$p : 3(\mathbb{S}^4) \cong \mathbb{P}^3_{\mathbb{C}} \longrightarrow \mathbb{S}^4 \, ,$$

with fibre $E_x \cong \mathbb{P}^1_{\mathbb{C}}$ the *twistor line* at each $x \in \mathbb{S}^4$; this is the set of all complex structures on $T_x\mathbb{S}^4$ which are compatible with the metric and orientation.

We think of $\mathbb{S}^4$ as being the space of right quaternionic lines in $\mathcal{H}^2$, as above. Recall that to get the twistor fibration in this way we observed that multiplication on the right by i determines a complex structure on $\mathcal{H}^2 \cong \mathbb{R}^8$, so that each *right* quaternionic line L_q in $\mathcal{H}^2$ becomes a 2-dimensional complex space in $\mathbb{C}^4 \cong \mathcal{H}^2$. Moreover, given any $\alpha \in \mathcal{H}$, multiplication by α by the right preserves $L_q :=$ $\{q\lambda : \lambda \in \mathcal{H} , q \in \mathcal{H}^2 - \{(0,0)\} \}$, so each line L_q is covered by the complex lines $l_{q\alpha} := \{q\alpha\lambda : \lambda \in \mathbb{C}, \} \subset L_q$.

If we identify each complex line $l_{q\alpha}$ to a point we obtain $\mathbb{P}^3_{\mathbb{C}} = 3(\mathbb{S}^4)$, and if we identify each quaternionic line L_q to a point we obtain $P^1_{\mathcal{H}} = \mathbb{S}^4$. This gives the 2-sphere bundle, $p : \mathbb{P}^3_{\mathbb{C}} \to \mathbb{S}^4$, where each fibre is a projective line $\mathbb{P}^1_{\mathbb{C}}$.

Now, given $h \in \mathrm{Conf}_+(\mathbb{S}^4)$, its canonical lifting to a holomorphic map

$$\widetilde{h} : 3(\mathbb{S}^4) = \mathbb{P}^3_{\mathbb{C}} \to \mathbb{P}^3_{\mathbb{C}} \, ,$$

can be defined through the above identification

$$\mathrm{Conf}_+(\mathbb{S}^4) \cong \mathrm{M\ddot{o}b}(2, \mathcal{H}) \cong \mathrm{PSL}(2, \mathcal{H}) \subset \mathrm{PSL}(4, \mathbb{C}) \, .$$

This also says that $\mathrm{Conf}_+(\mathbb{S}^4)$ actually lifts to $\mathrm{PSL}(4, \mathbb{C})$ as a group, carrying twistor lines into twistor lines.

Let us now say something about the way this lifting transforms the twistor lines.

We recall that $\mathbb{P}^1_{\mathbb{C}}$ has the Fubini-Study metric (see Chapter 2 or [232]), which coincides with the standard metric on $\mathbb{S}^2$. This metric is essentially the angle between the complex lines in $\mathbb{C}^2$. More precisely, we think of $\mathbb{S}^2 \cong \mathbb{P}^1_{\mathbb{C}}$ as being the quotient $\mathbb{S}^3/\, \mathrm{U}(1)$, which inherits a metric from the usual metric on $\mathbb{S}^3$, which is $\mathrm{U}(1)$-invariant.

For each line L_q, with the above complex structure, we consider the standard Hermitian metric. Then a transformation $A \in \mathrm{PSL}(2, \mathcal{H})$ sends the right line L_q isometrically into the right line $L_{q'}$, because the vectors $\{q, qi, qj, qk\}$ form a real orthonormal basis in L_q, and their image in $L_{q'}$ is the basis $\{q', q'i, q'j, q'k\}$. Therefore A preserves the angle between complex lines contained in the *same* right quaternionic line, so it preserves the Fubini-Study metric on the fibres of the twistor fibration. We thus arrive at the following theorem of [202]:

Theorem 10.2.2. *Let $h \in \mathrm{M\ddot{o}b}(2, \mathcal{H})$ and let $\widetilde{h}$ be its canonical lifting to an automorphism of $\mathbb{P}^3_{\mathbb{C}}$. Then $\widetilde{h}$ carries twistor fibres isometrically onto twistor fibres, with respect to the Fubini-Study metric on the fibres.*

10.2.2 The Canonical lifting in higher dimensions

We now discuss the generalisations of Theorems 10.2.1 and 10.2.2 to higher dimensions.

The Lie group $SO(2n+1)$, being compact, has a canonical bi-invariant metric, that we may think of as being the distance between two frames. This descends to a metric on $\mathfrak{Z}(\mathbb{S}^{2n}) \cong SO(2n+1)/U(n)$, which is invariant under the left action of $SO(2n+1)$ and restricts to the corresponding metric on each twistor fibre $\mathfrak{L}_x^{(2n)} := p^{-1}(x) \cong SO(2n)/U(n)$.

With this, the projection $p : \mathfrak{Z}(\mathbb{S}^{2n}) \to \mathbb{S}^{2n}$ becomes a Riemannian submersion. Furthermore, each element $\mathcal{F} \in SO(2n+1)$ can be regarded as being of the form $(x, \mathcal{F}_x^{2n})$, where x is a point in $\mathbb{S}^{2n}$ and $\mathcal{F}_x^{2n} = (v_1(x), \ldots, v_{2n}(x))$ is an orthonormal basis of the tangent space $T_x \mathbb{S}^{2n}$.

If $\gamma \in \mathrm{Conf}_+(\mathbb{S}^{2n})$ is a conformal diffeomorphism, then the derivative of $d\gamma$ carries $\mathcal{F}_x^{2n}$ into a basis $\mathcal{F}_{\gamma(x)}^{2n}$ of $T_{\gamma(x)}\mathbb{S}^{2n}$, which is orthonormal up to a scalar multiple. Thus γ lifts canonically to a diffeomorphism $\widetilde{\gamma}$ of the $\mathfrak{Z}(\mathbb{S}^{2n})$. There is another way of defining this lifting of γ to $\mathfrak{Z}(\mathbb{S}^{2n})$: at each point $x \in \mathbb{S}^{2n}$, the basis $\mathcal{F}_x^{2n}$ provides an identification $T_x(\mathbb{S}^{2n}) \cong \mathbb{C}^n$, so it endows $T_x(\mathbb{S}^{2n})$ with a complex structure J_x^1. Then $d\gamma$ determines the basis $d(\gamma)_x(\mathcal{F}_x^{2n})$ of $T_{\gamma(x)}(\mathbb{S}^{2n})$, hence an isomorphism $T_{\gamma(x)}(\mathbb{S}^{2n}) \cong \mathbb{C}^n$ and a complex structure $J_{\gamma(x)}^1$ on $T_{\gamma(x)}(\mathbb{S}^{2n})$. This gives the lifting $\widetilde{\gamma}$ of γ.

More generally, if N is a closed, oriented Riemannian $2n$-manifold, then its twistor space $\mathfrak{Z}(N)$ has a natural metric g, which turns it into an almost Hermitian manifold (following the notation in [189]): this metric is defined locally on $T_{\widetilde{x}}(\mathfrak{Z}(N)) = H_{\widetilde{x}} \oplus V_{\widetilde{x}}$ as the product of the metric on the horizontal subspace $H_{\widetilde{x}}$ and the above metric on the fibre $\mathfrak{Z}(\mathbb{S}^{2n-2})$. It is clear that the constructions above can be extended to this more general setting. Hence, whenever we have an orientation preserving conformal automorphism γ of N, we have a canonical lifting of γ to an automorphism $\widetilde{\gamma}$ of $\mathfrak{Z}(N)$ that carries twistor fibres isometrically into twistor fibres. Moreover, it is clear that one has: $(d\widetilde{\gamma})_{\widetilde{x}} \widetilde{J}_{\widetilde{x}} = \widetilde{J}_{\widetilde{\gamma}(\widetilde{x})} (d\widetilde{\gamma})_{\widetilde{x}}$ for every $\widetilde{x} \in \mathfrak{Z}(N)$, so that $\widetilde{\gamma}$ is in fact an "almost-holomorphic" automorphism of $\mathfrak{Z}(N)$, i.e., an automorphism that preserves the almost complex structure. (These maps are called holomorphic in [189].) If the almost complex structure on $\mathfrak{Z}(N)$ is integrable, then $\widetilde{\gamma}$ is actually holomorphic. One has the following theorem from [202]. The first statement in it is actually well-known; we include it here for completeness and to emphasise that the canonical lifting of conformal maps to the twistor space actually lifts the whole conformal group, as a group, a fact which is essential for us in the sequel.

Theorem 10.2.3. *Let N be as above, a closed, oriented, Riemannian $2n$-manifold, let $p: \mathfrak{Z}(N) \to N$ be its twistor fibration and endow $\mathfrak{Z}(N)$ with the metric g as above, i.e., it is locally the product of the metric on N, lifted to the horizontal distribution given by the Levi-Civita connexion on N, by the metric on the fibre induced by the bi-invariant metric on $SO(2n)$. Then:*

(i) *The group* $\mathrm{Conf}_+(N)$, *of conformal diffeomorphisms of N that preserve the orientation, lifts canonically to a subgroup* $\widetilde{\mathrm{Conf}}_+(N) \subset \mathrm{Aut}_{hol}(3(N))$ *of almost-holomorphic transformations of* $3(N)$. *Moreover, if the almost-complex structure* $\tilde{J}$ *on* $3(N)$ *is integrable, then these transformations are indeed holomorphic.*

(ii) *Each element in* $\widetilde{\mathrm{Conf}}_+(N)$ *carries twistor fibres isometrically into twistor fibres.*

10.3 Complex Kleinian groups on $\mathbb{P}^3_{\mathbb{C}}$

We now recall from Chapter 3 that given a subgroup $G \subset \mathrm{PSL}(4, \mathbb{C})$, its Kulkarni limit set $\Lambda_{\mathrm{Kul}} = \Lambda_{\mathrm{Kul}}(G)$ is the union $L_0(G) \cup L_1(G) \cup L_2(G)$, where $L_0(G)$ is the closure of the set of points in $\mathbb{P}^3_{\mathbb{C}}$ with infinite isotropy; $L_1(G)$ is the closure of the set of accumulation points of orbits in $\mathbb{P}^3_{\mathbb{C}} \setminus L_0$; and $L_2(G)$ is the closure of the set of accumulation points of orbits of compact sets in $\mathbb{P}^3_{\mathbb{C}} \setminus (L_0 \cup L_1)$. The complement $\widetilde{\Omega}_{\mathrm{Kul}}(G) := \mathbb{P}^3_{\mathbb{C}} \setminus \Lambda_{\mathrm{Kul}}$ is the (Kulkarni) region of discontinuity of G.

From now on we essentially follow [202].

Proposition 10.3.1. *Let* $\widetilde{G} \subset \mathrm{PSL}(4, \mathbb{C})$ *be the canonical lifting of a discrete subgroup* G *of* $\mathrm{Conf}_+(\mathbb{S}^4)$, *let* $p : \mathbb{P}^3_{\mathbb{C}} \to \mathbb{S}^4$ *be the twistor fibration, and let* $\Lambda(G) \subset \mathbb{S}^4$ *be the limit set of G. Then* $\Lambda_{\mathrm{Kul}}(\widetilde{G}) = p^{-1}(\Lambda(G))$, *and therefore* $\widetilde{\Omega}_{\mathrm{Kul}}(G) = p^{-1}(\Omega(G))$, *where* $\Omega(G) = \mathbb{S}^4 \setminus \Lambda(G)$ *is the discontinuity region of G. Hence, if G is Kleinian, then* $\widetilde{G}$ *is a complex Kleinian group.*

Proof. Since the action of $\widetilde{G}$ on $\mathbb{P}^3_{\mathbb{C}}$ is a lifting of the G-action on $\mathbb{S}^4$, it is clear that one has $\Lambda_{\mathrm{Kul}}(\widetilde{G}) \subset p^{-1}(\Lambda(G))$. To prove this we observe first that if $\tilde{x} \in \mathbb{P}^3_{\mathbb{C}}$ is not in $p^{-1}(\Lambda(G))$, then there is a neighbourhood $\mathcal{U}$ of $x = p(\tilde{x})$ totally contained in $\Omega(G)$ which meets at most finitely many of its orbits under the action of G on $\mathbb{S}^4$. Hence $\widetilde{\mathcal{U}} := p^{-1}(\mathcal{U})$ meets at most finitely many of its orbits under the action of $\widetilde{G}$ on $\mathbb{P}^3_{\mathbb{C}}$. Therefore $p^{-1}(\Lambda(G))$ contains $\Lambda(\widetilde{G})$.

Observe that since $\widetilde{G}$ acts discontinuously on $p^{-1}(\Omega(G))$ by Proposition 3.3.6, to prove $p^{-1}(\Lambda(G)) = \Lambda(\widetilde{G})$ it will be enough to show that $p^{-1}(\Lambda(G)) = L_0(\widetilde{G}) \cup L_1(\widetilde{G})$. For this, let $x \in \Lambda(G)$, thus there is a sequence of distinct elements $\gamma_m \subset G$ such that $\gamma_m(y) \xrightarrow[m \to \infty]{} x$, for each $y \in \Omega(G)$. Let $(\tilde{\gamma}_m) \subset \widetilde{G}$ be the sequence of the respective canonical liftings of the γ_m. Taking $\hat{\gamma}_m = \tilde{\gamma}_m |_{p^{-1}(y)}$, we will deduce that the family $\{\hat{\gamma}_m : m \in \mathbb{N}\}$ is equicontinuous and pointwise relatively compact; this claim is supported by the fact that each element in $\widetilde{\mathrm{Conf}}_+(N)$ carries twistor fibres isometrically into twistor fibres (see theorem 10.2.3) and $\mathbb{P}^3_{\mathbb{C}}$ is compact. Thus for the Arzelà-Ascoli theorem, there is a continuous function $\hat{\gamma} : p^{-1}(y) \to \mathbb{P}^3_{\mathbb{C}}$ and a subsequence, still denoted by $(\hat{\gamma}_m)$, such that $\hat{\gamma}_m \xrightarrow[m \to \infty]{} \hat{\gamma}$ uniformly, since $\hat{\gamma}_m(p^{-1}(y))$ is a twistor fibre for each m, we conclude that $\hat{\gamma}(p^{-1}(y))$ is contained in a twistor fibre. Moreover, since γ_m is an isometry we conclude that $\hat{\gamma}$ is an isometry and therefore $\hat{\gamma}(p^{-1}(y))$ is a twistor fibre. By the following commnutative diagram we conclude that this fibre must be $p^{-1}(x)$.

$$
\begin{array}{ccc}
p^{-1}(y) & \xrightarrow{\ \widehat{\gamma_m}\ } & \hat{\gamma}_m(p^{-1}(y)) \\
\downarrow{\scriptstyle p} & & \downarrow{\scriptstyle p} \\
\{y\} & \xrightarrow{\ \gamma_m\ } & \{\gamma_m(y)\}.
\end{array}
$$

Hence the limit set in $\mathbb{P}^3_{\mathbb{C}}$ is as stated. In particular, if G is Kleinian, then $\widetilde{G}$ is complex Kleinian. $\square$

Now consider a discrete subgroup G of $\mathrm{Conf}_+(\mathbb{S}^4)$; we recall that G is said to be *Fuchsian* if it leaves invariant a 3-dimensional round sphere $\mathbb{S}^3_\delta$ in $\mathbb{S}^4$ where by *round sphere* we mean, a sphere at infinity which is the boundary of a complete totally geodesic subspace of the hyperbolic space $\mathbb{H}^5$ (see Subsection 1.2.4). Every such group is automatically a Kleinian group in $\mathbb{S}^4$, because its limit set is contained in $\mathbb{S}^3_\delta$. The fundamental group of every complete hyperbolic n-orbifold with $n < 5$ is a Fuchsian group in $\mathbb{S}^4$ because the canonical inclusion $\mathrm{Iso}(\mathbb{H}^n) \hookrightarrow \mathrm{Iso}(\mathbb{H}^5)$ leaves invariant a hyperplane in $\mathbb{H}^5$, hence this group is also Kleinian.

Theorem 10.3.2. *Let $\widetilde{G}$ and G be as in Theorem* 10.3.1. *Then:*

(i) *If G is the fundamental group of a hyperbolic n-orbifold with $n < 4$, via the inclusion $\mathrm{Iso}(\mathbb{H}^n) \hookrightarrow \mathrm{Iso}(\mathbb{H}^5)$, then the action of $\widetilde{G}$ on $\mathbb{P}^3_{\mathbb{C}}$ leaves invariant a proper submanifold of $\mathbb{P}^3_{\mathbb{C}}$ and it is not minimal on the limit set.*

(ii) *If G is the fundamental group of a hyperbolic orbifold of dimension $n = 4, 5$ and $\Lambda(G)$ is the whole $\mathbb{S}^{n-1} \subset \mathbb{S}^4$, then the action of $\widetilde{G}$ is minimal on its limit set $\Lambda_{\mathrm{Kul}} \cong \mathbb{S}^{n-1} \times \mathbb{P}^1_{\mathbb{C}}$. Hence the action of $\widetilde{G}$ on $\mathbb{P}^3_{\mathbb{C}}$ is algebraically-mixing, i.e., there is no proper complex submanifold (nor sub-variety) of $\mathbb{P}^3_{\mathbb{C}}$ which is $\widetilde{G}$-invariant.*

We recall that the action of a group acting on a topological space is said to be *minimal* if each orbit is dense.

To prove Theorem 10.3.2 we need the following theorem, which is of independent interest.

Theorem 10.3.3. *Let G be a discrete subgroup of $\mathrm{Conf}_+(\mathbb{S}^4)$. Let $\widetilde{G}$ be the canonical lift of G to $\mathbb{P}^3_{\mathbb{C}}$ and let $H \subset \Lambda(\widetilde{G})$ be a nonempty minimal subset for the action of $\widetilde{G}$. Then:*

(i) *The restriction of $p : \mathbb{P}^3_{\mathbb{C}} \to \mathbb{S}^4$ to H is a locally trivial continuous fibre bundle over all of $\Lambda(G)$.*

(ii) *If* $H \neq \Lambda(\widetilde{G})$, *then each fibre* H_x *of* $p|_H$ *is either a point or a copy of the round circle* $\mathbb{S}^1$, *and there exists a* $\widetilde{G}$-*invariant continuous section of the bundle* $p : \mathbb{P}^3_{\mathbb{C}} \to \mathbb{S}^4$ *over the points in* $\Lambda(G) \subset \mathbb{S}^4$.

Proof of Theorem 10.3.3. We first note that, because H is compact, nonempty and the action of G on $\Lambda(G)$ is minimal, H intersects every twistor line over $\Lambda(G)$. Let $x \in \Lambda(G)$ and $H_x = H \cap p^{-1}(\{x\})$. Then $\widetilde{\gamma}(H_x) = H_{\gamma(x)}$ for every $\widetilde{\gamma} \in \widetilde{G}$, because $\widetilde{G}$ acts minimally on H and it carries twistor lines onto twistor lines . Moreover, the action on the twistor lines is by isometries. Thus for every $x, y \in \Lambda(G)$, H_x is isometric to H_y. Also by minimality, if for a sequence $\{x_i\}$ of $\Lambda(G)$ one had

$$\lim_{i \to \infty} x_i = x, \ but \ \lim_{x_i \to x} H_{x_i} \neq H_x,$$

where H_{x_i} converges to F_x in the Hausdorff metric, then $F_x \cup H_x$ would be isometric to H_x, which is not possible. Thus H_x depends continuously on x in the Hausdorff metric of compact subsets of $\mathbb{P}^3_{\mathbb{C}}$. Hence, for each $x \in \Lambda(G)$ there exists an open neighbourhood $U_x \subset \Lambda(G)$ and a continuous map $\psi : U_x \to SO(3)$, such that if we consider a trivialisation of the Calabi-Penrose fibration $p^{-1}(U_x) \cong U_x \times \mathbb{S}^2$, then $(y, H_y) = (y, \psi(y)(H_x))$. Thus we can trivialise $p|_H$ in U_x by the function $(y, w) \mapsto (y, \psi(y)(w))$, $w \in H_x \subset \mathbb{S}^2$, from $U_x \times H_x \subset U_x \times \mathbb{S}^2$ to $p|_H^{-1}(U_x)$. This proves statement (i).

Suppose that $H \neq p^{-1}(\Lambda(G))$, then we also have a fibration $p_1 : \mathbb{P}^3_{\mathbb{C}} - H \to \Lambda(G)$, where the fibres are $p_1^{-1}(\{x\}) = p^{-1}(\{x\}) - H_x := \Sigma_x$, and Σ_x is an open subset of the sphere $H_{\pi^{-1}(\{x\})}$. Thus Σ_x is isometric to Σ_y for all $x, y \in \Lambda(G)$ and $\widetilde{\gamma}$ sends Σ_x isometrically onto $\Sigma_{\gamma(x)}$. Suppose that for a fixed $x \in \Lambda(G)$ the function $y \mapsto d(y, H_x)$, from Σ_x to $\mathbb{R}$, attains its maximum at a *unique* point z_x, where d denotes the spherical distance in $p^{-1}(\{x\})$. Then, by minimality, the closure of the orbit of z_x under $\widetilde{G}$ meets every fibre of p_1, and it can not meet the fibre in more than one point because z_x is the unique point at maximal distance to H_x. Hence the closure of the $\widetilde{G}$-orbit of z_x is the graph of a continuous section of p_1. The image of this section is a closed set, it is $\widetilde{G}$-invariant, with a minimal action of $\widetilde{G}$.

Let us now show that, for each $x \in \Lambda(G)$, H_x is homogeneous. Let $w_1, w_2 \in H_x$. Then there exists a sequence $\{\widetilde{\gamma}_i\}$ in $\widetilde{G}$ such that $\widetilde{\gamma}_i(w_1)$ converges to w_2, by minimality, and we can obtain a subsequence $\widetilde{\gamma}_{i_j}$ such that the restriction $\widetilde{\gamma}_{i_j}|_{H_x}$ is convergent, because $SO(3)$ is compact. Hence the subgroup of $SO(3)$ that leaves invariant H_x is compact and it acts transitively on H_x. Then the connected component of this group is either trivial and H_x is a section of $p|_{\Lambda(\widetilde{G})}$, or else it is $SO(2)$ or $SO(3)$. If it is $SO(2)$, then H_x is a round circle and we can apply the previous argument to obtain an invariant section (for instance we could take the set of points which are centres of one of the discs in which the circle divides the fibre). If this group is $SO(3)$, then $H = \Lambda(\widetilde{G})$, which is a contradiction. This proves statement (ii). □

Proof of Theorem 10.3.2. Let us prove statement (i) first. This is done by using [35], [36] to show that if G is a conformal Kleinian group in $\mathbb{S}^4$ that leaves invariant a maximal round sphere $\mathbb{S}^2 \subset \mathbb{S}^4$, then $\mathbb{S}^2$ lifts to a holomorphic Legendrian curve in $\mathbb{P}^3_{\mathbb{C}}$, which is $\widetilde{G}$-invariant and it is transversal to all the twistor lines that this line meets.

We first recall that the bundle normal to the twistor lines in $\mathbb{P}^3_{\mathbb{C}}$, with respect to the Fubini-Study metric, is a complex two-dimensional (holomorphic) sub-bundle of the tangent bundle of $\mathbb{P}^3_{\mathbb{C}}$. This gives a holomorphic contact structure to $\mathbb{P}^3_{\mathbb{C}}$ (by [9] or [136, p. 204]). We recall that a complex structure on R^4 can be thought of as being a choice of an oriented 2-plane $P \subset \mathbb{R}^4$: the orientation determines a complex structure on P, and also an orientation and a complex structure on the orthogonal complement of P. Hence, if $\Sigma \looparrowright \mathbb{S}^4$ is an immersed oriented surface in $\mathbb{S}^4$, then Σ can be lifted canonically to $\mathbb{P}^3_{\mathbb{C}}$, and by [35], [36], this is a Legendrian (or *horizontal*) surface $\widehat{\Sigma}$ in $\mathbb{P}^3_{\mathbb{C}}$, i.e., it is tangent to the contact structure. Moreover, if Σ is the Riemann sphere, then every minimal immersion $\Sigma \looparrowright \mathbb{S}^4$ lifts to a holomorphic curve $\widehat{\Sigma}$ in $\mathbb{P}^3_{\mathbb{C}}$ (see [33, p. 466], also [35], [36], [136]).

Let us now consider a maximal round sphere $\mathbb{S}^2$ in $\mathbb{S}^4$ and consider a conformal Kleinian group G on $\mathbb{S}^4$ that leaves invariant this $\mathbb{S}^2$. Then $\mathbb{S}^2$ lifts to a holomorphic curve L in $\mathbb{P}^3_{\mathbb{C}}$, which is horizontal. The action of $\widetilde{G}$ on $\mathbb{P}^3_{\mathbb{C}}$ preserves L and it also preserves all lines in the Calabi-Penrose fibration that intersect L. Hence L is a proper complex submanifold of $\mathbb{P}^3_{\mathbb{C}}$ which is $\widetilde{G}$-invariant and the action on $\Lambda(\widetilde{G})$ is not minimal, because the action on the fibres is by isometries, so the points in $p^{-1}(\mathbb{S}^2) - L$ can never accumulate towards L. This proves statement (i).

To prove statement (ii) we first observe the standard fact that Zorn's lemma implies that there exists a subset $H \subset \Lambda(\widetilde{G}) \subset \mathbb{P}^3_{\mathbb{C}}$ where the action of $\widetilde{G}$ is minimal. We claim that one must have $H = \Lambda(\widetilde{G})$. Suppose $H \neq \Lambda(\widetilde{G})$ and $n = 4$, so that the limit set $\Lambda(G)$ is a round 3-sphere $\mathbb{S}^3 \subset \mathbb{S}^4$. By statement (ii) in Theorem 10.3.3, there exists a continuous family of almost complex structures $J_x : T_x\mathbb{S}^4 \to T_x\mathbb{S}^4$ for all $x \in \mathbb{S}^3 \subset \mathbb{S}^4$, which is compatible with the metric and the orientation of $\mathbb{S}^4$, and which is G-invariant. Consider the associated 2-plane field $\Pi := \{\Pi_x := T_x\mathbb{S}^3 \cap J_x(T_x\mathbb{S}^3) \subset T_x\mathbb{S}^4, x \in \mathbb{S}^3\}$. This plane field is G-invariant. Let $\mathcal{L}$ be the line field tangent to $\mathbb{S}^3$ which is orthogonal to Π, then $\mathcal{L}$ is also G-invariant by the conformality of the action, and this is not possible.

In fact, following the idea of the proof of Mostow's Rigidity Theorem [154], if α is a geodesic whose endpoints are in $\mathbb{S}^3$, then we can use parallel transport along α to transport the line at one end point of α at infinity, to a line at the other end point at infinity. The angle of these two lines is a continuous G-invariant function in $\mathbb{S}^3 \times \mathbb{S}^3$ which must be a constant because, under the hypothesis, G acts ergodically on pairs of points in $\mathbb{S}^3$. This is impossible by Theorem 5.9.10 in [222], in which Thurston gives a proof of Mostow's Rigidity Theorem [153, 154] (see also Chapter 1 above) using the nonexistence of G-invariant measurable line fields.

We can also use the following argument: Let $\mathfrak{H}$ be the family of *all* horocycles, of dimension 1, which are contained in $\mathbb{H}^4 \subset \mathbb{H}^5$ and which are tangent at infinity to the line field $\mathcal{L}$. Since $G = \pi_1(M^4)$, this family $\mathfrak{H}$ determines a proper, closed and invariant subset for a unipotent one-parameter subgroup of $SO(4,1)$ on the unit tangent bundle of M. But this is not possible because every such action is minimal, hence every orbit is dense (see [49]). Therefore $H = \Lambda(\widetilde{G})$ and $\widetilde{G}$ acts minimally on $\Lambda(\widetilde{G}) = \mathbb{S}^3 \times \mathbb{P}^1_{\mathbb{C}}$. Hence the action of $\widetilde{G}$ on $\mathbb{P}^3_{\mathbb{C}}$ is algebraically mixing, since any invariant algebraic sub-variety of $\mathbb{P}^3_{\mathbb{C}}$ must contain $\Lambda(\widetilde{G})$, which has real dimension 5, so it must have complex dimension 3. This proves statement (ii) when $n = 4$. If $n = 5$ and the action of $\widetilde{G}$ on $\Lambda(\widetilde{G})$ were not minimal, then, by Theorem 10.3.3, there would exist a section of the Calabi-Penrose fibration over all of $\mathbb{S}^4$. This is impossible since the sphere $\mathbb{S}^4$ does not have an almost complex structure. $\qquad\square$

Theorem 10.3.2 implies the following corollary:

Corollary 10.3.4. *There exist discrete subgroups of the projective group* $\mathrm{PSL}(4,\mathbb{C})$ *which act minimally on* $\mathbb{P}^3_{\mathbb{C}}$. *More precisely, let G be a discrete subgroup of* $\mathrm{Iso}_+(\mathbb{H}^5)$ *such that* $\mathbb{H}^5/G$ *has finite volume. Let $\widetilde{G}$ be its canonical lifting to* $\mathrm{PSL}(4,\mathbb{C})$. *Then* $\widetilde{G}$ *acts minimally on* $\mathbb{P}^3_{\mathbb{C}}$.

In fact we will prove later that these groups also act ergodically on $\mathbb{P}^3_{\mathbb{C}}$ with respect to the geometric measure.

We consider now subgroups $G \subset \mathrm{Conf}_+(\mathbb{S}^4)$ which are *geometrically finite*. Recall (see Chapter 1) that this implies that the group has a fundamental domain with finitely many faces. The group is *Zariski-dense* if its Zariski closure is the whole $\mathrm{Conf}_+(\mathbb{S}^4)$. The previous Theorem 10.3.2 can be generalised as follows:

Theorem 10.3.5. *Let G be a geometrically-finite discrete subgroup of* $\mathrm{Iso}_+(\mathbb{H}^m)$, $m = 4, 5$, *which is Zariski-dense. Let Λ be its limit set in* $\mathbb{S}^4$. *Let $\widetilde{G}$ be the lifting of G to* $\mathbb{P}^3_{\mathbb{C}}$. *Then, $\widetilde{G}$ acts minimally on its limit set* $\Lambda_{\mathrm{Kul}} = \Lambda \times \mathbb{P}^1_{\mathbb{C}} \subset \mathbb{P}^3_{\mathbb{C}}$.

Theorem 10.3.5 implies that if we consider a hyperbolic Schottky group acting on $\mathbb{H}^5$, whose Cantor limit set is not contained in any round sphere of dimension less than 4, then its twistorial lifting acts minimally on its limit set. This is a question that C. Series asked, motivating Theorem 10.3.5 and the equivalent statement in (10.5.3) below, both of them proved in [202]. To prove Theorem 10.3.5 we use a theorem of L. Flaminio and R. Spatzier (Theorem 1.3 in [59]) stated as Theorem 1.3.20 of this monograph.

Proof of Theorem 10.3.5. Suppose the action is not minimal. Then, by Theorem 10.3.3, there exists a continuous invariant section of the Calabi-Penrose fibration restricted to Λ. This section is therefore a G-invariant continuous family of almost complex structures $\{J_x\}_{x\in\Lambda}$. If $m = 5$, let E_x be the subspace of dimension 2 of $T_x\mathbb{S}^4$ which is the eigenspace corresponding to the eigenvalue i of J_x. Then, the family $\{E_x\}_{x\in\mathbb{S}^4}$ is a 2-dimensional G-invariant distribution. This contradicts

Theorem 1.3.20. If $m = 4$, for each $x \in \Lambda \subset \mathbb{S}^3 \subset \mathbb{S}^4$, let E_x be the 2-plane $E_x = T_x(\mathbb{S}^3) \cap J_x(T_x(\mathbb{S}^3))$. Then, the family $\{E_x\}_{x \in \mathbb{S}^3}$ is a 2-dimensional G-invariant distribution. This contradicts Theorem 1.3.20. $\qquad\square$

Remark 10.3.6. It is worth noting that results of [104] show that Ahlfors' Finiteness Theorem and Sullivan's Finite Number of Cusps Theorem fail for conformal groups in $\mathbb{S}^3$. More precisely, there exist finitely generated conformal Kleinian groups G on all $\mathbb{S}^n$, $n > 2$, such that $\Omega(G)$ is not of finite topological type. Also, M. Kapovich has examples where $\Omega(G)$ contains infinitely many connected components (personal communication). And there exist finitely generated conformal Kleinian groups in all spheres $\mathbb{S}^n$, $n > 2$, having infinitely many nonequivalent cusps. Thus the results above show that Ahlfors' Finiteness Theorem fails for complex Kleinian groups acting on $\mathbb{P}^n_{\mathbb{C}}$, for all $n > 1$. Similarly, Sullivan's Finite Number of Cusps Theorem also fails for complex Kleinian groups in $\mathbb{P}^n_{\mathbb{C}}$, but one has to make precise what a "cusp" means in this context.

10.4 Kleinian groups and twistor spaces in higher dimensions

We now consider a subgroup $G \subset \mathrm{Conf}_+(N)$ and its canonical lifting $\widetilde{G} \subset \mathrm{Aut}_{hol}(3(N))$. Its Kulkarni Limit set $\Lambda_{\mathrm{Kul}} = \Lambda_{\mathrm{Kul}}(G)$ is defined as in Chapter 3. This is the union of the three sets L_0, L_1, L_2. Its complement $3(N) \setminus \Lambda$ is the Kulkarni region of discontinuity.

Theorem 10.4.1. *Let N be as above, a closed, oriented, Riemannian $2n$-manifold, let $p\colon 3(N) \to N$ be its twistor fibration and endow $3(N)$ with the metric g as above, i.e., it is locally the product of the metric on N, lifted to the horizontal distribution given by the Levi-Civita connexion on N, by the metric on the fibre induced by the bi-invariant metric on $\mathrm{SO}(2n)$. Then:*

(i) *The group $\mathrm{Conf}_+(N)$, of conformal diffeomorphisms of N that preserve the orientation, lifts canonically to a subgroup $\widetilde{\mathrm{Conf}_+(N)} \subset \mathrm{Aut}_{hol}(3(N))$ of almost-holomorphic transformations of $3(N)$. Moreover, if the almost-complex structure $\widetilde{J}$ on $3(N)$ is integrable, then these transformations are indeed holomorphic.*

(ii) *Each element in $\widetilde{\mathrm{Conf}_+(N)}$ carries twistor fibres isometrically into twistor fibres.*

(iii) *If $G \subset \mathrm{Conf}_+(N)$ is a discrete subgroup acting on N with limit set Λ, then its canonical lifting $\widetilde{\mathrm{Conf}_+(N)}$ acts on $3(N)$ with limit set in the sense of Kulkarni $\Lambda_{\mathrm{Kul}}(\widetilde{G}) = p^{-1}(\Lambda(G))$, so $\widetilde{\Lambda}$ is a fibre bundle over Λ with fibre $3(\mathbb{S}^{2n-2})$.*

Proof. We already know that every $\gamma \in \mathrm{Conf}_+(N)$ lifts canonically to an element $\widetilde{\gamma} \in \mathrm{Aut}_{hol}(\mathfrak{Z}(N))$. So the only thing to prove for statement (i) is that $\mathrm{Conf}_+(N)$ lifts to $\mathrm{Aut}_{hol}(\mathfrak{Z}(N))$ as a group, i.e., that given any $\gamma_1, \gamma_2 \in \mathrm{Conf}_+(N)$, one has $\widetilde{\gamma}_1 \circ \widetilde{\gamma}_2 = \widetilde{\gamma_1 \circ \gamma_2}$, but this is evident because the derivative satisfies the chain rule. We next recall that at each $x \in N$, the derivative $d\gamma_x : T_x N \to T_{\gamma(x)} N$ takes orthonormal framings into orthogonal framings. In other words, dividing $d_{\gamma(x)}$ by some positive real number, we obtain an orthogonal automorphism $T_x N \to T_{\gamma(x)} N$. Hence, given the splitting $T\mathfrak{Z}(N) = V \oplus H$, into the vertical and horizontal components, one has that, for each $\widetilde{x} \in \mathfrak{Z}(N)$, the induced action of the derivative, $d\widetilde{\gamma}(\widetilde{x})|_{V_{\widetilde{x}}} : V_{\widetilde{x}} \to V_{\widetilde{\gamma}(\widetilde{x})}$, is by orthogonal transformations. Therefore statement (ii) follows from the fact that the metric on the fibres comes from the bi-invariant metric on $\mathrm{SO}(2n)$. The proof of (iii) is the same as that of Proposition 10.3.1, since $\widetilde{G}$ acts by isometries on the twistor fibres in $\mathfrak{Z}(N)$. $\qquad\square$

Let us restrict now our attention to the case $N = \mathbb{S}^{2n}$.

Definition 10.4.2. A *twistorial Kleinian group* is a discrete subgroup of the group $\mathrm{Aut}_{hol}(\mathfrak{Z}(\mathbb{S}^{2n}))$ of holomorphic automorphisms, which acts on $\mathfrak{Z}(\mathbb{S}^{2n})$ with non-empty region of discontinuity.

It follows from Theorem 10.2.3 that if $G \subset \mathrm{Conf}_+(\mathbb{S}^{2n})$ is Kleinian, then its lifting $\widetilde{G}$ to $\mathrm{Aut}_{hol}(\mathfrak{Z}(\mathbb{S}^{2n}))$ is also Kleinian. We have the following generalisation of Theorem 10.3.2:

Theorem 10.4.3. *Let $G \subset \mathrm{Conf}_+(\mathbb{S}^{2n})$, $n > 1$, be a conformal Kleinian group. We set $\mathfrak{Z} := \mathfrak{Z}(\mathbb{S}^{2n})$ and let $\widetilde{G}$ be the canonical lifting of G to $\mathrm{Aut}_{hol}(\mathfrak{Z})$, whose limit set we denote by $\widetilde{\Lambda}$. Then one has:*

(i) *If G leaves invariant an m-sphere $\mathbb{S}^m \subset S^{2n}$, $m < 2n - 1$, then the action of $\widetilde{G}$ on $\mathfrak{Z}$ leaves invariant a copy of each twistor space $\mathfrak{Z}(\mathbb{S}^{2r}) \subset \mathfrak{Z}$ for all $r \geq m/2$, which are all complex (algebraic) submanifolds of $\mathfrak{Z}$. Hence the action of $\widetilde{G}$ on $\widetilde{\Lambda} \subset \mathfrak{Z}$ is not minimal.*

(ii) *If G is a geometrically-finite discrete subgroup of $\mathrm{Iso}_+(\mathbb{H}^m)$, $m = 2n, 2n+1$, which is Zariski-dense, then $\widetilde{G}$ acts minimally on $\widetilde{\Lambda}$. Hence, there are no proper complex submanifolds (nor subvarieties) of $\mathfrak{Z}$ which are $\widetilde{G}$-invariant, i.e., the action of $\widetilde{G}$ on $\mathfrak{Z}$ is algebraically-mixing.*

(iii) *Let G be a geometrically-finite discrete subgroup of $\mathrm{Iso}_+(\mathbb{H}^{2m+1})$, $m < n-1$, which is Zariski-dense (so $\widetilde{G}$ leaves invariant $\mathfrak{Z}(\mathbb{S}^{2m})$). Then the action of $\widetilde{G}$ on $\mathfrak{Z}(\mathbb{S}^{2m}) \subset \mathfrak{Z}$ has no invariant complex submanifolds, the restriction of the projection $p : \mathfrak{Z} \to \mathbb{S}^{2n}$ to $\widetilde{\Lambda}_{2m} := \widetilde{\Lambda} \cap \mathfrak{Z}(\mathbb{S}^{2m})$ is a fibre bundle over $\Lambda(G)$, with fibre $\mathfrak{Z}(\mathbb{S}^{2m-2})$, and the action of $\widetilde{G}$ on $\widetilde{\Lambda}_{2m}$ is minimal.*

Proof. If G leaves invariant an m-sphere $\mathbb{S}^m$, then it leaves invariant, via the inclusion, a sequence of spheres $\mathbb{S}^m \subset \mathbb{S}^{m+1} \subset \cdots \subset \mathbb{S}^{2n}$. Hence, for every sphere $\mathbb{S}^{2r}$

in this sequence, G takes almost complex structures on $\mathbb{S}^{2r}$ into almost complex structures on $\mathbb{S}^{2r}$, so $\widetilde{G}$ preserves $3(\mathbb{S}^{2r}) \subset 3$. Since $\widetilde{G}$ takes twistor fibres isometrically into themselves, preserving $3(\mathbb{S}^{2r})$, one has that the action of $\widetilde{G}$ on $\widetilde{\Lambda}$ is not minimal, because the orbits can not get too close to $\widetilde{\Lambda} \cap 3(\mathbb{S}^{2r})$. This proves statement (i).

The proof of statement (ii) relies heavily on Theorem 1.3.21, which generalises the theorem in [59] (Theorem 1.3) about ergodicity of the action of G on the bundle of frames. In fact, notice that as in Section 10.3, there exists necessarily a compact set $H \subset \widetilde{\Lambda} \subset 3$, where $\widetilde{G}$ acts minimally. If $H = \Lambda_{\mathrm{Kul}}$, then there is nothing to prove. Assume $H \neq \Lambda_{\mathrm{Kul}}$, then by Theorem 10.3.3 we know that $p|_H \colon H \to \Lambda$ is a fibre bundle.

The set H is a closed subset of the set of pairs (x, J), where $x \in \mathbb{S}^{m-1} \subset \mathbb{S}^{2n}$ and J is an almost complex structure on $T_x \mathbb{S}^{2n}$ compatible with the metric and orientation. If $m = 2n$, then H determines a closed family $\mathfrak{F}$ of hyperplanes of dimension $2n - 2$ tangent to $\mathbb{S}^{2n-1}$, in the same way as in Theorem 10.3.2: $\mathfrak{F} := T_x \mathbb{S}^{2n-1} \cap J(T_x \mathbb{S}^{2n-1})$, $(x, J) \in H$. This contradicts Theorem 1.3.21, so the action is minimal on its limit set when $m = 2n$. If $m = 2n + 1$, we consider the families of n-planes: $\Pi^{\pm i} := \bigcup_{x \in \Lambda} \Pi_x^{\pm i}$, where $\Pi_x^{\pm i}$ is the eigenspace in $T_x(\mathbb{S}^{2n})$ corresponding to the multiplication by $\pm i$ given by the corresponding complex structure. These are disjoint, G-invariant families of n-planes over Λ, contradicting Theorem 1.3.21. Hence the action of $\widetilde{G}$ is minimal on its limit set.

The second statement in (ii) now follows easily: the minimality of the action implies that any invariant complex submanifold (or subvariety) of 3 must have the same dimension as 3, so it must be all of 3. Statement (iii) is an easy combination of Theorem 10.3.3 with statements (i) and (ii), so we leave the proof to the reader. $\square$

As mentioned before, the twistor space $3 := 3(\mathbb{S}^{2n})$ embeds in a projective space $\mathbb{P}_{\mathbb{C}}^N$, for some N (see for instance [34]). This can be proved in the usual way: showing that there exists a holomorphic line bundle $\mathcal{L}$ over 3 with "enough" sections, which provide a projective embedding of 3.

However, in order to state our next result, we will give a more precise description of such an embedding, following [99]. For this we first recall some facts about the spin representation. We refer to [11], [137] or [63, Ch.3] for details.

If V is a real vector space of dimension m, with the usual quadratic form q, then its Clifford Algebra $\mathcal{C}(V)$ is the quotient $\mathcal{C}(V) := \bigotimes^r T^*(V)/I$, of the complete tensor algebra of V by the ideal generated by elements of the form $(e * e + q(e) \cdot 1)$. As a vector space, $\mathcal{C}(V)$ has dimension 2^m and it is isomorphic to the exterior algebra of V (see [137] for a nice description of this isomorphism).

For $m = 2n$ even, the group $\mathrm{Spin}(2n)$ is defined to be the multiplicative subgroup of $\mathcal{C}(V)$ consisting of all the elements that can be expressed in the form $v_1 * \cdots * v_{2r}$, where each v_i is a vector in V of unit length. $\mathrm{Spin}(2n)$ acts orthogonally on V, so there is a canonical surjective homomorphism $\mathrm{Spin}(2n) \to \mathrm{SO}(2n)$, whose kernel is the centre of $\mathrm{Spin}(2n)$, which consists of ± 1. Hence, for all $n > 1$, $\mathrm{Spin}(2n)$

is simply connected and it is the universal cover of $SO(2n)$. This group acts linearly on $\mathcal{C}(V)$, so it also acts on the complexification $\mathcal{C}_{\mathbb{C}}(V) = \mathcal{C}(V) \otimes \mathbb{C}$, which is a complex representation space for the spin group, of complex dimension 2^{2n}. As a left module, $\mathcal{C}_{\mathbb{C}}(V)$ splits as the direct sum of 2^n copies of a left module Δ of dimension 2^n, which is the *the spin representation* of $Spin(2n)$, by definition.

This is in fact a complex representation space for the whole Clifford algebra $\mathcal{C}(V)$, and it is its unique irreducible complex representation, up to equivalence. However, as a representation space of the spin group, this is still reducible: Δ splits as the direct sum of two irreducible, nonequivalent representations $\Delta^{\pm}$ of dimension 2^{n-1}, called the *(positive and negative) half-spin representations*.

Let $\mathbb{P}_{\mathbb{C}}(\Delta^+) \cong \mathbb{P}_{\mathbb{C}}^{2^{n-1}-1}$ be the projectivisation of the positive half-spin representation space Δ^+. Then $Spin(2n)$ acts on $\mathbb{P}_{\mathbb{C}}(\Delta^+)$ inducing an action of $SO(2n)$, whose isotropy group at a preferred point $\theta_{\varnothing}$ is $U(n)$. This gives an $SO(2n)$-equivariant embedding of $3(\mathbb{S}^{2n-2}) = SO(2n)/U(n)$ in the projective space $\mathbb{P}_{\mathbb{C}}(\Delta^+)$, isomorphic to $\mathbb{P}_{\mathbb{C}}^{2^{n-1}-1}$, see [99], pages 108 and 114. Furthermore, from [99], Theorem 3.7, we know that this is the projective embedding of $3(\mathbb{S}^{2n-2})$ of smallest codimension.

It is clear that the first Betti number of 3 is 0 and $H^2(3; \mathbb{Z}) \cong \mathbb{Z}$. Hence, by [27] or [119, Ch. III-9], given the above embedding $3 \hookrightarrow \mathbb{P}_{\mathbb{C}}^{2^n-1}$, every holomorphic automorphism of 3 extends to this projective space and, moreover, the group $\text{Aut}_{hol}(3)$ can be identified uniquely with the group of holomorphic automorphisms of $\mathbb{P}_{\mathbb{C}}^{2^n-1}$ that preserve 3. Thus we arrive at the following theorem:

Theorem 10.4.4. *Let G be a conformal Kleinian group on $\mathbb{S}^{2n}$. Then G is a complex Kleinian group in $\mathbb{P}_{\mathbb{C}}^{2^n-1}$. More precisely, G lifts canonically to a Kleinian group $\widetilde{G}$ on the twistor space 3 and, given the natural embedding $3 \hookrightarrow \mathbb{P}_{\mathbb{C}}^{2^n-1}$ via the spin representation, $\widetilde{G}$ extends uniquely to a complex Kleinian group in $\mathbb{P}_{\mathbb{C}}^{2^n-1}$.*

We remark that the only Riemannian manifold which is not a sphere and whose twistor space is Kähler is $\mathbb{P}_{\mathbb{C}}^2$, by [95], [208]; its twistor space is the manifold $F_3(\mathbb{C})$ of flags in $\mathbb{C}^3$. The above discussion applies also in this case; however, the group $\text{Conf}_+(\mathbb{P}_{\mathbb{C}}^2)$ is $PU(3)$, which is compact, hence every discrete subgroup of this group is finite.

10.5 Patterson-Sullivan measures on twistor spaces

It follows from the previous discussion that each Riemannian metric g on $\mathbb{S}^{2n}$ defines canonically a Riemannian metric $\widetilde{g}$ on 3 via the twistor fibration: This is the unique metric $\widetilde{g}$ for which the differential of p, dp, is an isometry in each horizontal plane H_x and which coincides with the metric on each twistor fibre inherited from the bi-invariant metric on $SO(2n)$. Two conformally equivalent Riemannian metrics on $\mathbb{S}^{2n}$ lift to two Riemannian metrics on 3 which are *horizontally conformal*, i.e., they coincide in the twistor fibres and differ by a conformal factor in the horizontal distribution. A similar remark holds for other $2n$-dimensional Riemannian manifolds, and also for measures.

Let G be a *geometrically-finite* Kleinian group on $\mathbb{H}^{2n+1}$, $n > 1$. Let $y \in \mathbb{H}^{2n+1}$ and let μ_y be the Patterson-Sullivan measure on the sphere at infinity $\mathbb{S}^{2n}$ obtained from the orbit of y (See Chapter 1 or [172], [216], [218], [163]). Let $\widetilde{G}$ be the lifted group acting on $\mathfrak{Z}$. For each $y \in \mathbb{H}^{2n+1}$, let $\widetilde{\mu}_y$ be the measure on $\mathfrak{Z}$, supported in $\Lambda(\widetilde{G})$, which is the product of μ_y on $\mathbb{S}^{2n}$ and the measure on the twistor fibres determined by the metric. This is well defined and the family $\{\widetilde{\mu}_y\}$ is a *horizontally conformal density* in $\Lambda(\widetilde{G})$ of exponent δ, where $\delta := \delta(\Lambda(G))$ is the Hausdorff dimension of $\Lambda(G)$ (see Chapter 1 or [163] for the definition of conformal densities). These measures are all proportional for $y \in \mathbb{H}^{2n+1}$. Moreover, since the limit set of $\widetilde{G}$ is the cartesian product of $\Lambda(G)$ and a manifold of dimension $(n^2 - n)$, Theorem 2 of [24] (see also [142]) says that the Hausdorff dimension of $\Lambda(\widetilde{G})$ is $\delta(\Lambda(\widetilde{G})) = \delta(\Lambda(G)) + n^2 - n$. Thus one can apply known results of discrete hyperbolic groups to obtain results about the Hausdorff dimension of $\widetilde{G}$. In particular, by Theorem D in [226] (c.f., [218]) one has the following theorem:

Theorem 10.5.1. *Let G be a geometrically-finite conformal Kleinian group on $\mathbb{H}^{2n+1}$, with $n > 1$. Let $\widetilde{G}$ be its lifting to $\mathfrak{Z}$. Then $\delta(\Lambda(\widetilde{G})) = \delta(\Lambda(G)) + (n^2 - n) < n^2 + n$.*

Also, by results of R. Bowen [31] and D. Ruelle [188], we obtain:

Theorem 10.5.2. *Let $\{G_t\}$ be an analytic family of conformal Kleinian groups acting on $\mathbb{H}^{2n+1}$, which are geometrically-finite and without parabolic elements. Let $\widetilde{G}_t$ be their liftings to $\mathfrak{Z}$. Then $\delta(t) := \delta(\Lambda(\widetilde{G}_t))$ is a real analytic function of t.*

It would be interesting to find conditions under which this theorem holds for general complex Kleinian groups on $\mathbb{P}_{\mathbb{C}}^N$.

We now recall:

(i) If G is a subgroup of $\mathrm{Iso}(\mathbb{H}^m)$, $m \le 2n+1$, then G is a subgroup of $\mathrm{Iso}(\mathbb{H}^{2n+1})$ via the inclusion $\mathrm{Iso}(\mathbb{H}^m) \hookrightarrow \mathrm{Iso}(\mathbb{H}^{2n+1})$.

(ii) The Patterson-Sullivan density $\{\mu_y\}$, $y \in \mathbb{H}^{2n+1}$, associated to $G \subset \mathrm{Iso}_+(\mathbb{H}^{2n+1})$ is *ergodic* if for any G-invariant Borel subset A, either $\mu_y(A) = 0$ or $\mu_y(-A) = 0$, where $-A := \Lambda(G) - A$.

(iii) If a discrete subgroup $G \subset \mathrm{Iso}(\mathbb{H}^{2n+1})$ is geometrically-finite, then the densities $\{\mu_y\}$, $y \in \mathbb{H}^{2n+1}$, are all proportional, and so are their liftings $\{\widetilde{\mu}_y\}$ to $\mathfrak{Z}$. Hence, ergodicity for a fixed μ_y implies ergodicity for all μ_y, $y \in \mathbb{H}^{2n+1}$.

Theorem 10.5.3. *Let $\widetilde{G}$ be a group of holomorphic transformations of $\mathfrak{Z} := \mathfrak{Z}$ which is the lifting of a geometrically-finite discrete subgroup $G \subset \mathrm{Conf}_+(\mathbb{S}^{2n})$, $n > 1$. Assume G is indeed contained in $\mathrm{Iso}_+(\mathbb{H}^{m+1}) = \mathrm{Conf}_+(\mathbb{S}^m) \subset \mathrm{Conf}_+(\mathbb{S}^{2n})$, for some $m \le 2n$. Let $\widetilde{\Lambda} := \Lambda(\widetilde{G})$ be the limit set of $\widetilde{G}$.*

(i) *If $m < 2n - 1$, then the action of $\widetilde{G}$ on $\widetilde{\Lambda}$ is not ergodic with respect to the measures $\widetilde{\mu}_y$, $y \in \mathbb{H}^{m+1}$.*

(ii) *If m is either $2n - 1$ or $2n$ and $\Lambda(G) = \mathbb{S}^{m-1}$, then the action of $\widetilde{G}$ on $\widetilde{\Lambda}$ is ergodic with respect to the measures $\widetilde{\mu}_y$, $y \in \mathbb{H}^{m+1}$.*

(iii) *In fact, if m is either $2n - 1$ or $2n$ and G is Zariski-dense in $\mathrm{Iso}_+(\mathbb{H}^{m+1})$, then $\widetilde{\mu}$ is ergodic, $y \in \mathbb{H}^{m+1}$.*

(iv) *Let $m = 2r < 2n - 1$, so that (by Theorem 10.4.3) one has a $\widetilde{G}$-invariant twistor space $\mathfrak{Z}(\mathbb{S}^{2r})$ in $\mathfrak{Z}$, whose intersection with $\widetilde{\Lambda}$ is a fibre bundle over Λ with fibre $\mathfrak{Z}(\mathbb{S}^{2r-2})$. If G is Zariski-dense in $\mathrm{Iso}_+(\mathbb{H}^{m+1})$, then the action on $\widetilde{\Lambda} \cap \mathfrak{Z}(\mathbb{S}^{2r})$ is ergodic for $\widetilde{\mu}_y$, $y \in \mathbb{H}^{m+1}$.*

Notice that statement (iii) implies statement (ii), so we only prove statement (iii). We also notice that, by [216] and [218], the action of $G \subset \mathrm{Conf}_+(\mathbb{S}^m)$ on its limit set $\Lambda \subset \mathbb{S}^m$ is ergodic with respect to the Patterson-Sullivan densities (see Chapter 1). If $\Lambda = \mathbb{S}^m$, these measures are constant multiples of the Lebesgue measure on $\mathbb{S}^m$.

Proof. Assume $m < 2n - 1$ and suppose $m = 2r$ is even. Then, by Theorem 10.4.3, one has a $\widetilde{G}$-invariant twistor space $\mathfrak{Z}(\mathbb{S}^{2r})$ in $\mathfrak{Z}$, whose intersection with $\widetilde{\Lambda}$ is a fibre bundle over Λ with fibre $\mathfrak{Z}(\mathbb{S}^{2r-2})$. Furthermore, by Theorem 10.2.3, $\widetilde{G}$ takes twistor fibres isometrically into twistor fibres. This implies that for every $\varepsilon > 0$, the ε-neighbourhood of $\mathfrak{Z}(\mathbb{S}^{2r}) \cap \widetilde{G}$ in $\widetilde{G}$, is an invariant set of positive μ_y-measure, whose complement in $\widetilde{G}$ has also positive measure if ε is small. Hence these measures are not ergodic, proving i) when m is even. If $m < 2n - 1$ is odd, then $m + 1$ is even and $m + 1 < 2n - 1$, so we can apply the above arguments taking the inclusion $\mathrm{Conf}_+(\mathbb{S}^m) \hookrightarrow \mathrm{Conf}_+(\mathbb{S}^{m+1})$, thus proving (i).

Now let $m = 2n - 1$. Suppose there exists $A \subset \Lambda(\widetilde{G}) \subset \mathfrak{Z}$ which is a $\widetilde{G}$-invariant Borel subset such that $\widetilde{\mu}_y(A) \neq 0 \neq \widetilde{\mu}_y(-A)$, where $-A$ is the complement of A in $\Lambda(\widetilde{G})$. The set $p(A) \subset \Lambda(G) \subset \mathbb{S}^{2n-1} \subset \mathbb{S}^{2n}$ is G-invariant. Since the measure μ_y is ergodic, by [216], then either $\mu_y(p(A)) = 0$ or $\mu_y(-p(A)) = 0$, where $-p(A) := \mathbb{S}^{2n-1} - p(A)$. We can assume $\mu_y(-p(A)) = 0$, so that $p(A)$ has full measure, $\mu_y(p(A)) = 1$. Then, by Fubini's theorem applied to the fibration p, the set $p^{-1}(p(A))$ has full measure in $\Lambda(\widetilde{G})$. The set $p^{-1}(p(A))$ consists of A and $B = p^{-1}(p(A)) \cap (-A)$, which are disjoint sets of, necessarily, positive measure.

The limit set $\widetilde{\Lambda} \cong p^{-1}(\Lambda)$ is the set of all almost complex structures compatible with the orientation and the canonical metric of $T_x \mathbb{S}^{2n}$ for $x \in \Lambda \subset \mathbb{S}^{2n-1}$. An almost complex structure J_x of $T_x \mathbb{S}^{2n}$, at a point $x \in \Lambda \subset \mathbb{S}^{2n-1}$, determines the oriented $(2n-2)$-plane $\mathcal{P}_x := T_x \mathbb{S}^{2n-1} \cap J_x(T_x \mathbb{S}^{2n-1})$, tangent to $\mathbb{S}^{2n-1}$ at $x \in \Lambda$. Let $\mathcal{L}_x$ be the line in $T_x \mathbb{S}^{2n-1}$ orthogonal to $\mathcal{P}_x$; this line determines the family $\mathcal{H}_x$, consisting of all horocycles in $\mathbb{H}^{2n}$ which are tangent to $\mathcal{L}_x$.

Let $\mathfrak{H}$ be the space of *all* one-dimensional horocycles in $\mathbb{H}^{2n-1}$. It is clear that the group $\mathrm{Iso}_+(\mathbb{H}^{2n}) \cong \mathrm{Conf}_+(\mathbb{S}^{2n-1})$ acts transitively on $\mathfrak{H}$. So $\mathfrak{H}$ is a homogeneous space with a unique invariant measure class, which is clearly ergodic, because the action of $\mathrm{Iso}_+(\mathbb{H}^{2n})$ is transitive. Therefore the restriction of this action to G also acts ergodically on $\mathfrak{H}$, by Moore's Ergodicity Theorem (see [233, Th. 2.2.6] or Theorem 1.3.1 above).

Let $\mathcal{H}_A$ and $\mathcal{H}_B$ be the subsets of $\mathfrak{H}$ consisting of *all* horocycles in $\mathbb{H}^{2n} \subset \mathbb{H}^{2n+1}$ which are tangent to the lines determined by the points of $x \in \Lambda$ which are in A and B, respectively. Then $\mathcal{H}_A$ and $\mathcal{H}_B$ are two disjoint Borel subsets of $\mathfrak{H}$ which are G-invariant and of positive measure, because $p(A) = p(B)$ has full measure in $\mathbb{S}^{2n-1}$. This is a contradiction and the statement (ii) is proven when $m = 2n - 1$. Notice that this would also contradict Theorem 1.3.21 in Chapter 1.

Let now $m = 2n$. Then the limit set of G is contained in the sphere $\mathbb{S}^{2n}$. Suppose that the action of $\widetilde{G}$ is not ergodic. Then, as before, there exists a $\widetilde{G}$-invariant open set $A \subset 3$ such that both A and $-A$ have positive Lebesgue measure. If $z \in A$, then z corresponds to an almost complex structure J_z at the tangent space, $T_x(\mathbb{S}^{2n})$, of the point $x := p(z)$.

The tangent space decomposes as the direct sum: $T_x(\mathbb{S}^{2n}) = E_z^1 \oplus E_z^2$, where E_z^1 and E_z^2 are the eigenspaces which correspond to the eigenvalues i and $-i$, respectively. Now we use the same argument as before: the set of horocycles which are tangent to the family $\{E_z^1\}_{z \in \mathcal{U}}$ is a $\widetilde{G}$-invariant set in the space of all horocycles which has positive measure and whose complement has also positive measure. This contradicts both, Moore's ergodicity theorem and Theorem 1.3.21. This completes the proof of statement (iii).

The proof of statement (iv) is immediate from the above discussion. $\qquad\square$

10.6   Some remarks

Consider a discrete subgroup Γ of $\mathrm{Conf}_+(\mathbb{S}^4) \cong \mathrm{Iso}_+(\mathbb{H}_\mathbb{R}^5)$, the group of orientation preserving conformal automorphisms of the 4-sphere, which has real dimension 15. One has a canonical embedding of $\mathrm{Conf}_+(\mathbb{S}^4)$ in $\mathrm{PSL}(4,\mathbb{C})$, which is a Lie group of complex dimension 15, and we may have deformations of Γ in $\mathrm{PSL}(4,\mathbb{C})$ that do not come from $\mathrm{Conf}_+(\mathbb{S}^4)$.

If the group Γ is cocompact, then Mostow's rigidity theorem implies that Γ is rigid. In particular it has no infinitesimal deformations. Since the Lie algebra of $\mathrm{PSL}(4,\mathbb{C})$ is the complexification of the Lie algebra of $\mathrm{Conf}_+(\mathbb{S}^4)$, it follows that Γ is rigid also in $\mathrm{PSL}(4,\mathbb{C})$. Yet, when $\Gamma \subset \mathrm{Conf}_+(\mathbb{S}^4)$ is not a lattice, then it may have a rich deformation theory, and it can be interesting to study its deformations in $\mathrm{PSL}(4,\mathbb{C})$. For instance, we can start with a Schottky subgroup of $\mathrm{Conf}_+(\mathbb{S}^4)$, lift it to $\mathrm{PSL}(4,\mathbb{C})$ and deform it there. Among other things, this gives a new method for constructing compact complex 3-folds with a projective structure and an interesting Teichmüller-type theory.

We may consider also discrete subgroups of $\mathrm{PSL}(2,\mathbb{R}) \subset \mathrm{PSL}(2,\mathbb{C}) \cong \mathrm{Conf}_+(\mathbb{S}^2)$, include them in $\mathrm{Conf}_+(\mathbb{S}^4)$ and then lift them to $\mathrm{PSL}(4,\mathbb{C})$, where they can have interesting deformations.

Problem. *Study deformations in* $\mathrm{PSL}(4,\mathbb{C})$ *of discrete subgroups of* $\mathrm{Conf}_+(\mathbb{S}^4)$.

There are indeed many ways in which this problem can be turned into concrete questions and problems. Some of these are related with the behaviour of limit sets under deformations. For instance we know from the work of R. Bowen and others, that if we start with a cocompact Fuchsian group $\Gamma \subset \mathrm{PSL}(2, \mathbb{R})$ and we deform it "slightly" in $\mathrm{PSL}(2, \mathbb{C})$, then the limit set of the deformed group is a quasi-circle and the Hausdorff dimension of the limit set depends analytically on the parameters of the deformation. What if we deform it in $\mathrm{PSL}(4, \mathbb{C})$?, or in $\mathrm{PSL}(3, \mathbb{C})$? This is of course related to the work done by various authors, studying deformations of Fuchsian groups in $\mathrm{PU}(2, 1)$ (see for instance [171]).

Bibliography

[1] H. Abels, G. Margulis, G. Soifer. *Semigroups containing proximal linear maps.* Israel J. Math. **91** (1995), 1-30.

[2] A. Adem, J. Leida, Y. Ruan. *Orbifolds and stringy topology.* Cambridge Tracts in Mathematics 171. Cambridge University Press (2007).

[3] L. V. Ahlfors. *Finitely generated Kleinian groups.* Am. J. Math. **86** (1964), 415-429.

[4] L. V. Ahlfors. *Möbius transformations in several dimensions.* Lecture notes, School of Mathematics, University of Minnesota, Minneapolis (1981).

[5] L. V. Ahlfors. *Möbius transformations and Clifford numbers.* In "Differential Geometry and Complex Analysis", eds. I. Chavel and H. M. Farkas. Springer-Verlag, New York (1985), 65-74.

[6] B. N. Apanasov. *Geometrically finite hyperbolic structures on manifolds.* Ann. Glob. Anal. & Geom. **1** (1983), 1-22.

[7] B. N. Apanasov. *Discrete Groups in Space and Uniformization Problems.* Mathematics and Its Applications. Kluwer Academic Publishers. Dordrecht-Boston-London 1991.

[8] B. N. Apanasov. *Conformal geometry of discrete groups and manifolds.* De Gruyter Expositions in Mathematics, 32. Walter de Gruyter & Co., Berlin, 2000.

[9] V. I. Arnold. *Mathematical Methods of classical mechanics.* Springer-Verlag, 1978.

[10] M. F. Atiyah. *Geometry of Yang-Mills Fields.* Lezioni Ferminiane Accademia Nazionale dei Lincei, Scuola Normale Superiore, Pisa, Italia, 1979.

[11] M. F. Atiyah, R. Bott, A. Shapiro. *Clifford modules.* Topology **3**, Suppl. 1 (1964), 3-38.

[12] M. F. Atiyah, N. J. Hitchin, I. M. Singer. *Self duality in four dimensional Riemannian geometry.* Proc. Royal Soc. London, serie A (Math. and Phys. Sciences) **362** (1978), 425-461.

[13] W. Ballmann. *Lectures on spaces of nonpositive curvature*. DMV Seminar, **25**. Birkhauser, 1995

[14] W. Barrera, A. Cano, J. P. Navarrete. *The limit set of discrete subgroups of* PSL(3, $\mathbb{C}$). Math. Proc. Camb. Philos. Soc. **150** (2011), 129-146.

[15] W. Barrera, A. Cano, J. P. Navarrete, *Two dimensional Kleinian groups with four lines on its limit set*. Conf. Geom. Dyn., 15 (2011), 160–176.

[16] W. Barrera and J.P. Navarrete. *Discrete subgroups of* PU(2, 1) *acting on* $\mathbb{P}^2_\mathbb{C}$ *and the Kobayashi metric*. Bull. Braz. Math. Soc. New Series (1) **40** (2009), 99–106.

[17] W. Barrera and J.P. Navarrete. *Pappus' Theorem and Complex Kleinian Groups*. Preprint 2012.

[18] W. P. Barth, K. Hulek, C.A.M. Van Peters. *Compact Complex Surfaces*. Series: Ergebnisse der Mathematik und ihrer Grenzgebiete, Vol. 4, 2nd enlarged ed., 2004.

[19] A. F. Beardon. *The geometry of discrete groups*. Springer Verlag, 1983.

[20] Y. Benoist. *Convexes divisibles. II*. Duke Math. J. **120** (2003), 97-120.

[21] Y. Benoist. *Convexes divisibles. I*. In "Algebraic groups and arithmetic". Proceedings of the international conference, Mumbai, India, 2001. Dani, S. G. (ed.) et al. Tata Institute of Fundamental Research (2004) 339-374.

[22] Y. Benoist. *Convexes divisibles. III*. Ann. Sci. Éc. Norm. Supér. **38** (2005), 793-832.

[23] Y. Benoist. *Convexes divisibles. IV: Structure du bord en dimension 3*. Invent. Math. **164** (2006), 249-278.

[24] A. S. Besicovitch, P. A. P. Moran. *The measure of product and cylinder sets*. J. London Math. Soc. **20** (1945), 110-120.

[25] G. Besson, G. Courtois, S Gallot. *Entropies et rigidités des espaces localement symétriques de courbure strictement négative*. Geom. Funct. Anal. **5** (1995), no. 5, 731799.

[26] G. D. Birkhoff. *Dynamical systems*. First printing of revised edition. Providence R.I.: American Mathematical Society (AMS). XII. (1966).

[27] A. Blanchard. *Sur les variétés analytiques complexes*. Ann. Sci. ENS **63** (1958), 157-202.

[28] M. Boege, G. Hinojosa, A. Verjovsky. *Wild knots in higher dimensions as limit sets of Kleinian groups*. Conf. Geom. Dyn. **13** (2009), 197-216.

[29] A. Borel, Harish-Chandra. *Arithmetic subgroups of algebraic groups*. Ann. Math. **75** (1962), 485-535.

[30] B. H. Bowditch, *Geometrically finiteness for hyperbolic groups*. J. Functional Analysis 113 (1993), 245-317.

[31] R. Bowen. *Hausdorff dimension of quasi-circles*. Publ. Math. IHES **50** (1979), 11-25.

[32] M. Bridson, A. Haefliger. *Metric spaces of non-positive curvature*. Springer Verlag, Grundlehren der Mathematischen Wissenschaften 319 (1999).

[33] R. L. Bryant. *Conformal and minimal immersions of compact surfaces into the 4-sphere*. J. Diff. Geom. **17** (1982), 455-473.

[34] F. E. Burstall, J. H. Rawnsley. *Twistor Theory for Riemannian Symmetric Spaces*. Springer Verlag, Lect. Notes in Maths. **1424**, 1990.

[35] E. Calabi. *Quelques applications de l'analyse complexe aux surfaces d'aire minima*. In "Topics in Complex Manifolds", ed. H. Rossi. Les Presses de l'Université de Montreal (1967), 59-81.

[36] E. Calabi. *Minimal immersions of surfaces in Euclidean spheres*. J. Diff. Geom. **1** (1967), 111-125.

[37] C. Camacho, N. Kuiper, J. Palis. *The topology of holomorphic flows with singularity* Publ. Math., Inst. Hautes Étud. Sci. 48, 5-38 (1978).

[38] A. Cano. *Schottky groups can not act on P_C^{2n} as subgroups of* $\mathrm{PSL}_{2n+1}(C)$. Bull. Braz. Math. Soc. (N.S.) **39** (2008), 573-586.

[39] A. Cano, L. G. Loeza. *Projective Cyclic Groups in Higher Dimensions*. Preprint, 2011. arXiv:1112.4107

[40] A. Cano, J. Seade. *On the equicontinuity region of discrete subgroups of* $\mathrm{PU}(1,n)$. J. Geom. Anal. **20** (2010) , 291-305.

[41] A. Cano, J. Seade. *On discrete subgroups of automorphism of* $\mathbb{P}_\mathbb{C}^2$. Preprint, 2012. http://arxiv.org/abs/0806.1336.

[42] A. Cano, L. G. Loeza, J. Seade. *Towards Sullivan's dictionary in complex dimension two*. Tentative title, work in process, 2012.

[43] W. Cao and K. Gongopadhyay. *Algebraic characterization of isometries of the complex and the quaternionic hyperbolic planes*. Geometria Dedicata **157** (2012), Number 1, 23-39.

[44] W. Cao, J. Parker, X. Wang. *On the classification of quaternionic Möbius transformations*. Math. Proc. Cambridge Philos. Soc. 137 (2004), no. 2, 349–361.

[45] S. S. Chen and L. Greenberg. *Hyperbolic Spaces*. Contributions to Analysis, Academic Press, New York, pp. 49-87 (1974).

[46] S. Choi. *Geometric Structures on Orbifolds and Holonomy Representations*. Geometriae Dedicata, **104** (2004), 161-199.

[47] S. Choi, W. M. Goldman, *Convex real projective structures on closed surfaces are closed.* Proc. A.M.S. **118** (1993), 657-661.

[48] S. Choi, W. M. Goldman. *The classification of real projective structures on compact surfaces.* Bull. A. M. S. **34**, Number 2 (1997), 161-171.

[49] S. G. Dani. *Invariant measures and minimal sets of horospherical flows.* Inven. Math. **64** (1981), 357-385.

[50] P. Deligne, G. D. Mostow. *Commensurabilities among lattices in* $PU(n,1)$. Ann. Math. Study **132** (1993), Princeton Univ. Press.

[51] T.-C. Dinh, N. Sibony. *Dynamics in several complex variables: endomorphisms of projective spaces and polynomial-like mappings.* Gentili, Graziano (ed.) et al., Holomorphic dynamical systems. Lectures given at the C.I.M.E. summer school, Cetraro, Italy, July 7–12, 2008. Springer. Lecture Notes in Mathematics **1998**, 165-294 (2010).

[52] M. Dubois-Violette. *Structures complexes au-dessus des variétés, applications.* In "Mathématique et Physique", edit L. Boutet et al. Progress in Maths. **37**. Birkhäuser, 1983.

[53] F. Dutenhefner, N. Gusevskii, *Complex hyperbolic Kleinian groups with limit set a wild knot.* Topology **43** (2004), no. 3, 677-696.

[54] J. Eells, L. Lemaire. *Another report on harmonic maps.* Bull London Math. Soc. **20** (1988), 285-524.

[55] C. Ehresmann. *Sur les espaces fibrés differentiables.* C. R. Acad. Sci. Paris **224** (1947), 1611-1612.

[56] E. Falbel, P.-V. Koseleff. *Rigidity and flexibility of triangle groups in complex hyperbolic geometry.* Topology **41** (2002), 767-786.

[57] K. Falk, P. Tukia, *A note on Patterson measures.* Kodai Math. J. **29** (2006), no. 2, 227236.

[58] G. Faltings. *Real projective structures on Riemann surfaces.* Compositio. Math. **48** (1983), 223-269.

[59] L. Flaminio, R. J. Spatzier. *Geometrically finite groups, Patterson-Sullivan measures and Ratner's rigidity theorem.* Inv. Math. **99** (1990), 601-626.

[60] J. E. Fornæss, N. Sibony. *Complex Dynamics in Higher Dimension II.* In "Modern Methods in Complex Analysis", Annals of Mathematics Studies **137**. ed. T. Bloom, D. W. Catlin *et al* (1996), 134-182.

[61] H. Furstenberg. *A Poisson formula for semisimple Lie groups.* Ann. of Maths. **77** (1963), 335-383.

[62] I. Gelfand, S. Fomin. *Geodesic flows on manifolds of constant negative curvature*, Uspekhi Matem. Nauk, bf 7 (1952), 118-137.

[63] P. B. Gilkey. *Invariance theory, the Heat Equation and the Atiyah-Singer Index Theorem.* Publish or Perish, Math. Lecture Series No. **11**, 1984.

[64] G. Giraud. *Sur certaines fonctions automorphes de deux variables.* Ann. Ec. Norm. (3), **38** (1921), pp. 43-164.

[65] W. M. Goldman. *Projective structures with Fuchsian holonomy.* J. Differential Geom. **25** (1987), 297-326.

[66] W. M. Goldman, *Convex real projective structures on compact surfaces.* J. Differential Geom. **31** (1990), 791-845.

[67] W. Goldman. *Complex Hyperbolic Geometry.* Oxford Science Publications, 1999.

[68] W. M. Goldman, *Projective geometry on manifolds.* Notes 2009.

[69] W. M. Goldman. *Locally homogeneous geometric manifolds.* Proceedings of the International Congress of Mathematicians. Volume II, 717744. Edited by Rajendra Bhatia, Arup Pal, G. Rangarajan, V. Srinivas and M. Vanninathan. Hindustan Book Agency, New Delhi, 2010.

[70] W. Goldman, J. Parker. *Complex hyperbolic ideal triangle groups.* J. Reine Angew. Math. 425 (1992), 71–86.

[71] W. Goldman, J. Parker. *Dirichlet polyhedra for dihedral groups acting on complex hyperbolic space.* J. Geom. Anal. **2** (1992), 517-554.

[72] K. Gongopadhyay. *Algebraic characterization of the isometries of the hyperbolic 5-space.* Geom. Dedicata 144 (2010), 157–170.

[73] K. Gongopadhyay, S. Parsad. *Classification of quaternionic hyperbolic isometries.* Preprint 2012.

[74] K. Gongopadhyay, J. R. Parker, S. Parsad. *Classification of complex hyperbolic isometries.* Work in process, 2012.

[75] V. Goryunov, S. H. Man. *The complex crystallographic groups and symmetries of J_{10}.* Singularity theory and its applications, 55–72, Adv. Stud. Pure Math., 43, Math. Soc. Japan, Tokyo, 2006.

[76] L. Greenberg. *Discrete subgroups of the Lorentz group.* Math. Scand., **10** (1962), 85-107.

[77] L. Greenberg. *Fundamental polyhedra for Kleinian groups.* Annals of Maths. **84** (1966), 433-441.

[78] M. Gromov. *Hyperbolic groups.* In "Essays in Group Theory", MSRI Publ., Vol. **8** (1987), 75-263.

[79] M. Gromov, H. B. Lawson, W. Thurston. *Hyperbolic 4-manifolds and conformally flat 3-manifolds.* Publ. Math. I.H.E.S. Vol. **68** (1988), 27- 45.

[80] M. Gromov, I. I. Piatetski-Shapiro. *Non-arithmetic groups in Lobachevsky spaces.* Publ. Math., Inst. Hautes Étud. Sci. **66** (1988), 93-103.

[81] M. Gromov, R. Schoen. *Harmonic maps into singular spaces and p-adic superrigidity for lattices in groups of rank one.* Publ. Math., Inst. Hautes Étud. Sci. **76** (1992), 165-246.

[82] A. Grothendieck. *Sur quelques points d'algèbre homologique.* Tohoku Math. J., II. Ser. **9** (1957), 119-221.

[83] R. Gunning. *On uniformization of complex manifolds; the role of connections.* Princeton Univ. Press, Math. Notes **22**, 1978.

[84] N. Gusevskii, J. Parker. *Representations of free Fuchsian groups in complex hyperbolic space.* Topology 39 (2000), no. 1, 33–60.

[85] A. Haefliger. *Complexes of groups and orbihedra,* in "Group theory from a geometrical viewpoint, Trieste 1990", pages 504-540, World Scientific (1991) ed. É. Ghys, A. Haefliger and A. Verjovsky.

[86] R. Harvey. *Removable singularities of cohomology classes* Amer. Journal of Math. **96** (1974), 498-504.

[87] R. Hartshorne. *Algebraic geometry.* Graduate Texts in Mathematics, Springer-Verlag. New York - Heidelberg - Berlin **52** (1984).

[88] G. M. Henkin. *Integral representation of functions in strictly pseudoconvex domains and applications to the $\bar{\partial}$-problem.* Math. USSR, Sb. **11** (1970), 273-281.

[89] P. J. Hilton, U. Stammbach. *A Course in Homological Algebra.* Graduate Texts in Mathematics, Vol IV. Springer-Verlag, New York, 1971.

[90] G. Hinojosa. *Wild knots as limit sets of Kleinian groups.* In "Geometry and dynamics. International conference in honor of the 60th anniversary of Alberto Verjovsky". Eells, James (ed.) et al. Contemporary Mathematics 389 (2005) A.M.S., 125-139.

[91] G. Hinojosa. *A wild knot $\mathbf{S}^2 \hookrightarrow \mathbb{S}^4$ as limit set of a Kleinian group: Indra's pearls in four dimensions.* J. Knot Theory & Ramifications **16** (2007), 1083-1110.

[92] G. Hinojosa, A. Verjovsky. *Actions of discrete groups on spheres and real projective spaces.* Bull. Braz. Math. Soc. (N.S.) 39 (2008), no. 2, 157–171.

[93] F. Hirzebruch. *Arrangements of lines and algebraic surfaces.* Arithmetic and geometry, Pap. dedic. I. R. Shafarevich, Vol. II: Geometry. M. Artin and J. Tate, ed. Prog. Math. 36, 113-140 (1983).

[94] F. Hirzebruch. *Chern numbers of algebraic surfaces.* Math. Ann. 266, 351-356 (1984).

[95] N. J. Hitchin. *Kählerian twistor spaces.* Proc. London Math. Soc. **43** (1981), 133-150.

[96] N. J. Hitchin. *Lie groups and Teichmüller space.* Topology **31** (1992), 449-473.

[97] S. A. Huggett and K. P. Tod. *An introduction to twistor theory.* London Mathematical Student Texts **4** (New edition, 1994).

[98] M. Inoue. *On surfaces of Class* VII$_0$. Invent. Math. **24** (1974), pp 269-310.

[99] Y. Inoue. *Twistor spaces of even dimensional Riemannian manifolds.* J. Math. Kyoto Univ. **32** (1992), 101-134.

[100] M. Inoue, Sh. Kobayashi, T. Ochiai. *Holomorphic affine connections on compact complex surfaces.* J. Fac. Sci. Univ. Tokyo Sect. IA Math. 27 (1980), no. 2, 247-264.

[101] S. M. Ivashkovich. *Extension of locally biholomorphic mappings of domains into complex projective space.* (Translation from Izv. Akad. Nauk SSSR, Ser. Mat. **47**, No.1, 197-206 (Russian), (1983) Math. USSR, Izv. (1984), 181-189.

[102] S. Kamiya. *Notes on nondiscrete subgroups of $\hat{U}(1, n; F)$.* Hiroshima Math. J. *13*, pp. 501-506 (1982).

[103] S. Kamiya. *Notes on elements of* U$(1, n; \mathbb{C})$. Hiroshima Math. J. **21**, pp. 23-45 (1991).

[104] M. Kapovich, L. Potyagailo. *On the absence of Ahlfors finitness theorem for Kleinian groups in dimension three.* Topology and Applications **40** (1991), 83-91.

[105] M. Kapovich. *Kleinian groups in higher dimensions.* In "Geometry and dynamics of groups and spaces. In memory of Alexander Reznikov". Kapranov, Mikhail (ed.) et al. Birkhäuser. Progress in Mathematics **265**, 487-564 (2008).

[106] M. Kapovich. *Hyperbolic manifolds and discrete groups.* Reprint of the 2001 hardback edition. Modern Birkhäuser Classics. Birkhäuser (2009).

[107] M. Kato. *On compact complex 3-folds with lines.* Japanese J. Math. **11** (1985), 1-58.

[108] M. Kato. *Compact Complex 3-folds with Projective Structures; The infinite Cyclic Fundamental Group Case.* Saitama Math. J. **4** (1986), 35-49.

[109] M. Kato. *Factorization of compact complex 3-folds which admit certain projective structures.* Tôhoku Math. J. **41** (1989), 359-397.

[110] M. Kato. *Compact Quotient Manifolds of Domains in a Complex 3-Dimensional Projective Space and the Lebesgue Measure of Limit Sets.* Tokyo J. Math. **19** (1996), 99-119.

[111] M. Kato. *Compact quotients of large domains in a complex projective 3-space.* Tokyo J. Math. **29** (2006), 209-232.

[112] M. Kato. *Compact quotients with positive algebraic dimensions of large domains in a complex projective 3-space.* J. Math. Soc. Japan **62** (2010), 1317-1371.

[113] I. Kim, J. Parker. *Geometry of quaternionic hyperbolic manifolds.* Math. Proc. Cambridge Philosophical Society 135 (2003) 291-320.

[114] S. Kinoshita. *Notes on Covering Transformation Groups.* Proc. American Mathematical Society, Vol. 19, No. 2 (Apr., 1968), pp. 421 -424.

[115] A. A. Kirilov. *Elements of the Theory of Representations.* Grundlehren der mathematischen Wissenschaften, Springer-Verlag **220** (1976).

[116] B. Klingler. *Structures Affines et Projectives sur les Surfaces Complexes.* Annales de L' Institut Fourier, Grenoble, **48** 2 (1998), 441-447.

[117] B. Klingler. *Un théoreme de rigidité non-métrique pour les varietés localement symétriques hermitiennes.* Comment. Math. Helv. **76** (2001), no. 2, 200-217.

[118] B. Klingler. *Sur la rigidité de certains groupes fondamentaux, l'arithmticité des réseaux hyperboliques complexes, et les "faux plans projectifs".* Invent. Math. 153, No.1, 105-143 (2003).

[119] S. Kobayashi. *Transformation groups in differential geometry.* Springer Verlag, 1972.

[120] S. Kobayashi, T. Ochiai. *Holomorphic Projective Structures on Compact Complex Surfaces (I and II).* Math. Ann. **249** (255), 1980 (81), 75-94 (519-521).

[121] S. Kobayashi, *Hyperbolic Manifolds and Holomorphic Mappings,* Marcel Dekker, 1970.

[122] K. Kodaira. *Complex Manifolds and Deformation of Complex Structures.* Springer Verlag, 1985.

[123] K. Kodaira, L. Nirenberg, D. C. Spencer. *On the existence of deformations of complex analytic structures.* Ann. of Maths. **68** (1958), 450-459.

[124] P. Koebe. *Über die Uniformisierung der Algebraischen Kurven II.* Math. Ann. 69 (1910),1-81.

[125] A. I. Kostrikin, Yu I. Manin. *Linear Algebra and Geometry.* CRC Press (1989).

[126] R. K. Kovacheva: An analogue of Montels theorem to rational approximating sequences. Comp. Ren. Acad. Bulg. Scien. 50 (1997), 912.

[127] I. Kra. *On Lifting Kleinian Groups to* $SL(2,\mathbb{C})$. In "Differential Geometry and Complex Analysis", ed. I. Chavel and H. M. Farkas. Springer-Verlag, 1985, 181-194.

[128] I. Kra, B. Maskit. *Remarks on projective structures*. In "Riemann surfaces and related topics: Proceedings of the 1978 Stony Brook Conference", ed. I. Kra and B. Maskit. Princeton Univ. Press, Annals of Maths. Study **97** (1981), 343-360.

[129] S. G. Krantz, *Function Theory of Several Complex Variables*, Pure and Applied Mathematics, John Wiley and Sons, New York, 1982.

[130] H. Kriete (1998). Progress in Holomorphic Dynamics. Pitman Research Notes in Maths. CRC Press, 1998.

[131] S. L. Krushkal, B. N. Apanasov, and N. A. Guseviskii, *Kleinian groups and uniformization in examples and problems*. Translated from Russian by H. H. McFaden, Translations of Mathematical Monographs, vol. 62, American Mathematical Society, Providence, R.I., 1986.

[132] R. S. Kulkarni. *Groups with domains of discontinuity*. Math. Ann. **237** (1978), 253-272.

[133] R. S. Kulkarni. *Conformal structures and Möbius structures*. In "Aspects of Mathematics", eds. R.S. Kulkarni and U. Pinkhall, Max Planck Institut fur Mathematik, Vieweg 1988, p. 1-39.

[134] M. Kuranishi. *On the locally complete families of complex analytic structures*. Ann. of Maths. **75** (1962), 536-577.

[135] S. Lang. *Introduction to Complex Hyperbolic Spaces*. Springer Verlag, 1987.

[136] H. B. Lawson. *Les surfaces minimales et la construction de Calabi-Penrose*. (Séminaire Bourbaki, 624, 1984, 59-81) Asterisque **121-122** (1985), 197-211.

[137] H. B. Lawson, M. L. Michelshon. *Spin geometry*. Princeton Univ. Press, 1989.

[138] F. Ledrappier. *Harmonic measures and Bowen-Margulis measures*. Israel Journal of Mathematics **71** (1990), 275-287.

[139] A. Marden. *Schottky groups and circles*. Contributions to Analysis, (1974) 273-278.

[140] A. Marden. *The geometry of finitely generated Kleinian groups*. Annals of Maths. **99** (1974), 383-462.

[141] G. A. Margulis. *Discrete subgroups of semisimple Lie groups* . Springer Verlag, 1991.

[142] J. M. Marstrand. *The dimension of cartesian product sets*. Proc. Cambridge Phil. Soc. **50** (1954), 198-202.

[143] B. Maskit. *A characterization of Schottky groups*. J. d'Analyse Math. **19** (1967), 227-230.

[144] B. Maskit. *Kleinian groups*. Springer-Verlag, 1972.

[145] F. I. Mautner. *Geodesic flows on symmetric Riemann spaces*. Ann. of Math. **65** (1957), 416-431.

[146] K. Matsuzaki, M. Taniguchi Hyperbolic manifolds and Kleinian groups, Oxford University press. 1998.

[147] C. T. McMullen. *Rational maps and Kleinian groups*. In "Proceedings International Congress of Mathematicians", Kyoto, 1990, p. 889-900, Springer Verlag 1991.

[148] C. T. McMullen. *Complex Dynamics and Renormalization*. Annals of Mathematics Studies **135** (1994), Princeton University Press. NJ.

[149] C. T. McMullen. *Hausdorff dimension and conformal dynamics. II: Geometrically finite rational maps*. Comment. Math. Helv. **75** (2000), 535-593.

[150] C. T. McMullen. *Riemann Surfaces, Dynamics and Geometry*. Published by Harvard University 2008. Electronic book:
www.e-booksdirectory.com/details.php?ebook=3048.

[151] Y. Miyaoka. *Algebraic surfaces with positive indices*. In "Classification of algebraic and analytic manifolds", K. Ueno, ed. Proc. Symp. Katata/Jap. 1982, Prog. Math. 39, 281-301 (1983).

[152] C.C. Moore. *Ergodicity of flows on homogeneous spaces*. Amer. J. Math. **88** (1966), 154-178.

[153] G. D. Mostow. *Quasi-conformal mappings in n-space and the rigidity of the hyperbolic space forms*. Publ. Math. IHES **34** (1968), 53-104.

[154] G. D. Mostow. *Strong rigidity of locally symmetric spaces*. Annals of Mathematics Studies, Princeton University Press. NJ **78** (1973).

[155] G. D. Mostow. *On a remarkable class of polyhedra in complex hyperbolic space*. Pac. J. Math. 86, 171-276 (1980).t

[156] D. Mumford. *Abelian Varieties*. With appendices by C. P. Ramanujam and Yuri Manin. Corrected reprint of the second (1974) edition. Tata Institute of Fundamental Research Studies in Mathematics, 5. Published for the Tata Institute of Fundamental Research, Bombay; by Hindustan Book Agency, New Delhi, 2008.

[157] D. Mumford, J. Fogarty, F. Kirwan. *Geometric invariant theory*. 3rd enl. ed. Ergebnisse der Mathematik und ihrer Grenzgebiete. 3. Folge. **34**. Berlin: Springer-Verlag, 1993.

[158] D. Mumford, C. Series, D. Wright. *Indra's Pearls*. Cambridge University Press, 2002.

[159] P. J. Myrberg. *Beispiele von Automorphen Funktionen*. Annales Acad. Sci. Fennice **89** (1951), 1-16.

[160] J.-P. Navarrete. *On the limit set of discrete subgroups of* PU(2, 1). Geom. Dedicata 122, 1-13 (2006).

[161] J.-P. Navarrete. *The Trace Function and Complex Kleinian Groups in* $\mathbb{P}^2_\mathbb{C}$. Int. Jour. of Math. Vol. 19, No. 7 (2008) 865-890.

[162] J.-P. Navarrete and W. Barrera. *Discrete subgroups of* PU(2, 1) *acting on* $\mathbb{P}^2_\mathbb{C}$ *and the Kobayashi metric*. Bull. Braz. Math. Soc. (N.S.) **40** (2009), 99-106.

[163] P. J. Nicholls. *The Ergodic theory of discrete groups*. L. M. S. Lecture Notes Series vol. **143** (1989).

[164] M. V. Nori. *The Schottky groups in higher dimensions*. Proceedings of Lefschetz Centennial Conference, Mexico City. Edited by A. Verjovsky et al. AMS Contemporary Maths.**58**, part I (1986), 195-197.

[165] N. R. O'Brien, J. H. Rawnsley. *Twistor spaces*. Ann. Global Anal. Geom. **3** (1985), 29-58.

[166] J.-P. Otal. *Le théorem d'hyperbolisation pour les variétés fibrées de dimension trois*. Astérisque 235 (1996).

[167] J. Palis and W. de Melo. *Geometric theory of dynamical systems: an introduction*. Springer-Verlag. (1982)

[168] J. R. Parker. *Complex hyperbolic lattices*. In "Discrete Groups and Geometric Structures". Contemporary Mathematics 501 AMS, (2009), 1-42.

[169] J. R. Parker. *Complex hyperbolic Kleinian Groups*. Cambridge Univ. Press (to appear).

[170] J. R. Parker, I. Short. *Conjugacy classification of quaternionic Möbius transformations*. Comput. Methods Funct. Theory 9 (2009), no. 1, 1325.

[171] John R. Parker, I. D. Platis. *Complex hyperbolic quasi-Fuchsian groups*. In "Geometry of Riemann surfaces", 309–355, London Math. Soc. Lecture Note Ser., 368, Cambridge Univ. Press, Cambridge, 2010.

[172] S. J. Patterson. *The limit set of a Fuchsian group*. Acta Math. **136** (1976), 241-273.

[173] S. J. Patterson. *Lectures on measures on limit sets of Kleinian groups*. In "Analytic and geometric aspects of hyperbolic space". London Math. Soc., Lecture Notes **111** (1987), 281-323.

[174] R. Penrose. *The twistor programme*. Rep. Math. Phys. **12** (1977), 65-76.

[175] R. Penrose. *The complex geometry of the natural world.* In Proceedings of the International Congress of Mathematicians. Held in Helsinki, August 1523, 1978. p. 189-194. Edited by Olli Lehto. Academia Scientiarum Fennica, Helsinki, 1980.

[176] R. Penrose. *Pretzel twistor spaces.* In "Further advances in twistor theory", ed. L.J. Mason and L. P. Hughston, Pitman Research Notes in Maths. vol. **231** (1990), 246-253.

[177] H. Poincaré. *Papers on Fuchsian Functions.* In Collected articles on Fuchsian and Kleinian groups, translated by J. Stillwell. Springer Verlag, 1985.

[178] V. L. Popov. *Discrete complex reflection groups.* Commun. Math. Inst., Rijksuniv. Utr. 15 (1982).

[179] G. Prasad. *Strong rigidity of Q-rank 1 lattices.* Invent. Math. **21** (1973), 255-286.

[180] J.-F. Quint. *Mesures de Patterson-Sullivan en rang supérieur.* Geom. Funct. Anal. **12** (2002), 776-809.

[181] J.-F. Quint. *An overview of Patterson-Sullivan theory,* Notes for the Workshop The barycenter method, FIM, Zurich, May 2006.

[182] M. S. Raghunathan. *Discrete subgroups of Lie groups.* Springer Verlag, 1972.

[183] E. Ramirez de Arellano. *Ein Divisionsproblem und Randintegraldarstellungen in der Komplexen Analysis.* Math. Ann. *184* (1970), 172-187.

[184] A. I. Ramirez-Galarza, J. Seade. *Introduction to classical geometries.* Birkhäuser Verlag 2007.

[185] J. Ratcliffe. *The foundations of hyperbolic manifolds.* Graduate texts in Math. **149**, Springer Verlag, 1994.

[186] M. Ratner. *Rigidity of horocycle flows.* Ann. of Maths. **115** (1982), 597-614.

[187] J. J. Rotman. *An introduction to the theory of groups.* Springer-Verlag, 1994.

[188] D. Ruelle. *Bowen's formula for the Hausdorff dimension of self-similar sets.* Progr. Phys. **7** (1983), 351-358.

[189] S. Salamon. *Harmonic and holomorphic maps.* Springer Verlag, Lecture Notes in Maths. No. **1164**, by M. Meschiri et al (1985), 162-224.

[190] I. Satake: *On a generalisation of the notion of manifold.* Proc. Nat. Acad. Sciences 42 (1956), 359-363.

[191] G. Scheja. *Riemannsche Hebbarkeitsätze für Cohomoligieklassen.* Math. Ann. **144** (1961), 345-360.

[192] F. Schottky , *Über die conforme Abbildung mehrfach zusammenhängender ebener Flächen.* Crelle, Journal f. d. Reine und Angewandte Mathematik **83** (1877), 300–351.

[193] H. Schubart. *Über normal-discontinuierliche lineare Gruppen in 2 komplexen Variablen.* Comm. Math. Helv. **12** (1939-1940), 81-129.

[194] I. Schur. *Über die darstellung der endlichen Gruppen durch gebrochen lineare Substitutionen.* Crelle J. Reine Angew. Math. **127** (1904), 20-50.

[195] O. V. Schvartsman. *Discrete reflection groups in the complex ball.* Funct. Anal. Appl. 18 (1984), 88-89.

[196] R. Schwartz, *Pappus' theorem and the modular group,* Inst. Hautes tudes Sci. Publ. Math. No. 78 (1993), pp. 187-206.

[197] R. E. Schwartz. *Ideal triangle groups, dented tori, and numerical analysis.* Ann. of Math. (2) 153 (2001), no. 3, 533–598.

[198] R. E. Schwartz, *Complex hyperbolic triangle groups.* Proceedings of the International Congress of Mathematicians, Vol. II (Beijing, 2002), 339-349, Higher Ed. Press, Beijing, 2002.

[199] R. E. Schwartz. *A better proof of the Goldman-Parker conjecture.* Geom. Topol. 9 (2005), 1539–1601.

[200] R. E. Schwartz. *Spherical CR geometry and Dehn surgery.* Annals of Mathematical Studies 165. Princeton, NJ: Princeton Univ. Press (2007).

[201] J. Seade, A. Verjovsky. *Actions of Discrete Groups on Complex Projective Spaces.* Contemporary Math., **269** (2001),155-178.

[202] J. Seade, A. Verjovsky. *Higher Dimensional Kleinian groups.* Math. Ann., **322** (2002), 279-300.

[203] J. Seade, A. Verjovsky. *Complex Schottky groups.* Geometric methods in dynamics. II. Astérisque **287** (2003), xx, 251-272.

[204] I. E. Segal, J. von Neumann. *A theorem on unitary representations of semisimple Lie groups.* Ann. of Math. **52**, (1950). 509-517.

[205] C. L. Siegel. *Discontinuous groups.* Annals of Maths. **44**, pp. 674-689 (1943).

[206] M. A. Singer. *Flat twistor spaces, conformally flat manifolds and four-dimensional field theory* Commun. Math. Phys. **133** (1990), 75-90.

[207] G. Sienra. *Complex Kleinian Groups and Limit Sets in P^2.* Complex Variables, *49* No. 10, pp. 689-701 (2004).

[208] M. Slupinski. *Espaces de twisteurs Kählériens en dimension $4k$, $k > 1$.* J. London Math. Soc. 33 (1986), 535-542.

[209] S. Smale. *Differentiable dynamical systems* . Bull. Amer. Math. Soc. **73**, (1967) 747-817.

[210] J. Smillie, G. Buzzard. *Complex Dynamics in several Complex Variables.* In "Flavors in Geometry", Silvio Levy editor, M.S.R.I. Publ. **31**, Camb. Univ. Press, 1977.

[211] E. H. Spanier, *Algebraic topology*, McGraw Series in Higher Mathematics, McGraw-Hill, New York, 1966.

[212] E. Sperner, *Ueberdie fixpunktfreien Abbildungen Ebene*, Abh. Math. Sem. der Univ. Hamburg 10 (1934), pp. 1-47.

[213] D. P. Sullivan. *The density at infinity of a discrete group of hyperbolic motions*. Publ. Math. I. H. E. S. **50** (1979), 171-202.

[214] D. P. Sullivan. *A finiteness theorem for cusps*. Acta Math. **147** (1981), 289-294.

[215] D. P. Sullivan. *On the ergodic theory at infinity of an arbitrary discrete group of hyperbolic motions*. In "Riemann surfaces and related topics: Proceedings of the 1978 Stony Brook Conference", ed. I. Kra and B. Maskit. Ann. Maths. Study 97, 465-496 (1981).

[216] D. P. Sullivan. *Seminar on conformal and hyperbolic geometry*; notes by M. Baker and J. Seade. Institut des Hautes Études Scientifiques, Bures-sur-Yvette, France, 1982.

[217] D. P. Sullivan. *Conformal Dynamical Systems*. Lecture Notes in Mathematics, Springer Verlag **1007** (1983), 725-752.

[218] D. P. Sullivan. *Entropy, Hausdorff measures old and new, and limit sets of geometrically-finite Kleinian groups*. Acta Math. **153** (1984), 259-277.

[219] D. P. Sullivan. *Quasiconformal homeomorphisms and dynamics I*. Ann. of Maths. **122** (1985), 401-418.

[220] D. P. Sullivan. *Quasiconformal homeomorphisms and dynamics II*. Acta Math. **155** (1985) 243-260.

[221] T. Suwa. *Compact quotients of $\mathbb{C}^2$ by affine transformation groups*. J. Diff. Geometry, **10** (1975), 239-252.

[222] W. P. Thurston. *The Geometry and topology of 3-manifolds*. Princeton University Notes 1980.

[223] W. P. Thurston. *Three-dimensional Geometry and Topology* . Princeton Mathematical Series **35**, Princeton University Press, 1997.

[224] W. P. Thurston. *Shapes of polyhedra and triangulations of the sphere*. I. Rivin et al, ed.In "The Epstein Birthday Schrift dedicated to David Epstein on the occasion of his 60th birthday". Warwick: University of Warwick, Institute of Mathematics, Geom. Topol. Monogr. **1** (1998), 511-549.

[225] S. Tokunaga, M. Yoshida. *Complex crystallographic groups. I*. J. Math. Soc. Japan 34 (1982), no. 4, 581–593.

[226] P. Tukia. *The Hausdorff dimension of the limit set of a geometrically-finite Kleinian group* . Acta Math. **152** (1984), 127-140.

[227] P. Tukia. *Differentiability and rigidity of Möbius groups.* Invent. Math. **82** (1985), 557-578.

[228] P. Tukia. *Mostow-rigidity and noncompact hyperbolic manifolds.* Quart. J. Math. Oxford Ser. **42** (1991), no. 166, 219-226.

[229] V. S. Varadajan. *Lie groups, Lie Algebras and Their Representations.* Springer -Verlag, 1974.

[230] A. Vitter. *Affine structures on compact complex manifolds.* Invent. Math. 17 (1972), 231–244.

[231] C. T. C. Wall. *Geometric Structures on Compact Complex Analytic Surfaces.* Topology **25**, No. 2, pp. 119-156, 1986.

[232] R. O., Jr. Wells. *Differential Analysis on Complex Manifolds.* Prentice-Hall, Englewood Cliffs, NJ, 1973.

[233] R. J. Zimmer. *Ergodic theory and semisimple groups.* Monographs in Mathematics, Birkhäuser **81** 1984.

Index

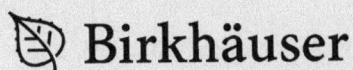
Progress in Mathematics (PM)

Edited by
Hyman Bass, University of Michigan, USA
Joseph Oesterlé, Institut Henri Poincaré, Université Paris VI, France
Alan Weinstein, University of California, Berkeley, USA
Yuri Tschinkel, Courant Institute of Mathematical Sciences, New York, USA

Progress in Mathematics is a series of books intended for professional mathematicians and scientists, encompassing all areas of pure mathematics. This distinguished series, which began in 1979, includes research level monographs, polished notes arising from seminars or lecture series, graduate level textbooks, and proceedings of focused and refereed conferences. It is designed as a vehicle for reporting ongoing research as well as expositions of particular subject areas.

PM 302: Mastrolia, P.; Rigoli, M.; Setti, A.G.
Yamabe-type Equations on Complete, Noncompact Manifolds (2012).
ISBN 978-3-0348-0375-5

PM 301: Ruzhansky, M.; Sugimoto, M.; Wirth, J. (Eds.)
Evolution Equations of Hyperbolic and Schrödinger Type. Asymptotics, Estimates and Nonlinearities (2012).
ISBN 978-3-0348-0350-2

PM 300: Bump, D.; Friedberg, S.; Goldfeld, D. (Eds.)
Multiple Dirichlet Series, L-functions and Automorphic Forms (2012).
ISBN 978-0-8176-8333-7

PM 299: Müller-Hoissen, F.; Pallo, J. M.; Stasheff, J. (Eds.)
Associahedra, Tamari Lattices and Related Structures. Tamari Memorial Festschrift (2012).
ISBN 978-3-0348-0404-2

PM 298: Getz, J.; Goresky, M.
Hilbert Modular Forms with Coefficients in Intersection Homology and Quadratic Base Change (2012).
ISBN 978-3-0348-0350-2

PM 297: Dai, X.; Rong, X. (Eds.)
Metric and Differential Geometry. The Jeff Cheeger Anniversary Volume (2012).
ISBN 978-3-0348-0256-7

PM 296: Itenberg, I.; Jöricke, B.; Passare, M. (Eds.)
Perspectives in Analysis, Geometry, and Topology. On the Occasion of the 60th Birthday of Oleg Viro (2012).
ISBN 978-0-8176-8276-7

PM 295: Joseph, A.; Melnikov, A.; Penkov, I. (Eds.)
Highlights in Lie Algebraic Methods (2012).
ISBN 978-0-8176-8273-6

PM 294: Barreira, L.
Thermodynamic Formalism and Applications to Dimension Theory (2011).
ISBN 978-3-0348-0205-5

PM 293: Mazzucchelli, M.
Critical Point Theory for Lagrangian Systems (2011).
ISBN 978-3-0348-0162-1

PM 292: van den Ban, E. P.; Kolk, J.A.C. (Eds.)
Geometric Aspects of Analysis and Mechanics. In Honor of the 65th Birthday of Hans Duistermaat (2011).
ISBN 978-0-8176-8243-9

PM 291: Greene, R.E.; Kim, K.-T.; Krantz, S.G.
The Geometry of Complex Domains (2011).
ISBN 978-0-8176-4139-9

PM 290: Mantegazza, C.
Lecture Notes on Mean Curvature Flow (2011).
ISBN 978-3-0348-0144-7

PM 289: Colombo, F.; Sabadini, I.; Struppa, D. C.
Noncommutative Functional Calculus (2011).
ISBN 978-3-0348-0109-6

PM 288: Neeb, K.-H.; Pianzola, A. (Eds.)
Developments and Trends in Infinite-Dimensional Lie Theory (2011).
ISBN 978-0-8176-4740-7

PM 287: Cattaneo, A.S.; Giaquinto, A.; Xu, P. (Eds.)
Higher Structures in Geometry and Physics (2011).
ISBN 978-0-8176-4734-6

PM 286: Abbes, A.
Éléments de Géométrie Rigide. Volume 1: Construction et Étude Géométrique des Espaces Rigides (2011).
ISBN 978-3-0348-0011-2